A REAL-TIME APPROACH TO PROCESS CONTROL

A REAL-TIME APPROACH TO PROCESS CONTROL

William Y. Svrcek
University of Calgary
Calgary, Canada

Donald P. Mahoney
AEA Technology Engineering Software, Hyprotech Ltd
Calgary, Canada

Brent R. Young
University of Calgary
Calgary, Canada

JOHN WILEY & SONS, Ltd
Chichester · New York · Weinheim · Brisbane · Singapore · Toronto

Baffins Lane, Chichester,
West Sussex PO19 1UD, England

National 01243 779777
International (+44) 1243 779777

e-mail (for orders and customer service enquiries: cs-books@wiley.co.uk
Visit our Home Page on http://www.wiley.co.uk
or http://www.wiley.com

Other Wiley Editorial Offices

John Wiley & Sons, Inc., 605 Third Avenue,
New York, NY 10158-0012, USA

WILEY-VCH Verlag GmbH, Pappelallee 3,
D-69469 Weinheim, Germany

Jacaranda Wiley Ltd., 33 Park Road, Milton,
Queensland 4064, Australia

John Wiley & Sons (Asia) Pte Ltd., 2 Clementi Loop #02–01
Jin Xing Distripark, Singapore 129809

John Wiley & Sons (Canada) Ltd., 22 Worcester Road,
Rexdale, Ontario M9W 1L1, Canada

Library of Congress Cataloging-in-Publication Data
Svrcek, William Y.
A real-time approach to process control/William Y. Svrcek,
Donald P. Mahoney, Brent R. Young.
p. cm.
Includes bibliographical references and index.
ISBN 0-471-80363-4 (alk. paper) – ISBN 0-471-80452-5 (pbk. : alk. paper)
1. Process control–Data processing. 2. Real-time control.
I. Mahoney, Donald P. II. Young, Brent R. III. Title.

TS156.8. S86 2000
670.42′75433–dc21 99-089624

British Library Cataloguing in Publication Data
A catalogue record for this book is available from the British Library

ISBN 0–471–80363–4 (cloth)
0–471–80452–5 (pbk)

Typeset in 11/13 Times Ten by Florence Production Ltd, Stoodleigh, Devon
Printed and bound in Great Britain by Biddles Ltd, Guildford and King's Lynn
This book is printed on acid-free paper responsibly manufactured from sustainable forestry, in which at least two trees are planted for each one used for paper production.

To our wives – Johann, Susan and Suzanne – for their unending support, and in memory of Ruby Mahoney

CONTENTS

ABOUT THE AUTHORS

William Svrcek is Professor of Chemical and Petroleum Engineering at the University of Calgary, Alberta, Canada. He received his B.Sc. (1962) and Ph.D. (1967) degrees in Chemical Engineering from the University of Alberta, Edmonton. Prior to joining the University of Calgary he worked for Monsanto Company as a senior systems engineer and as an Associate Professor (1970–75) in the Department of Biochemical and Chemical Engineering at the University of Western Ontario, London, Ontario. Dr. Svrcek's teaching and research interests center on process simulation control and design. He has authored or co-authored over 150 technical articles/reports and has supervised over 30 graduate students. He has been involved for many years in teaching the continuing education course titled "Computer Aided Process Design – Oil and Gas Processing" that has been presented worldwide. Most recently this course has been modified to include not only steady-state simulation but also dynamic simulation and control strategy development and verification. Dr. Svrcek was also a senior partner in Hyprotech from its incorporation in 1976. As a Principal, Director, and President (1981–1993) he was instrumental in establishing Hyprotech as a leading international process simulation software company. He is a registered Professional Engineer in both Alberta and Ontario, and a member of professional societies that include The Canadian Society for Chemical Engineering, American Institute for Chemical Engineers, Canadian Gas Processors Association and the Instrument Society of America.

Donald Mahoney is Vice President with AEA Technology Engineering Software/Hyprotech. Mr. Mahoney earned a Bachelor's Degree in Mechanical Engineering from Penn State, a Master's Degree in Control Theory from Purdue University, and an MBA from the University of Delaware. Mr. Mahoney has held research and teaching positions at the US Navy's Applied Research Lab and Purdue University, where he was awarded the staff's "Outstanding Teaching Award". He has also lectured extensively on the topic of process control, and has published a number of journal articles on process simulation and control. Prior to joining AEA Technology/Hyprotech, Mr. Mahoney held industrial positions at General Motors and DuPont as a control systems engineer and process modelling and control consultant. While at DuPont, he was involved in the development and support of the company's object-oriented dynamic simulation package, TMODS.

Brent Young is Associate Professor of Chemical and Petroleum Engineering at the University of Calgary, Alberta, Canada. He received his B.E. (1986) and Ph.D. (1993) degrees in Chemical and Process Engineering from the University of Canterbury, New Zealand. Prior to his graduate studies, he worked as a chemical engineer for Ravensdown Fertilizer Coop's Super Phosphate Plant in Christchurch and developed a process model for the simulation of a rock phosphate grinding circuit. In 1991, he joined the University of Technology in Sydney, Australia as a lecturer, received tenure in 1994 and was promoted to Senior Lecturer in 1996, continuing his research in the areas of modelling and control of processes, particularly industrial processes. He joined the University of Calgary as an Associate Professor in late 1998. His current research is centred on two major areas, process simulation and control, and process development – particularly the processing of carbonaceous substances.

FOREWORD

"As plants are pushed beyond nameplate, it is increasingly obvious that the importance of process control has grown to the point where it is the single biggest leverage point for increasing manufacturing capacity and efficiency. The process engineer, who is best posed to use his process knowledge for getting the most from better control, typically has had just a single course in control. Furthermore, the approach was based on theory rather than on practice, and was immersed in the frequency domain. Real processes are diverse and complex and the view into their behavior is by means of real time trend recordings. This book provides a building block real time approach to understanding and improving process control systems. Practical examples and workshops using models drive home the points and make the principles much more accessible and applicable."

Gregory K. McMillan, Senior Fellow,
Solutia Inc.

"At the undergraduate chemical engineering level, the traditional, highly mathematical approach misses the point of what knowledge of control and dynamics the practicing process engineer requires. If BS graduates in chemical engineering simply understood the basics of time based process dynamics and control (capacitance, dead time, PID control action and controller tuning, inventory, throughput, and distillation control), the impact on process design and plant operations throughout the CPI would be immense. Today, these skills are among the least developed in BS chemical engineering graduates, despite having taken the requisite traditional process control course. This text is particularly suitable for any college, university, or technical training program seeking to provide its graduates with a truly practical and applied background in process dynamics and control. With today's widespread commercial availability of high fidelity process simulation software, the understanding gained from this text can be immediately and directly applied."

Thomas C. Hanson, Senior Engineering Associate,
Praxair, Inc.

"Several years ago, a recruiter from a major chemical company told me that his company was hesitant to interview students that indicated a first preference in the area of process control because his company 'did not have any jobs that made use of Laplace transforms and frequency domain skills'. This was an excellent example of the mismatch between what is frequently taught in universities, and what often gets applied in industry. After teaching chemical process control for over 30 years, I feel strongly that good process control is synonymous with good chemical engineering. Industry would be well served if all chemical engineering graduates, regardless of career paths, had a better, more practical working knowledge of process dynamics and control. I think the approach taken in this text is right on target, and is consistent with how we teach at the University of Tennessee. It provides a good hands-on feel for process dynamics and process control, but more importantly, it presents these concepts as fundamentals of chemical engineering. For undergraduate programs looking to transition away from the traditional mathematical-based approach to a more applied, hands-on approach, this text will be an invaluable aid."

Charles F. Moore, Professor of Chemical Engineering,
University of Tennessee

"What BS degree chemical engineers need is a base level understanding of differential equations, process dynamics, dynamic modeling of the basic unit operations (in the time domain), basic control algorithms (such as PID), cascade structures and feed forward structures. With these basic tools and an understanding of how to apply them, they can solve most of their control problems themselves. What they do not need is the theory and mathematics that usually surround the teaching of process control such as frequency domain analysis. Graduate education in process control is the place to introduce these concepts."

James J. Downs, Senior Engineering Associate,
Eastman Chemical Company

PREFACE

For decades, the subject of control theory has been taught using transfer functions, frequency-domain analysis, and Laplace transform mathematics. For linear systems – like those from the electromechanical areas from which these classical control techniques emerged – this approach is well suited. As an approach to the control of chemical processes, which are often characterised by nonlinearity and large doses of dead time, classical control techniques have some limitations.

In today's simulation-rich environment, the right combination of hardware and software is available to implement a "hands-on" approach to process control system design. Engineers and students alike are now able to experiment on virtual plants that capture the important non-idealities of the real world, and readily test even the most outlandish of control structures without resorting to non-intuitive mathematics or to placing real plants at risk.

Thus, the basis of this text is to provide a practical, hands-on introduction to the topic of process control by using only time-based representations of the process and the associated instrumentation and control. We believe this book is the first to treat the topic without relying at all upon Laplace transforms and the classical, frequency-domain techniques. For those students wishing to advance their knowledge of process control beyond this first, introductory exposure, we highly recommend understanding, even mastering, the classical techniques. However, as an introductory treatment of the topic, and for those chemical engineers not wishing to specialise in process control, but rather to extract something immediately practical and applicable, we believe our approach hits the mark.

This text is organised into a framework that provides relevant theory, along with a series of hands-on workshops that employ computer simulations that test and allow for exploration of the theory. Chapter 1 provides a historical overview of the field. Chapter 2 introduces the very important and often overlooked topic of instrumentation. In Chapter 3, we ground the reader in some of the basics of single input/single output (SISO) systems. Feedback control, the elements of control loops, system dynamics including capacitance and dead time, and system modelling are introduced here. Chapter 4 highlights the various PID control modes and provides a framework for understanding control loop design and tuning. Chapter 5 focuses specifically on tuning. Armed with an understanding of feedback control, control loop structures, and tuning, Chapter 6 introduces some more advanced control configurations including feed forward, cascade, and override control. Chapter 7 provides some practical rules of thumb for designing and tuning the more common control loops found in industry. In Chapter 8, we tackle a more complex control problem: the control of distillation columns. As with the rest of this text, a combination of theory and applied methodology is used to provide a practical treatment to this complex topic. Chapter 9 introduces the concept of multiple loop controllers. In Chapter 10, we take a look at some of the important issues relating to the plant-wide control problem. Finally, up-to-date information on computer simulation for the workshops can be found on the book's Web site, **http://www.ench.ucalgary.ca/~realtime**

While this text is designed as an introductory course on process control for senior university students in the chemical engineering curriculum, we believe it will also serve as a valuable desk reference for practising chemical engineers and as a text for technical colleges.

We believe the era of real-time, simulation-based instruction of chemical process control has arrived. We hope you'll agree! We wish you every success as you begin to learn more about this exciting and ever-changing field. Your comments on and suggestions for improving this textbook are most welcome.

ACKNOWLEDGEMENTS

It would be impossible to mention all of the individuals who contributed to the ideas that form the background of this text. Over the past five years, we have interacted with many students, academics, and perhaps most importantly, practitioners in the field of process control. This, combined with the more than 50 years of cumulative experience among the authors, has led to what we believe is a uniquely practical first encounter with the discipline of chemical process control.

Some who deserve special mention for their influence include Björn Tyréus and Ed Longwell from DuPont, and Paul Fruefauf from Applied Control Engineering. These gentlemen share a passion for the field and a commitment to the practical approach to both teaching and practicing process control.

As with any text, many more names were involved in its creation than the three printed on the cover. To those who put in such generous effort to help make this text a reality, we express our sincerest of thanks.

Special thanks are due to:

Joanna Williams, consulting engineer, for her many helpful suggestions. In particular, her careful editing of the original text and enhancements to the workshops is most appreciated.

Dr Wayne Monnery, consulting engineer, for preparing the section on control valve sizing.

Dr Martin Sneesby, consulting engineer, for the excellent effort in reviewing, testing, as well as suggested changes to the original group of workshops.

Ken Trumble and Darrin Kuchle of Spartan Controls for facilitating the provision of the detailed hardware schematics and photographs shown in the book. In particular, Ken's many helpful comments on the text are much appreciated.

Kel Stillman, University of Technology, Sydney, Australia; Professor Daniel Lewin, Technion, Israel; Dr Ingmar Tollet, Espoo University, Finland; Dr Seamus McMonagle, University of Limerick, Ireland; and Professor Michael Henson, Louisiana State University, USA, for the time they spent reading through an earlier draft of the book, and their helpful comments and suggestions, without which the final version would have been poorer.

Finally, the 1997, 1998, and 1999 fourth year chemical engineering students at the University of Calgary, Canada, for their constructive comments on the book, especially the workshops.

"Tell me and I forget,
Show me and I may remember,
Involve me and I understand."

Benjamin Franklin
Scientist and statesman.

1 A BRIEF HISTORY OF CONTROL AND SIMULATION

In order to gain an appreciation for process control and simulation it is important to have some understanding of the history and driving force behind the development of both control and simulation. Rudimentary control systems have been used for centuries to help humans use tools and machinery more efficiently and effectively. However, only in the last century has more time and effort been devoted to developing a greater understanding of controls and more sophisticated control systems. The expansion of the controls field has aided the growth of process simulation from relative obscurity to the indispensable and commonplace tool that it is today.

1.1 Control

Feedback control can be traced back as far as the early third century BC [1]. During this period, Ktesibios of Alexandria employed a float valve similar to the one found in today's automobile carburetors to regulate the level in the water clocks of that time [2]. Three centuries later, Heron of Alexandria described another float valve water level regulator similar to that used in toilet water tanks [1]. Arabic water clock builders used this same control device as late as 1206. The Romans also made use of this first control device in regulating the water levels in their aqueducts. The level regulating device or float valve remained unknown to Europeans and was reinvented in the eighteenth century to regulate the water levels in steam boilers and home water tanks.

The Europeans did however invent a number of feedback control devices; namely the thermostat or bimetallic temperature regulator, the safety relief valve, and the windmill fantail. In 1620, Cornlis Drebbel [2], a Dutch engineer, used a bimetallic temperature regulator to control the temperature of a furnace. Denis Papin [2], in 1681, used weights on a pressure cooker to regulate the pressure in the vessel. In 1745, Edmund Lee [1] attached a fantail at right angles to the main sail of a windmill, thus always keeping the main windmill drive facing into the wind. It was not until the Industrial Revolution, particularly in England, that feedback devices became more numerous and varied.

One-port automata (open loop) evolved as part of the Industrial Revolution and focused on a flow of commands that mechanised the functions of a human operator. In 1801, Joseph Farcot [3] fed punched cards past a row of needles to program patterns on a loom, and in 1796, David Wilkinson [4] developed a copying lathe with a cutting tool positioned by a follower on a model. Oliver Evans [3] built a water-powered flourmill near Philadelphia, in 1784, using bucket and screw conveyors to eliminate manual intervention. Similarly, biscuit making was automated for the Royal Navy in 1833, and meat processing was mechanised in America during the late 1860s. Henry Ford used the same concept for his 1910 automobile assembly plant automation. Unit operations, pioneered by Allen Rogers of the Pratt Institute [5] and Arthur D. Little of MIT [5], led to continuous chemical processing and extensive automation during the 1920s.

The concept of feedback evolved along with the development of steam power and steam-powered ships. The valve operator of Humphrey Potter [6] utilised piston displacement on a Newcomen engine to perform a deterministic control function. However, the flyball governor designed by James Watt [6] in 1769 modulated steam flow to overcome unpredictable disturbances and became the archetype for single loop regulatory controllers. Feedback was accompanied by a perplexing tendency to overshoot the desired operating level, particularly as controller sensitivity increased. The steam-powered steering systems of the ships of the mid-1800s used a human operator to supply feedback, but high rudder-positioning gain caused the ship to zigzag along its course. In 1867, Macfarlane Gray [1] corrected this problem with a linkage that closed the steering valve as the rudder approached the desired point. In 1872, Leon Farcot [1] designed a hydraulic system such that a displacement representing rudder position was subtracted from the steering position displacement, and the difference was used to operate the valve. The helmsman could then indicate a rudder position which would be achieved and maintained by the servo motor.

Subsequent refinements of the servo principle were largely empirical until Minorsky [7] in 1922 published an analytical study of ship steering which considered the use of proportional, derivative, and second-derivative controllers for steering ships and demonstrated how stability could be determined from the differential equations. In 1934 Hazen [8] introduced the term "servomechanism" for position control devices and discussed the design of control systems capable of close tracking of a changing set point. Nyquist [9], based on a study of feedback amplifiers, developed a general and relatively simple procedure for determining the stability of feedback systems from the open-loop response.

Experience with and theory in mechanical and electrical systems were therefore available when World War II created a massive impetus for weapon controls. While the eventual social benefit of this and subsequent military efforts is not without merit, the nature of the incentives emphasises the irony seen by Elting Morison [10]. Just as we attain a means of "control over our resistant natural environment we find we have produced in the means themselves an artificial environment of such complexity that we cannot control it."

Although the basic principles of feedback control can be applied to chemical processing plants as well as to amplifiers or mechanical systems, chemical engineers were slow to adapt the wealth of control literature from other disciplines for the design of process control schemes. The unfamiliar terminology was one major reason for the delay, but there was also the basic difference between chemical processes and servomechanisms, which delayed the development of process control theory and its implementation. Chemical plants normally operate with a constant set point, and large capacity elements help to minimise the effect of disturbances, whereas these would tend to slow the response of servomechanisms. Time delay or transport lag is frequently a major factor in process control, yet it is rarely mentioned in the literature on servomechanisms. In process control systems, interacting first order elements and distributed resistances are much more common than second order elements found in the control of mechanical and electrical systems. These differences made many of the published examples of servomechanism design of little use to those interested in process control.

A few theoretical papers on process control did appear during the 1930s. Notable among these was the paper by Grebe, Boundy and Cermak [11] that discussed the problem of pH control and showed the advantages of using derivative action to improve controller response. Callender, Hartree and Porter [12] showed the effect of time delay on the stability and speed of response of a control system. However, it was not until the mid-1950s that the first texts on process control were published by Young, in 1954 [13] and Ceaglske, in 1956 [14]. Between the late 1950s and the 1970s many texts appeared, notably those by EcKman [15], Campbell [16], Coughanowr and Koppel [17], Luyben [18], Harriott [19], Murrill [20] and Shinskey [21]. Process control became an integral part of every chemical engineering curriculum.

1.2 Simulation

Prior to the 1950s, calculations had been done manually[1] on mechanical or electronic calculators. In 1950, Rose and Williams [22] wrote the first steady-state, multistage binary distillation tower simulation program. The total simulation was written in machine language on an IBM 702, a major feat

[1] Using a slide rule

with the hardware of the day. The general trend through the 1950s was steady-state simulation of individual units. The field was moving so rapidly that by 1953 the American Institute of Chemical Engineers (AIChE) had the first annual review of Computers and Computing in Chemical Engineering. The introduction of FORTRAN by IBM in 1954, provided the impetus for the chemical process industry to embrace computer calculations. The 1950s can be characterised as a period of discovery [23].

From the early 1960s to the present day, steady-state process simulation has moved from a tool used only by experts to a software tool used daily to perform routine calculations. This was made possible by the advances in computing hardware, the most significant of which has been the proliferation of powerful desktop computers (PCs), the development of Windows-based systems software, and the development of object-oriented programming languages. This combination of inexpensive hardware and system tools have led to the proliferation of exceptionally user-friendly and robust software tools for steady-state process simulation and design. Dynamic simulation naturally developed along with the steady-state simulators [24]. Figure 1.1 presents a summary of the growth of dynamic process simulation.

During the 1960s, the size of the analog computer controlled the size of the simulation. These analog computers grew from a few amplifier systems to large systems of a hundred or more amplifiers and finally in the late 1960s to hybrid computers [25]. It was recognised very early that the major disadvantages of analog computers were problem size and dynamic range, both of which were limited by hardware size. Hybrid computers were an attempt to mitigate some of these problems. However, hybrid computers of the late 1960s and early 1970s still had the following problems that limited their general acceptance [25]:

1. Hybrid computers required detailed knowledge of operating both analog and digital computers. This translated into long training periods[2] before an engineer was able to work with the hybrid computer.
2. Hybrid computer simulations were composed of two parts, the analog and digital computer portions. This made debugging complicated since both parts had to be debugged and then integrated.
3. Documentation was required for both parts of the hybrid simulation, analog and digital. The analog part was documented by using wiring diagrams. These wiring diagrams quickly became outdated as changes were made to the analog board that were not always added to the wiring diagram[3].

---

[2] One week or more

[3] Human nature

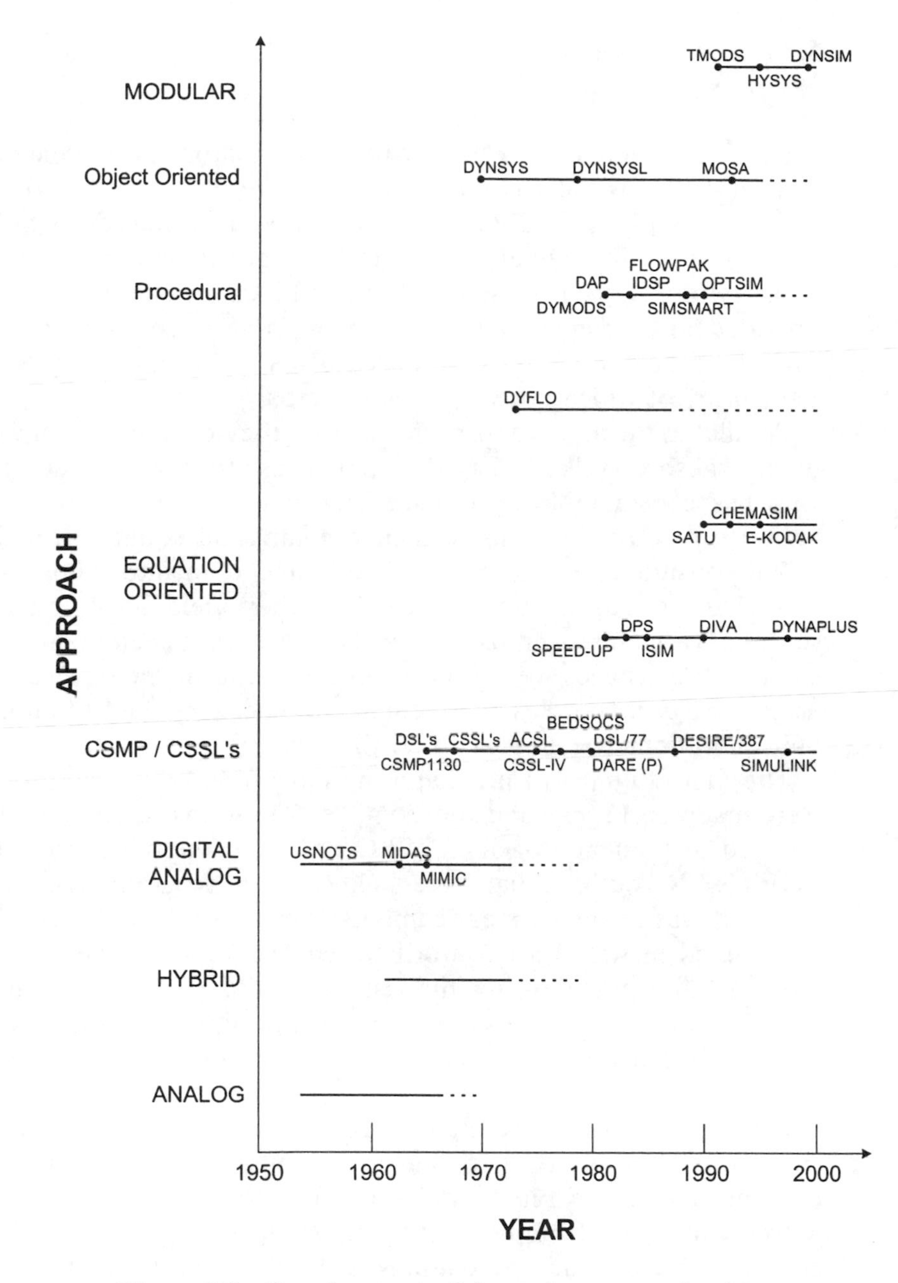

Figure 1.1 Development of dynamic process simulators.

4. Simulations using hybrid computers were extremely time-consuming. An engineer had to reserve time in the hybrid simulation lab and work in this lab in order to solve the problem. This time was devoted entirely to solving one problem and removed the engineer from other effective work.

5. For the majority of simulations, hybrid computers were more expensive to use than digital computers.

Engineers were searching for a dynamic simulator that paralleled steady state simulators being developed during the late 1960s and early 1970s. Early attempts simply moved the analog to a digital formation (CSMPs, Pactolus, etc.) by providing numerical integration algorithms and a suitable programming syntax. Latter versions of these block oriented dynamic simulators provided more functionality and an improved programming methodology. This approach resulted in various CSSLs (Continuous System Simulation Language) of which ACSL [26] is the most widely used.

Parallel to the above approach has been the development of equation based numerical solvers like SPEEDUP [27]. These tools are aimed at the specialist who has considerable experience in using the tool, knows how to model various processes in terms of their fundamental equations, and is willing to spend considerable time entering code and data into input files, which are compiled, edited and debugged before they yield results of time plots for selected variables over fixed time periods. These equation based dynamic simulation packages were very much the realm of the expert. Concepts such as ease of use, complex thermodynamic packages, and libraries of reusable unit operations had not migrated to these dynamic simulators.

The first attempts to provide a modular based dynamic process simulator were made by Franks and Johnson [28]. These two early modular simulators differed in their approach. DYFLO provided the simulator with a suite of FORTRAN routines that were linked via a program written by the user. Hence it was to some extent cumbersome but useable. DYNSYS [28] on the other hand provided a key word structure much like the steady state simulators of the era allowing the user to build a dynamic simulation. Both simulators found limited success due to the difficulty of producing a simulation and the actual run times on the computer hardware of the time were often greater than real-time.

During the late 1970s and throughout the 1980s only equation oriented simulators were used. There was a continuing effort to develop and extend dynamic models of plants and use these for control system development. Many companies, from necessity, had groups using this approach to develop specific plant dynamic simulations and subsequently used these simulations for control design and evaluation. Marquardt at CPCIV [29] presented a paper summarising key developments and future challenges in dynamic process simulation.

The key benefits of dynamic simulation [30] are related to the improved process understanding that it provides; plants are by their nature dynamic. By understanding the process more fully, several benefits follow naturally. These include improvements in control system design, improvements in the basic operation of the plant, and improvements in training for both operators and engineers.

Control system design is unfortunately still often left to the end of the design cycle. This practice frequently requires an elaborate control strategy in order to make the best of a poor design. Dynamic simulation, when involved early in the design phase, can help to identify the important operability and control issues and influence the design accordingly. Clearly the ideal is not just to develop a working control strategy, but also to design a plant that is inherently easy to control.

Using a rigorous dynamic model, control strategies can be designed, tested and even tuned prior to start-up. With appropriate hardware links, dynamic models may even be used to check out distributed control system (DCS) or other control system configurations. All of these features make dynamic simulation ideally suited to control applications.

Another benefit involves reconciling trade-offs between steady-state optimisations and dynamic operability. To minimise capital expenditures and operating utility costs, many plant designers have adopted the use of steady-state optimisation techniques. As a result, plant designs have become more complex and much more highly integrated and interactive. Examples include extensive heat-exchange networks, process recycles and minimum-holdup designs. While such designs may optimise the steady-state flowsheet, they present particular challenges to plant control and operations engineers, usually requiring advanced control strategies and a well-trained operating staff. This trade-off between steady-state optimisation and dynamic operability is classic and can only be truly reconciled using dynamic simulation.

Once a plant is in operation, manufacturing personnel are continually looking for ways to improve quality, minimise waste, maximise yield, reduce utility costs, and often to increase capacity. It is in this area of process improvements that dynamic simulation has perhaps the most value-adding impact. This is also the area where it is most important to minimise the usage barriers for dynamic simulation. Since plant operating personnel are typically busy with the day-to-day operation of the plant, simulation tools that are difficult to understand and use will never see any of the truly practical and value-adding applications. By allowing plant engineers to quickly and easily test theories, illustrate concepts, or compare alternative control strategies or operating schemes, dynamic simulation can have a tremendous cumulative benefit.

Over the past several years, the industry has begun to focus a great deal of attention and interest on dynamic simulation for training purposes. As mentioned earlier, the increased complexity of the plants being designed today requires well-trained operating personnel. In order to be effective, the training simulator should be interactive, realistic and run real-time. By running a relatively high-fidelity model operators can test "what if" scenarios, practise start-up and shutdown sequences, and respond to failure and alarm situations.

More recently, training simulators have provided links to a variety of DCS platforms. By using the actual control hardware to run a dynamic model of the plant, operators have the added benefit of training on the same equipment that will be used to operate the real process.

It is important at this point to introduce the notion of breadth-of-use for a model. We have discussed the use of dynamic simulation for design, control, operations, and now for training. Indeed, it would be beneficial if the same model used to design the plant, develop its controls, and study its operation could be used as the on-line training simulator for the DCS. While this may seem obvious, it is difficult to find examples of such applications. This is primarily due to the absence of commercial simulation tools that provide sufficient breadth of functionality – both engineering functionality and usability.

With all the benefits of dynamic simulation, why is it that this technology has only recently begun to see more widespread use? To answer this question, it is helpful to continue with the history of simulation and to consider the unique set of skills required to develop a dynamic simulation from first principles.

First, an understanding of and access to the basic data relating to the physical properties of the chemical system is needed. This includes the vapour-liquid equilibrium (VLE) and any reaction equations involved. Second, a detailed understanding of the heat and material balance relationships in the process equipment is required. Third, knowledge of appropriate numerical techniques to solve the sets of differential and algebraic equations is needed. Finally, experience in striking a balance between rigour and performance is needed in order to build a model that is at the same time useful and useable. Thus there is indeed a unique set of skills required to design a first-principles dynamic simulator.

Because of the computational load, dynamic simulations have been reserved for large mainframe or mini-computers. An unfortunate feature of these large computer systems was their often cumbersome user interfaces. Typically, dynamic simulations were run in a batch-mode where the model was built with no feedback from the program, then submitted to the computer to be solved for a predetermined length of simulation time. Only when the solution was reached could the user view the results of the simulation study.

With this approach, 50% to 80% of the time dedicated to a dynamic study was used up by the model-building phase. Roughly 20% was dedicated to running the various case studies, and 10% to documentation and presentation of results. This kind of cycle made it difficult for a casual user to conduct a study or even to run a model that someone else had prepared. While the batch-style approach consumed a disproportionate amount of time setting up the model, its real drawback was the lack of any interaction between the user and the simulation. By preventing any real interaction with the model as it is being solved, batch-style simulation sessions are much less effective. Additionally, since more time and effort are spent building model structures, submitting and waiting for batch input runs a smaller fraction of time is available to gain the important process understanding through "what if" sessions.

Thus between the sophisticated chemical engineering, thermodynamics, programming and modeling skills, the large and expensive computers, and cumbersome and inefficient user interfaces it is not surprising that dynamic simulation has not enjoyed widespread use. Normally, only the most complex process studies and designs justified the effort required to develop a dynamic simulation. We believe that the two most significant factors in increased use of dynamic simulation are [24]:

- the growth of computer hardware and software technology; and
- the emergence of new ways of packaging simulation.

As indicated previously, there has been a tremendous increase in the performance of personal computers (PCs) accompanied by an equally impressive drop in their prices. For example, it is not uncommon for an engineer to have a PC with memory of up to 128 Mb, a 5 to 10 Gb hard drive, and a Super VGA graphics monitor on his desktop costing less than $1,500. Furthermore, a number of powerful and interactive window environments have been developed for the PC and other inexpensive hardware platforms. Windows (95, 98, NT), X-Windows, and Mac Systems are just some examples.

The growth in the performance and speed of the PC has made the migration of numerically intense applications to personal computer platforms a reality. This, combined with the flexibility and ease of use of the window environments has laid the groundwork for a truly new approach to simulation.

There are literally thousands of person-years of simulation experience in the industry. With the existing computer technology providing the framework, there are very few reasons why most engineers should have to write and compile code in order to use dynamic simulation. Model libraries do not provide the answer since they do not eliminate the build–compile–link sequence that is often troublesome, prone to errors, and intimidating to many potential users. Given today's window environments and the new programming capabilities that languages such as object-oriented C++ provide, there is no need for batch-type simulation sessions.

It is imperative that a dynamic simulation is "packaged" in a way that makes it easy to use and learn, yet still be applicable to a broad range of applications and users. The criteria include the following:

- Easy to use and learn – must have an intuitive and interactive graphical environment that involves no writing of code or any compile–link–run cycle.

- Configurable – must provide reuseable modules which can easily be linked together to build the desired model.
- Accurate – must provide meaningful results.
- Fast – must strike a balance between rigour and performance so as not to lose the interactive benefits of simulation.
- Broadly applicable – must provide a broad range of functionality to span different industrial applications, as well as varying levels of detail and rigour.
- Desktop computer based – must reside on a convenient desktop computer environment such as a PC, Mac or workstation.

With these attributes, dynamic simulation becomes not only available, but also attractive to a much larger audience than ever before. While dynamic simulation is clearly a valuable tool in the hands of seasoned modelers, only when process engineers, control engineers and plant operating personnel feel comfortable with it will dynamic simulation deliver its most powerful and value-adding benefits.

Even with this emphasis on control system design, chemical plant design used the results of steady-state performance to size the equipment while heuristic methods rather than dynamic systems analysis chose the control schemes. Instruments were field-adjusted to give performance as good as or better than manual control. When the control schemes, sensing devices, valves and the process itself produced poor results, trial and error was used to find an acceptable level of performance. The lengthy analysis required for an accurate control system design using the equation based approach could not be justified for a very few critical loops. Vogel [31] states that even as late as 1991, only the most challenging and troublesome processes were modeled dynamically with the aim of developing process dynamic behaviour understanding and testing alternative control configurations.

For complex processes that required close control, the weakest link in the control scheme design was usually the dynamic description or model of the process. The response of the sensor, valve, and controller could easily be modeled to within 5%. The modeling error in predicting the dynamic behaviour of the process was generally two to three times greater. The lack of reliable, robust, reusable dynamic process models and suitable software [32] limited the acceptance and use of process control theory.

In summary, the traditional approach to control loop analysis has been through the use of frequency domain techniques such as Bodé diagrams, transfer functions, Nyquist plots, etc. Most of these analysis methods require a working knowledge of Laplace transforms and were developed as pencil and paper techniques for solving linear sets of different equations. Although these frequency domain techniques are useful for single control loops they are not easily applicable to real multiloop and nonlinear systems which comprise the actual plants that must be controlled in the fluid processing industries.

In the real-time[4] approach the same set of algebraic and differential equations are encountered as in the frequency domain. However, the major advantage of solving these equations in real time is the ability to observe the interactions of the process, control scheme and load variables much as the operator of a plant observes the behaviour of the plant. Dynamic simulation allows for the comparison of several candidate control strategies and assesses the propagation of variation through a process/plant. In other words, dynamic simulation allows for the evaluation of plant-wide versus single loop control schemes.

References

1. Mayr, O. *Feedback Mechanisms: In the Historical Collections of the National Museum of History and Technology*. Smithsonian Institution Press, City of Washington, 1971, p. 96.
2. Mayr, O. *The Origins of Feedback Control*. Cambridge, Massachusets, 1970.
3. Williams, T.I. *The History of Invention: From Stone Axes to Silicon Chips*. MacDonald & Co. Ltd, 1987, pp. 148–61.
4. Woodbury, R.S. *Studies in the History of Machine Tools*. M.I.T. Press, 1972, pp. 90–3.
5. Hougen, Olaf A. "Seven Decades of Chemical Engineering". *Chem. Eng. Prog.*, 1977, **73**: 89–104.
6. Bennett, S. "The Search for 'Uniform and Equable Motion': A Study of the Early Methods of Control of the Steam Engine". *International Journal of Control*, 1975, **21**: 115–16.
7. Minorsky, N. "Directional Stability of Automatically Steered Bodies". *J. Am. Soc. Naval Engrs*, 1922, **34**: 280.
8. Hazen, H.L. "Theory of Servomechanisms". *J. Franklin Inst.*, 1934, **218**: 279.
9. Nyquist, H. "Regeneration Theory". *Bell System Tech. J.*, 1932, **11**: 126.
10. Krigman, A. "ICON, the Helmsman". *Instruments and Control Systems*, 1970, **Dec.**: 4.
11. Grebe, J.J., Boundy, R.H. and Cermak, R.W. "The Control of Chemical Processes". *Trans. Am. Inst. Chem. Engrs*, 1933, **29**: 211.
12. Callender, A., Hartree, D.R. and Porter, A. "Time Lag in a Control System – I". *Phil. Trans. Roy. Soc. London, ser. A*, 1936, **235**: 415.
13. Young, A.J. *Process Control*. Instruments Publishing Company, Pittsburgh, 1954.
14. Ceaglske, N.H. *Automatic Process Control for Chemical Engineers*. John Wiley & Sons, 1956.
15. Eckman, D. *Automatic Process Control*. John Wiley & Sons, New York, 1958.
16. Campbell, D.P. *Process Dynamics*. John Wiley & Sons, New York, 1958.
17. Coughanowr, D.R. and Koppel, L.B. *Process System Analysis and Control*. McGraw-Hill, 1965.
18. Luyben, W.L. *Process Modeling, Simulation, and Control for Chemical Engineers*. McGraw-Hill, New York, 1973.
19. Harriott, P. *Process Control*. McGraw-Hill, New York, 1964.
20. Murrill, P.W. *Automatic Control of Processes*. International Textbook, 1967.
21. Shinskey, F.G. *Process Control Systems*. McGraw-Hill, 1967.
22. Rose, A. and Williams, T.J. "Punched Card Devices for Distillation Calculations". *Ind. Eng. Chem.*, 1950, **42**: p. 2494.

---

[4] Time domain dynamics

23. Lacey, J.W. and Svrcek, W.Y. "Computers and Chemical Engineers". Presented at 40th Canadian Chemical Engineering Conference, Halifax, NS, Canada, 1990.
24. Svrcek, W.Y., Sim, W.D. and Satyro, M.A. "From Large Computers and Small Solutions to Small Computers and Large Solutions". Proceedings of Chemeca '96, Sydney, Australia, vol. 2, 1996, pp. 11–18.
25. Svrcek, W.Y. and Sandholm, D.P. "An Experience in the Selection of a Computer Simulation Technique". *Simulation*, 1971, **17(6)**: 245–6.
26. Strauss, J.C., *et al.* "The SCi Continuous System Simulation Language (CSSL)". *Simulation*, 1967, **9(6)**: 281–92.
27. Perkins, J.D. "Survey of Existing Systems for the Dynamic Simulation of Industrial Processes". *Modeling, Identification and Control*, 1986.
28. Ausain, A. *Chemical Process Simulation*. John Wiley & Sons, New York, 1986. pp. 201–6.
29. Marqurdt, W. "Dynamic Process Simulation – Recent Progress and Future Challenges". Proceedings of CPCIV, Ed: Arkun, Y., Ray, W.H., AIChE, New York, 1991, pp. 131–80.
30. Fruehauf, P.S. and Mahoney, D.P. "Improve Distillation Column Control Design". *CEP*, 1994, **March**: 75–83.
31. Vogel, E.F. "An Industrial Perspective on Dynamic Flowsheet Simulation". Proceedings of CPCIV, Eds: Arkun, Y., Ray, W.H., AIChE, New York, 1991, pp. 181–208.
32. Tyreus, B.D., "Object-Oriented Simulation", Chapter 5, in Luyben, W.L. (Ed.) *Practical Distillation Control*. Van Nostrand Reinhold, 1992.

2 PROCESS CONTROL HARDWARE FUNDAMENTALS

In order to analyse a control system, the individual components that make up the system must be understood. Only with this understanding can the workings of a control system be fully comprehended. The rest of this book deals extensively with controller and process characteristics. It is therefore appropriate and necessary that hardware fundamentals for the primary elements and final control elements be studied first in this chapter. Discussion of controller hardware is delayed until Chapter 4 where the control equations governing the controllers are covered. Several of the concepts introduced in this chapter are discussed in further detail in later sections of this book.

2.1 Control system components

A control system is comprised of the following components:

1. Primary elements (or sensor/transmitters)
2. Controllers
3. Final Control Elements (usually control valves)
4. Processes

Figure 2.1 illustrates a level control system and its components. The level in the tank is read by a level sensor device, which passes the information on to the controller. The controller compares the level reading with the desired level or set point and then transmits this information on to the control valve, referred to as the final control element. The valve percent opening is adjusted to correct for any deviations from the set point.

Figure 2.2 is an information flow diagram that corresponds to the physical process flow diagram in Figure 2.1. The information is transmitted between the different control system elements as either pneumatic, electronic or digital signals. These signals often use a live zero. Typical levels are 20–100 kPa (or 3–15 psi) for pneumatic signals and 4–20 mA current loop that is often converted to 1–5 V for analog electronic signals and binary digits or bits for digital signals.

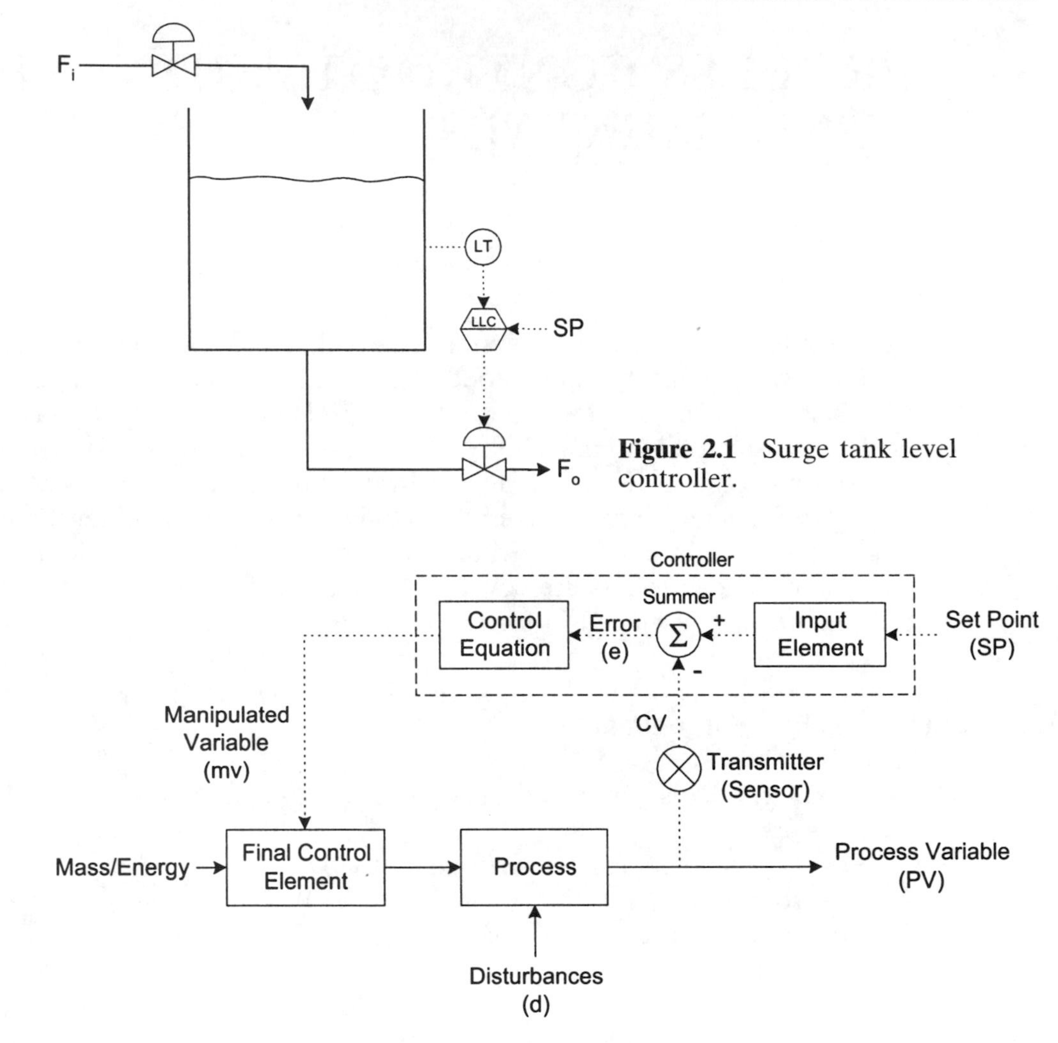

Figure 2.1 Surge tank level controller.

Figure 2.2 Single input/single output block diagram.

2.2 Primary elements

Primary elements, also known as sensor/transmitters, are the instruments used to measure variables in a process such as temperature, pressure, etc. A full listing of the types of primary elements available on the market would be very long, but these sensor types can be broadly classified into groups including the following:

1. Pressure and level
2. Temperature
3. Flow rate and total flow

4. Quality or analysis instruments (e.g. electrolytic conductivity, pH, pION, moisture, oxidation-reduction potential, gas analysers (O_2, CO_2, H_2) thermal conductivity, GLC, heat of reaction, calorific value).
5. Transducers (working with the above or as individual units).

Some specific examples of instruments from the more common groups listed above will be examined, including pressure, level, temperature, and flow. It is important to note that this list is not complete or fully representative of the complex developments in this area. Further information and details can be found in the references.

2.2.1 PRESSURE MEASUREMENT

There are numerous types of primary elements used for measuring pressure that could be studied; however, this discussion will be limited to some of the most common types encountered. These include manometers, Bourdon tubes, and differential pressure cells.

Manometers

Manometers are simple, rugged, cheap, and give reliable static measurements. They are therefore very popular as calibration devices for pressure measurement. The working concept of a manometer is simple. A fluid with a known density, ρ, is used to measure the pressure difference between two points, $P_1 - P_2$, based on Equation 2.1, where H is the height difference in the fluid level:

$$P_1 - P_2 = \rho g H \tag{2.1}$$

Figure 2.3 illustrates some of the different manometer types.

The Bourdon tube pressure gauge

The Bourdon tube pressure gauge, named after Eugene Bourdon (circa 1852) and shown in Figure 2.4, is probably the most common gauge used in industry. The essential feature of the Bourdon tube is its oval shaped cross-section. The operating principle behind the gauge is that when pressure is applied to the inside of the tube the tip is moved outward. This pulls up the link and causes the quadrant to move the pinion to which the pointer is attached. The resultant movement is indicated on a dial. A hairspring is also included (not shown) to take up any backlash that exists between quadrant and pinion; this has no effect on calibration.

The accuracy of the gauge is ±0.5% of full range for commercial models. When specifying it is usual to choose a value of 60% of full scale for normal working pressure.

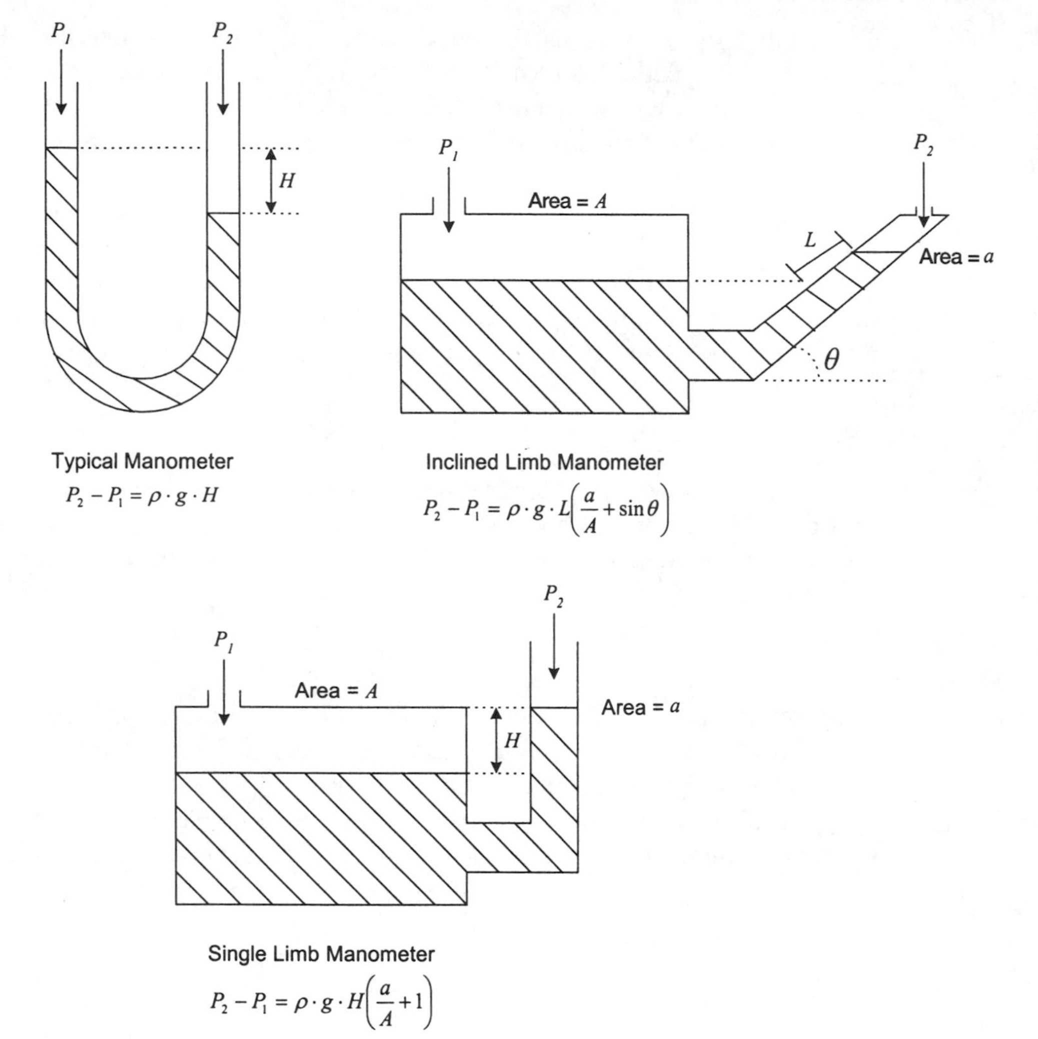

Figure 2.3 Various manometer types.

Other types of these gauges include the twist tube, spiral tube and helical tube. Diaphragm and bellows gauges are two other types of pressure sensors that were developed later. For more details on Bourdon tube materials and design refer to Giacobbe and Bounds [1], Goitein [2] and Considine [3].

The differential pressure cell

The differential pressure (DP) cell is considered by many as the start of modern automation. The DP cell was developed at the outbreak of World War II by Foxboro in Massachusetts, USA on a government grant provided

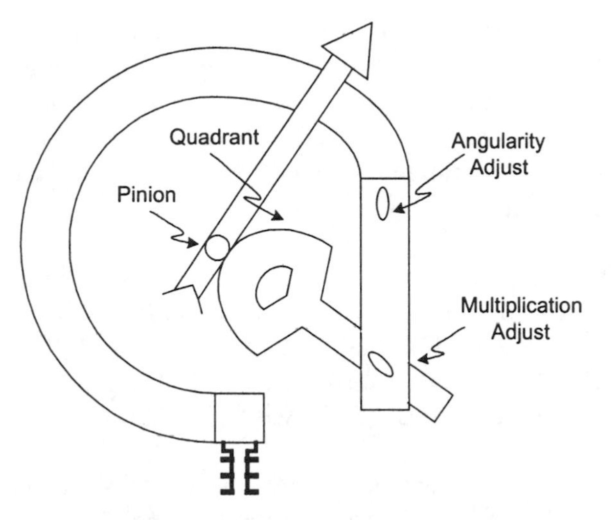

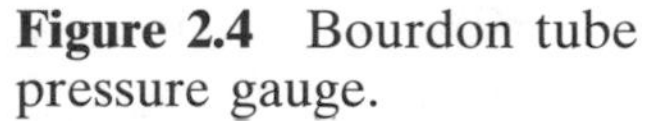
Figure 2.4 Bourdon tube pressure gauge.

that it was not patented. The idea was that competition would bring down the price of the instrument.

DP cells allow remote transmission to central control rooms where a small number of operators can control large, complex plants. For example, a typical petroleum refinery processing around 80,000 barrels/day (530 m^3/h) might have 2000 DP cells throughout the refinery.

Some of the major benefits of DP cells are that their maintenance is practically zero and no mercury is used in the operation of the transducer.

The pneumatic DP cell Figure 2.5 shows a schematic of a pneumatic DP cell.

Pressure is applied to the opposite sides of a silicone-filled twin diaphragm capsule. The pressure difference applies a force at the lower end of the force

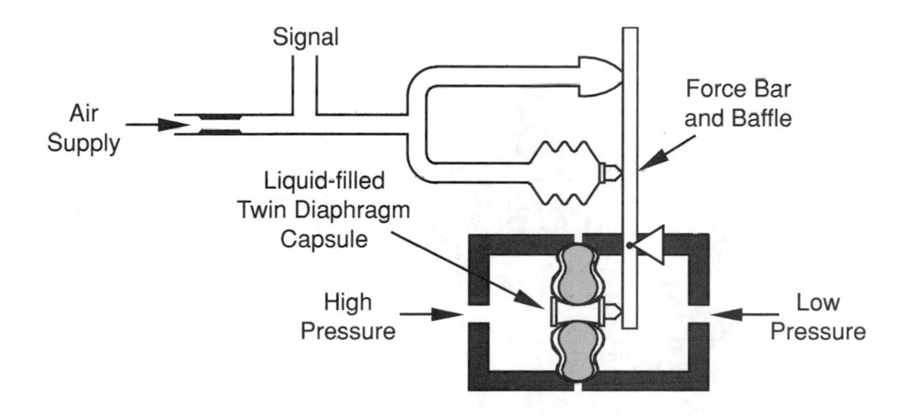

Figure 2.5 Pneumatic DP cell (courtesy of Rosemount Inc.).

bar, which is balanced through a simple lever system consisting of the force bar and baffle. This force exerted by the capsule is opposed through the lever system by the feedback bellows. The result is a 3 psi (or 20 kPa if calibrated in SI units) to 15 psi (or 100 kPa if calibrated in SI units) signal proportional to the differential pressure. The range of the cell is 1200 Pa to 210 kPa differential pressure with an accuracy of ±0.5% of the range.

Modern DP cells E-type electronic transducers, strain gauge, capacitive cell transducers and, most recently, digital electronics have replaced the pneumatic type DP cell. Figure 2.6 shows a schematic of an electronic DP cell.

Figure 2.7 is a picture of a classic electronic DP cell. Figure 2.8 is a picture of a modern electronic DP cell, Model 3051 with Foundation Fieldbus from Rosemount Inc.

The features of the modern electronic DP cell, such as the Rosemount Model 3051 or Honeywell's "smart" transmitter [4], include remote range change, diagnostics that indicate the location and type of any system faults, easy self-calibration, local digital display, reporting and interrogation functions, and local and remote reporting. The modern DP cell can also be directly connected to a process computer and has the ability to communicate with the computer indicating problem analysis that is then displayed on the computer screen.

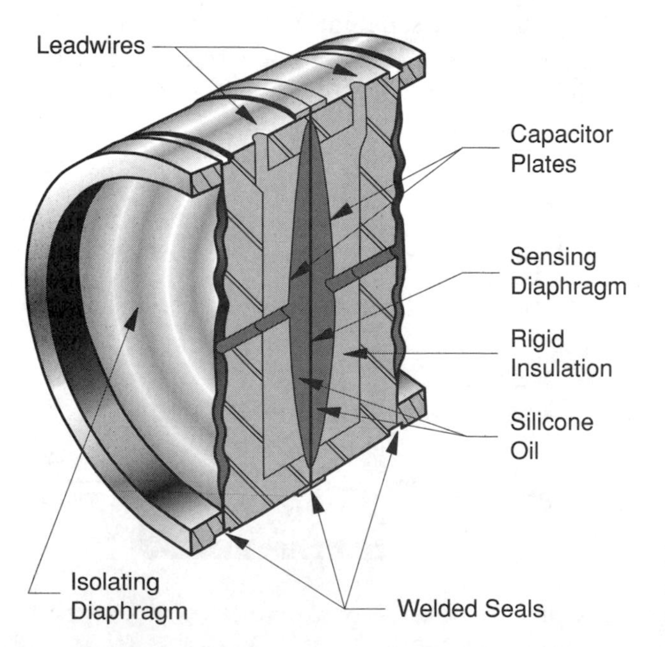

Figure 2.6 Electronic DP cell (courtesy of Rosemount Inc.).

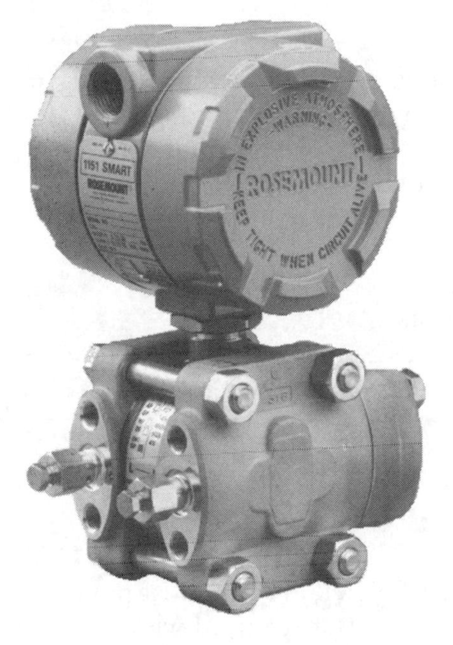

Figure 2.7 Model 1151 electronic DP cell (courtesy of Rosemount Inc.).

Figure 2.8 Model 3051 electronic DP cell (courtesy of Rosemount Inc.).

2.2.2 LEVEL MEASUREMENT

Level measurement is the determination of the location of the interface between two fluids which separate by gravity, with respect to a fixed plane. The most common level measurement is between a liquid and a gas. Methods of level measurement include the following [5]:

1. Float-actuated devices, such as:
   (a) Chain or tape float gauge
   (b) Lever and shaft mechanisms
   (c) Magnetically coupled devices
2. Pressure/head devices, i.e. DP cells or manometers:
   (a) Bubble tube systems
   (b) Electrical methods
3. Thermal methods.
4. Sonic methods.

Pressure/head devices such as the DP cell are the most popular of all level measurement devices. The DP cell can often be used where manometers are impracticable and floats would cause problems. Figure 2.9 demonstrates a typical setup for level measurement using a Rosemount Model 3095 level controller, which is essentially a combined DP cell and proportional controller.

2.2.3 TEMPERATURE MEASUREMENT

Methods of measuring temperature include [4]:

1. Change of state.
2. Expansion:
   (a) Bimetal thermostats
   (b) Liquid in glass
   (c) Liquid in metal
3. Pressure type:
   (a) Gas filled
   (b) Vapour pressure filled
4. Electrical:
   (a) Resistance
   (b) Thermocouple
5. Radiation pyrometers:
   (a) Total radiation
   (b) Optical

Bimetal thermostats, thermocouples and resistance thermometers will now be discussed in detail.

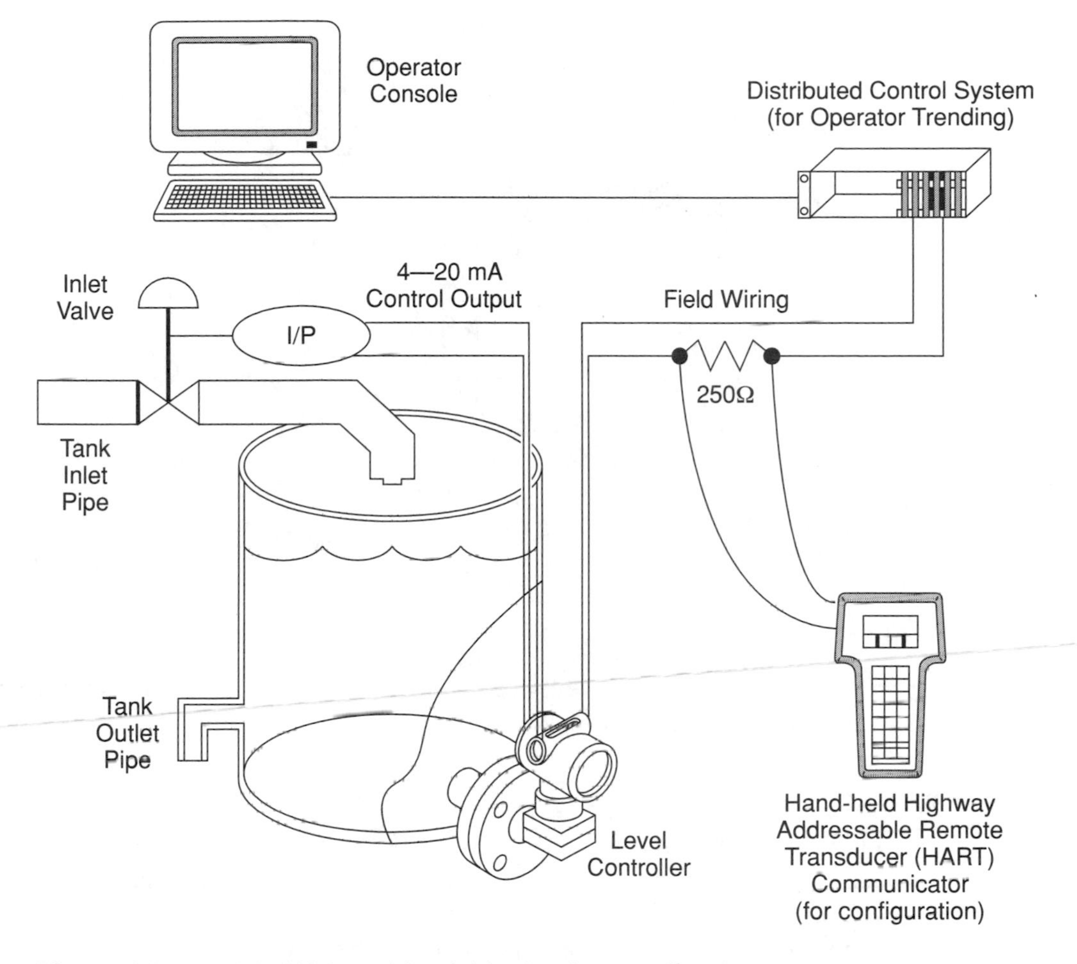

Figure 2.9 Model 3095 multivariable level controller in a liquid level process (courtesy of Rosemount Inc.).

Bimetal thermostats

The bimetal thermostat works on the concept that different metals expand by different amounts if they are subject to the same temperature rise. If two metals are fixed rigidly together, then a differential expansion takes place when the metals are heated, causing the composite bar to bend. The thermostat employs the bimetal bar to switch on or off a control device depending on the temperature. An illustration of a bimetal thermostat is given in Figure 2.10.

The temperature range for bimetal thermostats is 0 to 400°C with an accuracy of ±5%, although the accuracy can be increased to ±1% [5]. The deflection/temperature relationship is linear for many metal combinations over a particular temperature range only, and the materials must be chosen

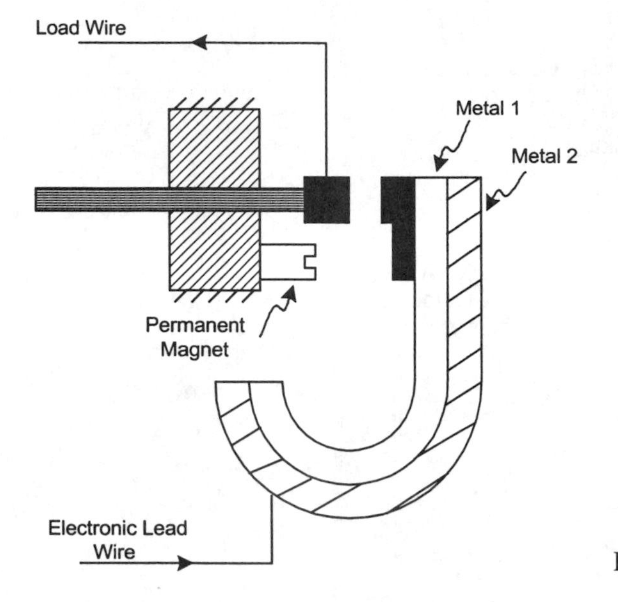

Figure 2.10 Bimetal thermostat.

with care. These instruments are rugged, cheap, offer direct reading, and can work under conditions of vibration.

Thermocouples

When two dissimilar metal or alloy wires are joined together at both ends to form a loop, and a difference in temperature exists between the ends, a difference in junction potentials exists resulting in a thermoelectric electromagnetic field (emf). This is known as the Seebeck Effect, after Seebeck's 1821 discovery of this phenomenon. The magnitude of the emf will depend on the types of materials used and the temperature difference. This is the concept behind a thermocouple for measuring temperature.

If one junction temperature, the reference or cold junction, is maintained at a constant and known value and the characteristics of the thermocouple are known, then the magnitude of the emf generated will be a measure of the temperature of the other junction. This other junction is called the hot junction.

The emf generated for any two particular metals at a given temperature will be the same regardless of the size of the wires, the areas in contact or the method of joining them together. The relationship between temperature and generated emf is nonlinear except over limited ranges. On the steep part of the curve, the relationship is:

$$e = a(T_1 - T_2) + b(T_1^2 - T_2^2) \tag{2.2}$$

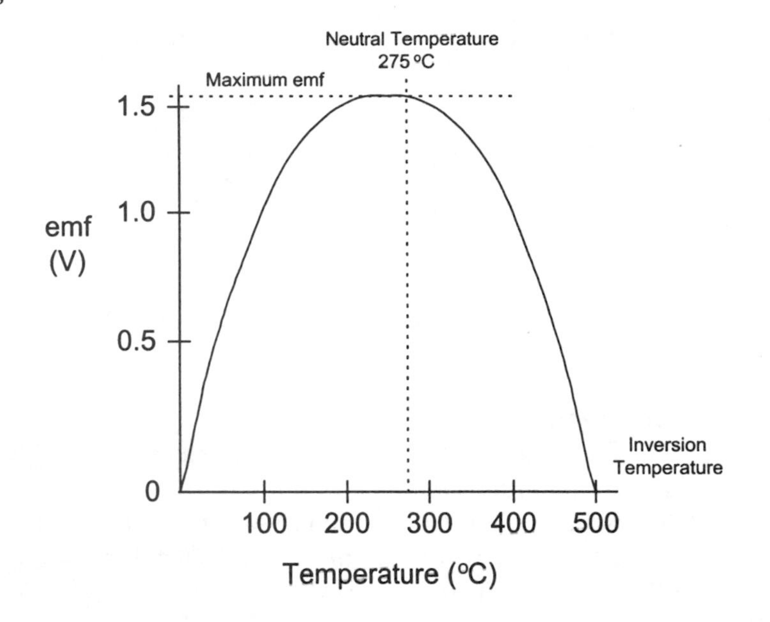

Figure 2.11 Relationship between temperature and emf for a Cu/Fe system.

In Equation 2.2, e is the generated emf, T_1 and T_2 are the hot and cold junction temperatures in degrees Kelvin, and a and b are constants for the given material. An example of the relation is given in Figure 2.11 for a Cu/Fe thermocouple system.

Peltier effect Of interest to note here is the phenomenon of the Peltier effect (1834). When a current is flowing through a junction, absorption or liberation of heat takes place depending on the direction of the current flow. When the current flows in the same direction as the Seebeck current, heat is absorbed at the hotter junction and liberated at the colder junction. This effect is defined as the change in heat content when one Coulomb of charge crosses the junction. This effect is the basis for thermoelectric refrigeration and heating [3].

Thermocouple types There are many thermocouple types. Common systems and their ranges are as follows:

Base metal thermocouple types

1. Constantan/Copper – 0 to 400°C (continuous) or to 500°C (intermittent), Type T.
2. Constantan/Chromel – 0 to 700°C (continuous) or to 1000°C (intermittent), Type E.
3. Constantan/Iron – 0 to 850°C (continuous) or to 1100°C (intermittent), Type J.

4. Alumel/Chromel – 0 to 1100°C (continuous) or to 1300°C (intermittent), Type K

Noble metal thermocouple types

1. Platinum/Platinum 13% Rhodium – 0 to 1400°C (continuous) or 0 to 1600°C (intermittent), Type R.
2. Platinum 5% Rhodium/Platinum 20% Rhodium – 0 to 1500°C (continuous) or 0 to 1700°C (intermittent).
3. Tungsten 20% Rhemum/Tungsten – 0 to 2500°C (continuous) or 0 to 2800°C (intermittent).

Poisons to thermocouples

- Iron (Fe) deteriorates quickly due to scaling in oxidising atmospheres at high temperatures.
- Chromel and alumel thermocouples are poisoned by gases that are carbon based, sulphurous, or contain cyanide groups. These thermocouples are better in an oxidising atmosphere than a reducing atmosphere.
- Tungsten should only to be used in reducing or inert atmospheres.
- Platinum must be protected from hydrogen and metallic vapours.

Resistance thermometer detectors (RTDs)

Resistance thermometer detectors are made of either metal or semiconductor materials as resistive elements that may be classed as follows [3]:

1. Wire wound – range −240°C to 260°C, accuracy 0.75%.
2. Photo etched – range −200°C to 300°C, accuracy 0.5%.
3. Thermistor beads – range 0°C to 400°C, accuracy 0.5%.

An example is the platinum resistance thermometer detector, which is the most accurate thermometer in the world.

RTD temperature sensors exhibit a highly linear and stable resistance versus temperature relationship. However, resistance thermometers all suffer from a self-heating effect that must be allowed for, and I^2R must be kept below 20 mW, where I is defined as the electrical current and R is the resistance.

When compared to thermocouples, RTDs have higher accuracy, better linearity, long term stability, do not require cold junction compensation or extension lead wires and are less susceptible to noise. However they have a lower maximum temperature limit and are slower in response time in applications without a thermal well (a protective well filled with conductive material in which the sensor is placed).

Selecting temperature sensors

Getting the right operating data is crucial in selecting the proper sensor. A good article on selecting the right sensor is by Johnson [6].

Figure 2.12 shows a selection of thermocouples, resistance thermometers (RTDs), and temperature accessories, such as thermal wells, that are typically available from instrument suppliers (in this case Rosemount Inc.). Figure 2.13 shows a picture of a typical temperature sensor and transmitter assembly.

2.2.4 FLOW MEASUREMENT

Flow measurement techniques can be divided up into the following categories [13]:

1. Obstruction type meters, such as:
   (a) Orifice plates
   (b) Flow nozzles
   (c) Venturi tubes
   (d) Pitot tubes
   (e) Dall tubes

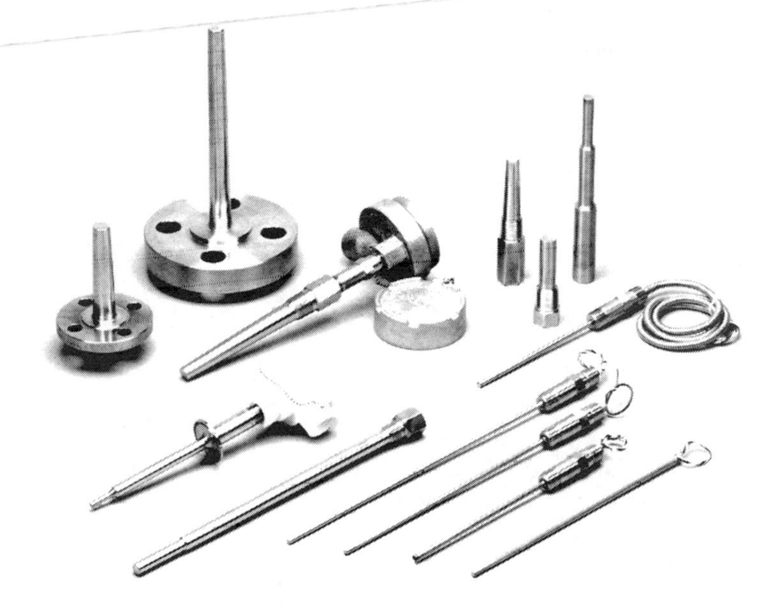

Figure 2.12 Selection of thermocouples, RTDs, and accessories (courtesy of Rosemount Inc.).

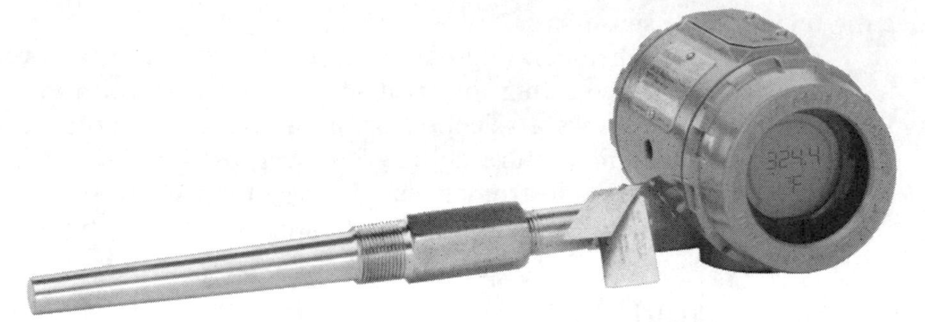

Figure 2.13 Temperature sensor and transmitter assembly (courtesy of Rosemount Inc.).

   (f) Combinations of (a) to (e)
   (g) Elbow and target meters
2. Rotational or turbine meters.
3. Variable area meters/rotameters.
4. Ultrasonic and thermal type meters.
5. Square root extractors for obstruction type meters.
6. Quantity or total flow meters, such as:
   (a) Positive displacement
   (b) Sliding vane
   (c) Bellows type
   (d) Nutating disc
   (e) Rotating piston
   (f) Turbine type
7. Magnetic flowmeters.
8. Vortex meters.
9. Mass flowmeters, such as:
   (a) Coriolis effect flowmeters
   (b) Thermal dispersion flowmeters

Of these, orifice plates and magnetic flowmeters will be discussed since they are the two most commonly found types in the fluid processing industry.

Orifice plates

The concentric orifice plate is the least expensive and the simplest of the head meters. The orifice plate is a primary device that constricts the flow of a fluid to produce a differential pressure across the plate. The result is proportional to the square of the flow. Figure 2.14 shows a typical thin plate orifice meter.

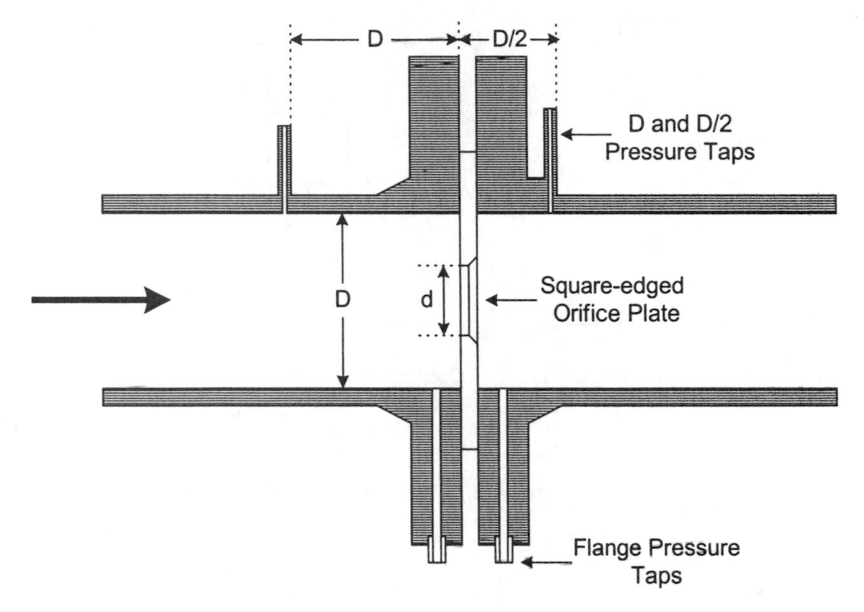

Figure 2.14 Thin plate orifice meter.

An orifice plate usually produces a larger overall pressure loss than other primary devices. A practical advantage of the orifice plate is that cost does not increase significantly with pipe size. They are used widely in industrial applications where line pressure losses and pumping costs are not critical.

The thin concentric orifice plate can be used with clean homogenous fluids, which include liquids, vapours, or gases, whose viscosity does not exceed 65 cP at 15°C. In general the Reynolds number (Re) should not exceed 10,000. The plate thickness should be 1.5 to 3.0 mm, or in certain applications up to 4.5 mm [7].

Many variations for orifice plates have been suggested, especially during the 1950s when oil companies and universities in North America and Europe sponsored numerous PhD studies on orifice plates [3]. Of these only a few have survived, which were the ones that incorporated cheaply some of the features of the more expensive devices. Figure 2.15 shows some of these designs. Other designs that are utilised include eccentric and segmental orifice configurations.

Magnetic flowmeters

The magnetic flowmeter is a device that measures flow using a magnetic field, as implied by the name. The working relationship for magnetic flowmeters is based upon Faraday's law (see Equation 2.3) which states that a voltage will be induced in a conductor moving through a magnetic field.

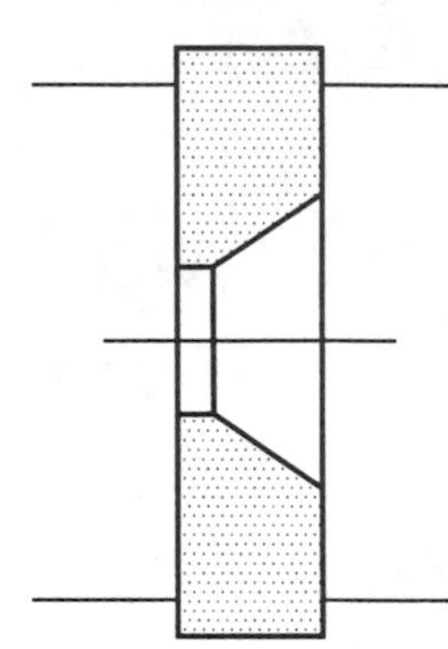
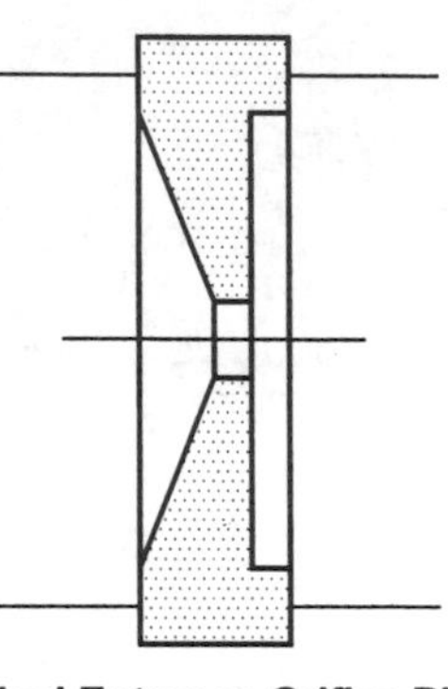
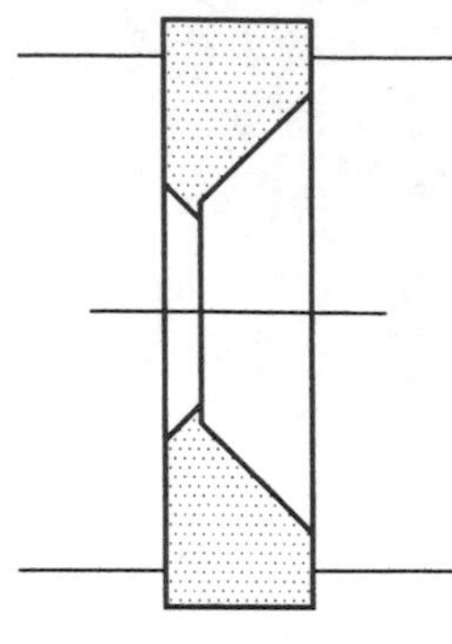
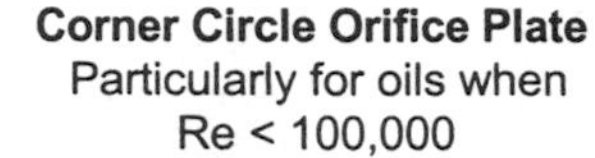

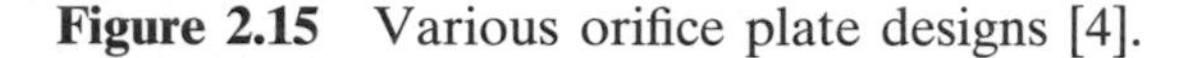
Figure 2.15 Various orifice plate designs [4].

$$E = k \cdot B \cdot D \cdot V \tag{2.3}$$

In Equation 2.3, E is the generated emf, B is the magnetic field strength, D is the pipe diameter, V is the average velocity of the fluid, and k is a constant of proportionality. As seen in Equation 2.3, when k, B, and D are kept constant, V is proportional to E.

Figure 2.16 illustrates the principle of operation of a magnetic flowmeter.

In the past, magnetic flowmeters have been very expensive compared to orifice plates and DP cells. However, now the cost is very competitive, and in fact magnetic flowmeters are replacing orifice plates where possible.

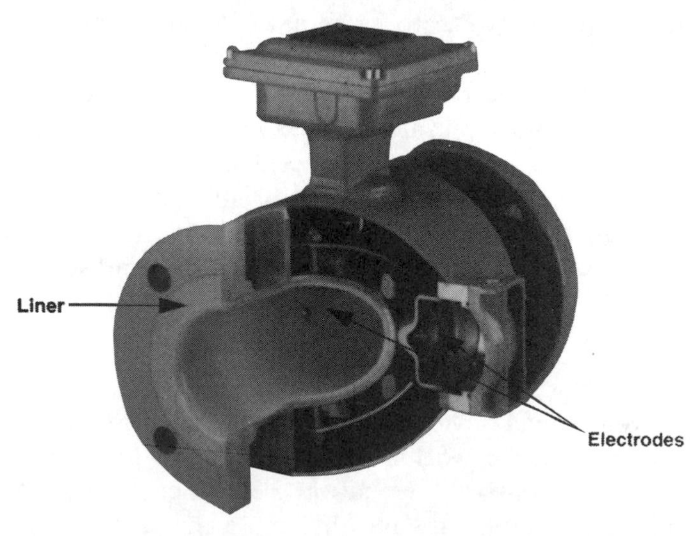

Figure 2.16 Cutaway view of a magnetic flowmeter flowtube (courtesy of Rosemount Inc.).

There are numerous benefits to using a magnetic flowmeter. With polytetrafluoroethylene (PTFE), fibreglass or rubber liners, the magnetic flowmeter can handle almost any corrosive liquid. The electrodes can be made from very corrosive-resistant metals. Gold, titanium and tantalum have been used in the past. The magnetic flowmeters are virtually maintenance free, and there is no flow obstruction to the stream being measured. Also, they can be readily connected to an electronic controller and they can give out a digital signal that can be directly fed to a computer.

When using a magnetic flowmeter it is necessary for the liquid to be conductive, although low conductivities are acceptable. Also, outside capacitance can create a big problem. Calibration should be done carefully initially, with accurate readings done on the liquid conductivity to ensure accurate set-up of the magnetic flowmeter.

2.3 Final control elements

Pneumatic, or air operated, diaphragm control valves are the most common final control element in process control applications. They are used to regulate the flow of material or energy into a system. Variable speed pumps are also possible but are often costly as motor control is expensive, they are less efficient, break down more often, and maintain maximum pressure if they fail. Electric valves are seen but only for large applications above 25 cm pipe/valve diameters. Variable electric power control elements such as rheostats are used in small applications such as laboratory water bath temperature control.

Since control valves are the most common final control element we will now devote our discussion to control valves.

2.3.1 CONTROL VALVES

The sliding stem control valves are the most common control valve configuration and have at least half the market in control valves. Figures 2.17a and b show a typical, modern sliding stem control valve assembly.

The pneumatic diaphragm-operated control valve is the commonly specified final control element in existence. Pneumatic valves have many advantages over electrically activated valves, but the main ones are initial lower purchase cost, relative ease of maintenance, speed of response, and developed power of the valve plug. This last reason has become less relevant since valve bodies have changed from contoured and ported styles to plug and cage styles in order to avoid unbalanced forces in single ported designs, especially for high pressure liquids. Figure 2.18 shows some of the newer styles of valve cages.

Figure 2.17a Typical modern sliding stem control valve assembly (courtesy of Fisher Controls International, Inc.).

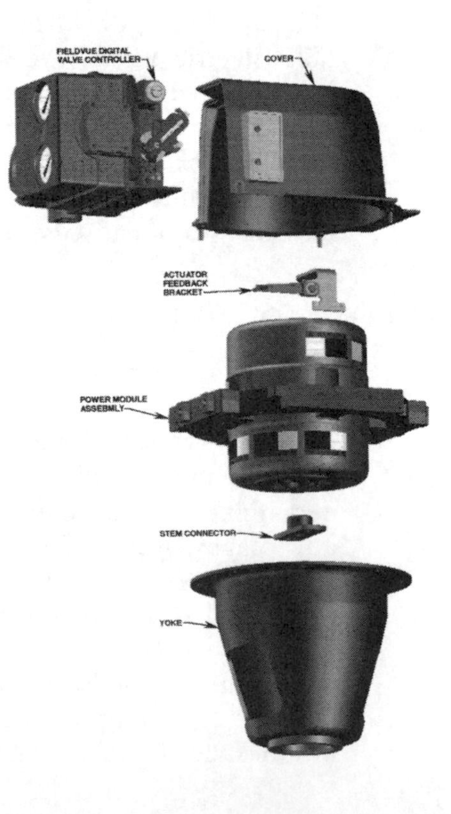

Figure 2.17b Single-acting spring return actuator and digital valve positioner for a modern sliding stem control valve (courtesy of Fisher Controls International, Inc.).

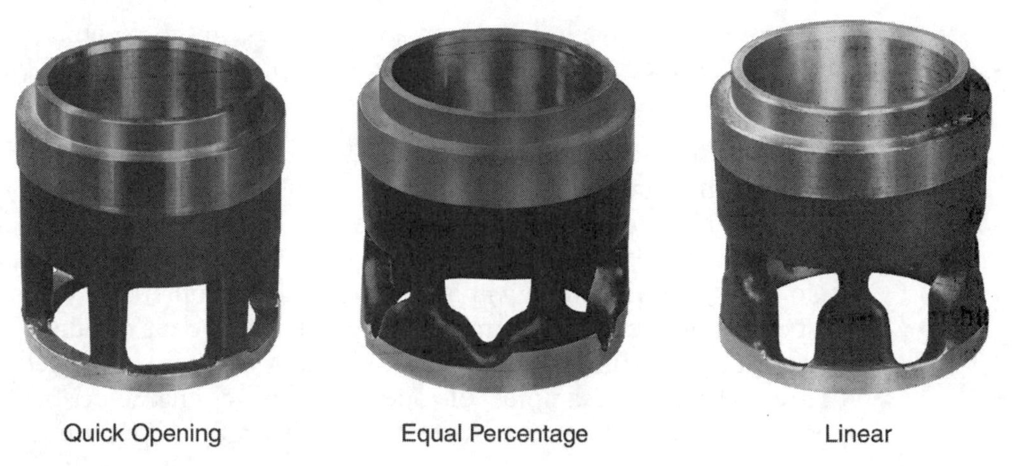

Figure 2.18 Characterised cages for globe style valve bodies (courtesy of Fisher Controls International, Inc.).

Control valve sizing

The common equation for the flow of a non-compressible fluid through a control valve is given in Equation 2.4 [8], which can be derived from Bernoulli's equation:

$$Q = C_v \sqrt{\left(\frac{\Delta P}{SG}\right)} \tag{2.4}$$

In Equation 2.4, Q is the volumetric flow rate, ΔP is the pressure drop across the valve, SG is the relative density compared to water, and C_v is the valve coefficient. C_v is defined by convention in field or imperial units as the number of US gallons that will pass through a control valve in one minute, when the pressure differential across the valve is 1 psi.

C_v varies negligibly with Re for most valve applications. Even in cases where the Reynolds number is low, the Reynolds number at the valve will be high due to the valve restricting flow, and so the valve is normally operating in a region where C_v is independent of the Reynolds number. For valves with a streamlined shape such as those used for slurries or very viscous liquids, the Reynolds number can be low so C_v becomes dependent. In these cases a correction factor for the low C_v is usually supplied in manufacturers' catalogues under the heading "Viscosity Correction Factors for C_v" [8, 9, 10, 11, 12].

The value of C_v is also a function of A, which is the flow area of the valve. For a given valve, this value of A varies extensively with valve opening. The curve giving the variation of C_v at high Reynolds number with valve opening is called the "inherent characteristic of the valve". The maximum value of C_v

occurs when the valve is wide open and depends on the design and size of the valve. For geometrically similar valves, C_v is proportional to the valve size.

Inherent valve characteristic

The inherent characteristic of a valve is a plot of C_v versus valve opening. This curve is usually plotted as C_v in percentage of maximum flow (or C_v). Inherent characteristics are usually plotted in this way rather that actual C_v versus actual lift so that the same curve will apply to a set of geometrically similar valves, irrespective of size. If the characteristic curve and the maximum C_v are known then the C_v at any intermediate lift or opening can be determined.

Three common examples of operating valve characteristics are quick opening, linear, and equal percentage, as illustrated in Figure 2.19.

Operating characteristic

The operating characteristic is a plot of flow versus lift for a particular installation. This is not an inherent property of the valve and is usually plotted as flow versus lift in a similar way to the inherent characteristic, with both the flow and the lift being plotted as a percentage of the maximum.

If the pressure drop did not change across the valve with valve opening, then the flow would vary proportionally with C_v, and thus the operating characteristic would be the same as the inherent characteristic curve if both were plotted as a percentage of the maximum. However, as the valve closes, the pressure drop across it increases. This increased ΔP is due to the fact that as the valve resistance increases, the valve's resistance becomes a larger fraction of the total system resistance. This is because as the valve closes, flow through the system decreases and the system ΔP other than the valve decreases but the valve ΔP increases. This means that as the valve closes, its C_v falls but its ΔP increases. The result is that the flow does not fall as fast as the C_v, and so the installed or operating flow characteristic differs from the inherent characteristic.

When the valve is shut its resistance is infinite and the whole available pressure drop occurs across it. Thus the ΔP across the valve varies from a maximum when closed, to a minimum when 100% open. The greater this variation, the more the operating characteristic varies from the inherent characteristic. A measure of this deviation is the parameter, β, defined by:

$$\beta = \frac{\Delta P_v \text{ (max. flow)}}{\Delta P_v \text{ (min. flow)}} \tag{2.5}$$

The smaller the value of β, the greater the deviation of the operating characteristic from the inherent characteristic. For a small β, an operating characteristic like that shown in Figure 2.20 is obtained.

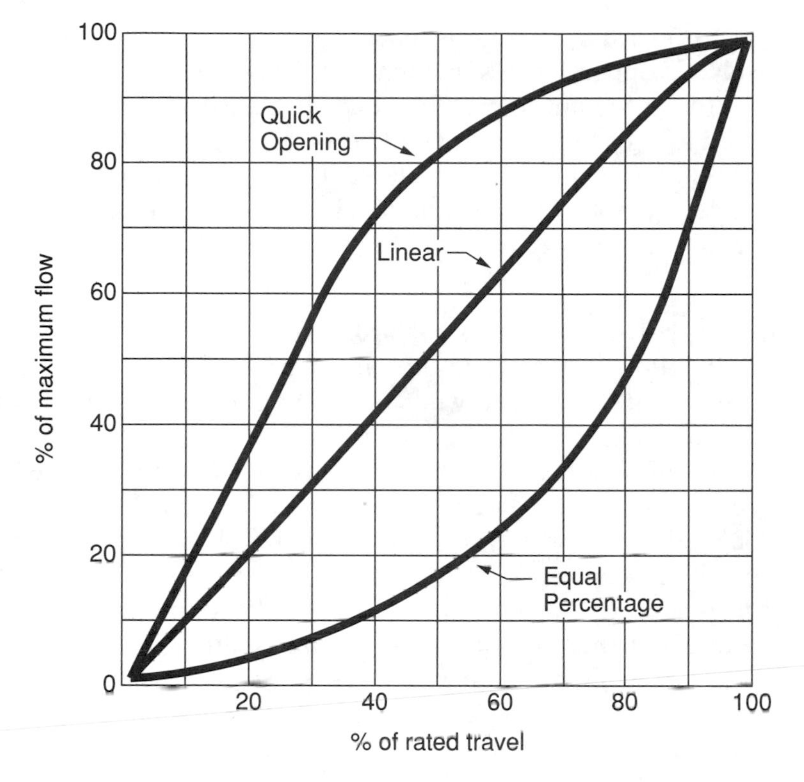

Figure 2.19 Examples of inherent valve characteristic curves (courtesy of Fisher Controls International, Inc.).

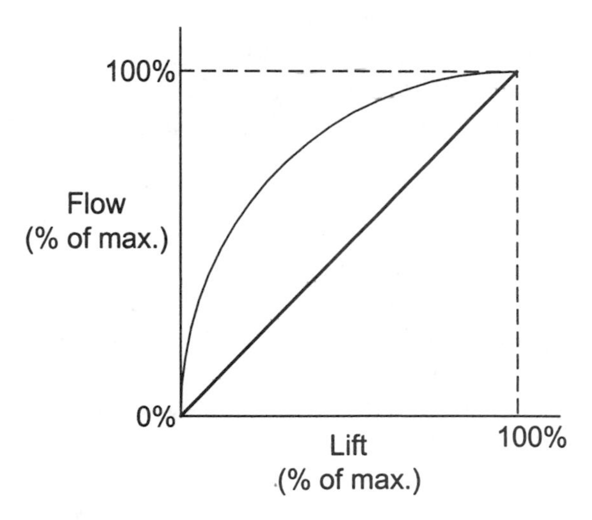

Figure 2.20 Operating characteristic for a small value of β.

With such a characteristic as a small β, nearly full flow is obtained when the valve is opened by a small amount. Thus, the effective stem movement range for throttling flow is greatly reduced and erosion of the valve is increased since the plug is nearly closed at all flows. Therefore, a small β is undesirable and is caused by a valve that is too large. Decreasing the valve size increases the value of ΔP_v (open). This, of course, increases the required pumping costs because of the extra power required to overcome the greater resistance of the valve.

Valve selection based on control performance

From the point of view of control performance, there are two aspects that need to be considered when selecting a valve. These are valve size and valve inherent characteristic.

As previously stated, these two aspects are not independent since the flow characteristics obtained depend on both. Ideally, the valve size should be decided during the design phase of the plant in conjunction with the choice of pipe and pump size. In this way, it can be ensured that the valve pressure drop is a satisfactory proportion of the total pressure drop, and thus will produce a satisfactory operating characteristic.

It is usually recommended that β be at least 30%. In satisfying this relationship the valve is seldom the same size as the pipe, and the valve will usually be one size smaller, with the minimum recommended size being 50% of the line size. When the control valve is added to an existing system, it is often sized to handle the maximum required flow with the available pressure drop. However, this generally results in an oversized valve since the pump was not originally designed for losses in the valve. This in turn then leads to a poor valve characteristic and unsatisfactory control. For this reason, an equal percentage characteristic should be selected so that the operating characteristic tends towards linear. In addition, the equal percentage characteristic is more forgiving when sized incorrectly or if there is insufficient pressure drop across the valve. The equal percentage valve plug has become the standard type of plug when valves are put into an existing system.

If the control system is included as part of the original design, an equal percentage may not be the best choice, and a linear flow characteristic may also not be suitable. If the process is not linear, then the required open loop gain to obtain a given degree of stability will vary with the operating point. With a nonlinear operating characteristic the valve gain will also vary with the operating point, and so it may be possible to match the valve's characteristic to that of the rest of the control system to produce roughly a constant system gain at all operating points.

This matching of the valve characteristic to that of the process is only relevant if the operating point of the process does not vary over the whole range, i.e. it is not subject to large disturbances. In this case, there may be no way

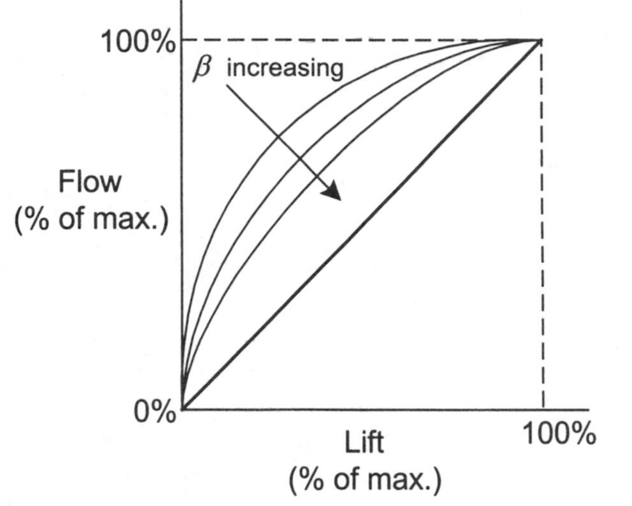

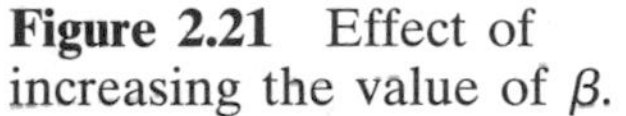
Figure 2.21 Effect of increasing the value of β.

of matching the valve to the process if there is more than one variable that produces appreciable changes to the operating point. This is because the best characteristic for the compensation of the effects of one load variable may be different from that required for another. However, if there is only one load variable, it is often possible to determine the best shape of the valve's inherent characteristic by the use of process dynamics.

Valve selection based on process dynamics

Gain is defined as the change in output divided by the change in input. Each component in the control loop has a gain term associated with it. The control valve has a very clearly defined gain term that depends on valve type, size, pressure drop, etc. The process gain term depends upon the process response to a change in input and the various load disturbances imposed upon it. A control system should be designed such that a controller produces an effect equal to the disturbance but 180° out of phase, to bring about cancellation. Thus, for good control the loop gain should be unity as shown in equation 2.6. If the gain is less than unity then the disturbance is not fully cancelled, and if the gain is greater than unity then the corrective action is excessive. This concept of gain is explained in greater detail in Chapter 3.

$$K_{controller} \cdot K_{transmitter} \cdot K_{process} \cdot K_{valve} = 1 \quad (2.6)$$

For a given set of controller settings, the controller gain and the sensor/transmitter gains may be considered as constants, resulting in the relationship of equation 2.7.

$$K_{process} \cdot K_{valve} = \text{Constant} \quad (2.7)$$

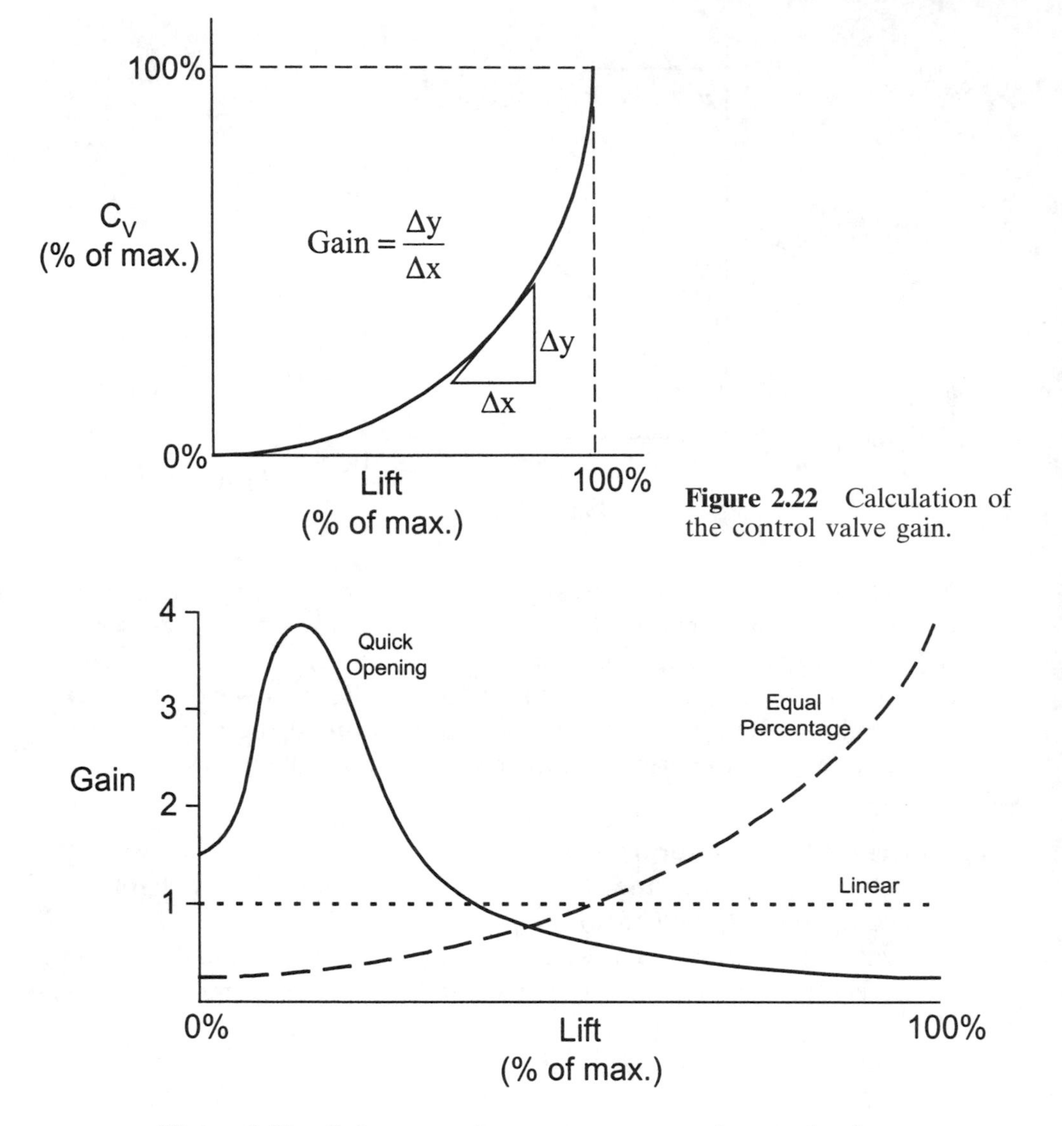

Figure 2.22 Calculation of the control valve gain.

Figure 2.23 Gain curves for common types of control valves.

The gain of the control valve can be computed from the C_v versus lift curves at the slope of the tangent to the curve, as shown in Figure 2.22.

If this is done for the common types of control valves over their whole range then the gain curves shown in Figure 2.23 are obtained.

Quick opening The gain increases to a maximum at about 20% of the lift from where it decreases exponentially resulting in decreasing effectiveness as the valve approaches a fully open position. This indicates that the valve would be good for a process whose gain increases with the variable used to control it.

Linear The gain of the truly linear valve is a constant not depending on lift at all. The flow is proportional to the valve position. For example, at 50% open the flow is 50% of the maximum. This characteristic would suit a plant whose gain was independent of the operating point.

Equal percentage The gain increases exponentially for equal percentage lifts between 10 and 100%. The use of this inherent characteristic is clearly best used for a process whose gain decreases as the load increases.

There are however rules of thumb for selecting control valves and matching inherent valve characteristics to common process control loops or processes where the valve pressure drop is fairly constant. The equal percentage characteristic is the most common and is used where variations in pressure drop are expected or in systems where a small percentage of the total system pressure drop is taken across the valve, such as in pressure and flow control. More detailed recommendations are available from control valve vendors (eg. [9, 10]).

Quantitatively, β is used to recommend a valve characteristic type. If $\beta > 0.5$, implying relatively less variation in pressure drop as the valve operates, then a linear characteristic is recommended. If $\beta < 0.5$, implying relatively more variation in pressure drop as the valve operates, then an equal percentage is recommended. Ultimately, however, valve gain is used to check the valve selection. For a selected valve, the procedure is as follows:

1. Calculate ΔP across the valve for several flows from the minimum to the maximum flow.
2. Calculate the corresponding C_v values (C_g for gases/vapour, or C_s for steam).
3. Obtain the % open data corresponding to the C_v (or C_g etc.) values calculated in Step 2 (vendor data C_v versus % open).
4. Calculate the control valve gain as: $K_{cv} = \%\Delta Flow/\%\Delta Opening$
5. Calculate the average gain.
6. Are the minimum and maximum control valve gains within ±50% of the average value calculated in Step 6? If the answer is yes then there will be good stable control. If the answer is no then check other valves' characteristics.

Control valve rangeability

The required range of C_v should fall between 20% and 80% of valve opening. This tends to provide a relatively constant gain and the most stable range of control. This can be checked by doing an analysis of the valve gain as discussed previously and in Purcell [13]. At less than 10% open, it can be difficult for the control valve to stabilise and it tends to oscillate, while at greater than 80% opening the gain begins to vary too much.

Control valve pressure drop

Control valves control flow by absorbing a pressure drop, which must be specified. This pressure drop is an economic loss to the system since, typically, it must be supplied by a pump or compressor. As such, economics might dictate sizing a control valve with a low pressure drop but this results in a larger valve which may have a decreased range of control. Often the pressure drop to be taken across the control valve is specified when detailed plant hydraulics are not complete and so it needs to be estimated. As such, there are several rules of thumb. Typically, the control valve pressure drop is estimated as 50% of the friction pressure drop taken across the equipment plus piping or 33% of the total system pressure drop (excluding the valve). Minimum pressure drops have been stated as 10% of the system pressure drop for equal percentage valves or 25% of the system pressure drop for linear valves [14] or 35 kPa for rotary valves and 69 kPa for globe valves [13].

As stated previously, the key to sizing control valves properly is to specify the range across which they have to function. Specifying a pressure drop with the above rule of thumb at one condition means that the pressure drop required at other conditions must be checked. The required pressure drops at other conditions are governed by system hydraulics. For example, assume that we have a system pressure drop of 50 kPa. Based on taking 33% of the total system drop at design flow, excluding the valve, a pressure balance results in the control valve taking 25% of 50 kPa i.e. 12.5 kPa, while the system takes 37.5 kPa. For simplicity, assume there is no elevation component. What happens if the flow increases by 20% above the design value? Hydraulics states that the system pressure drop will increase to 72 kPa. Where is this pressure drop going to come from? The answer is the control valve but in this case we do not have sufficient pressure drop to supply it and so this system could not have an increase of 20% flow above design. The rule of thumb pressure drop should be at maximum flow and then you should check what happens at minimum flow. A more engineered approach is derived by Connell [15] with the results given below. In this approach, the pressure drop across a control valve can be estimated using Equation 2.8:

$$\Delta P_{cv} = 0.05P_s + 1.1\left[\left(\frac{Q_m}{Q_d}\right)^2 - 1\right]\Delta P_f + \Delta P_b \tag{2.8}$$

where: ΔP_{cv} = pressure drop across a control valve, kPa
P_s = upstream or supply pressure, kPa
Q_m = maximum anticipated flow rate

Table 2.1 Base (minimum) pressure drops for control valve types

Control valve type	ΔP_b (kPa)
Single plug (globe)	75.8
Double plug (globe)	48.3
Cage	27.6
Butterfly	1.4
Ball	6.9

Q_d = design flow rate
ΔP_f = friction pressure drop at the design flow rate, kPa
ΔP_b = the base (minimum) control valve pressure drop, kPa

The first term accounts for a falloff in overall system pressure drop by using 5% of the system start pressure. The second term accounts for an increase in system flow from design to maximum and the corresponding friction pressure drop. The last term is the base (minimum) control valve pressure drop from [15], given in Table 2.1.

It should be noted that the values for butterfly and ball valves in Table 2.1 appear to be somewhat lower than what others recommend. As such, a minimum pressure drop of 27.6 kPa should be used unless experience indicates a specific low pressure drop application, for example a sulfur plant air control valve.

Practical control valve sizing

Note that to calculate the range of size required, the following sizing procedures require the pressure drop at minimum flow and at maximum flow, not just at design flow or an arbitrary multiple thereof. However, conditions at design flow can also be incorporated and helpful. Note that the procedures use the equations developed by Fisher [9, 10] but the sizing procedures are generic. In order to size a control valve properly, the following process information must be known:

- fluid type and viscosity;
- range of controlled flows (minimum and maximum);
- range of inlet and outlet pressures (minimum and maximum pressure drops corresponding to flows);
- specific gravity (SG); and
- temperature.

In addition, the following data need to be specified before a valve is purchased:

- shut-off pressure;
- leakage rate (ANSI/IEC Leakage Class [8, 11]); and
- noise tolerance (ANSI/IEC Standard [8, 11]).

Currently manufacturers worldwide are implementing the IEC Valve Sizing Code [11]. This is a procedure that allows tighter noise prediction, particularly in gas service.

Liquid control valve sizing

The basic procedure for sizing is, for a given flow rate and pressure drop, to calculate the required C_v as per a rearrangement of Equation 2.4:

$$C_v = \frac{Q}{\sqrt{\Delta P / SG}} \tag{2.9}$$

This calculation should be performed at minimum and maximum conditions to obtain $C_{v,\min}$ and $C_{v,\max}$. Subsequently, these required values are compared to a C_v range for a particular valve. Typically, the required C_v values should fall in the range of about 20 to 80% of the valve opening.

A plot of Equation 2.4 implies that the relationship of flow is linear with respect to the square root of pressure drop with the slope equal to C_v, and

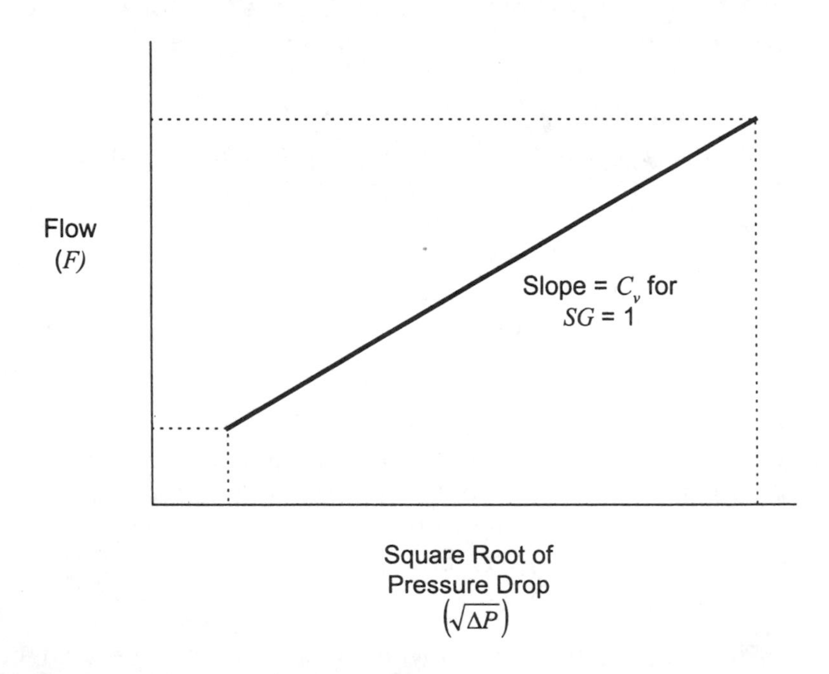

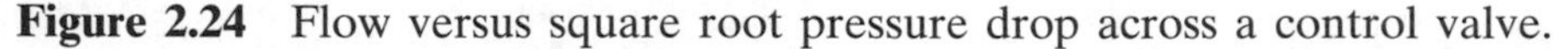

Figure 2.24 Flow versus square root pressure drop across a control valve.

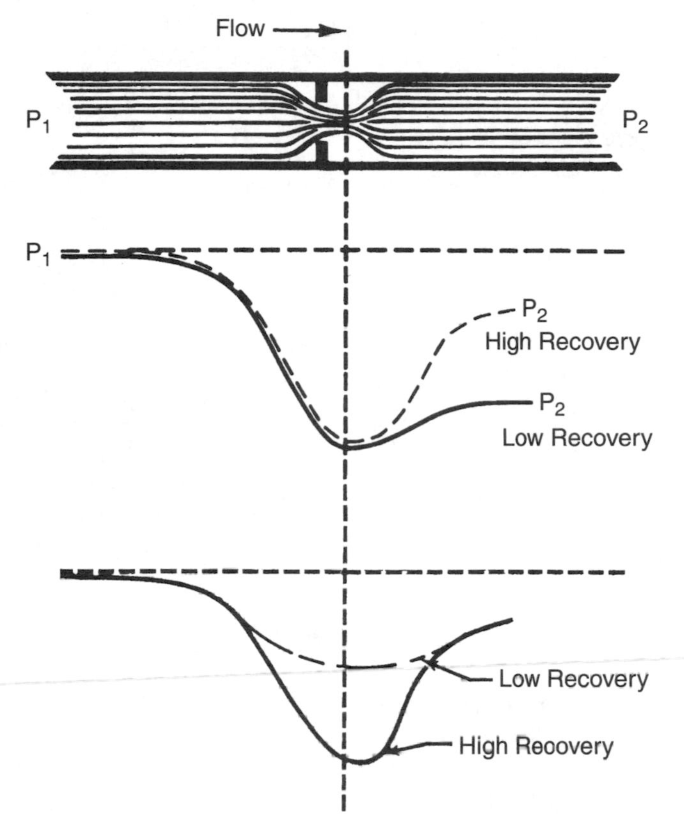

Figure 2.25 Pressure profiles across a control valve (courtesy of Fisher Controls International, Inc.).

that the flow can be continually increased with pressure drop.

Actually, a limit is reached where this is no longer true, known as choked flow. As liquid passes through a reduced cross-sectional area, the velocity increases and the pressure decreases. The point of minimum pressure and maximum velocity is the *vena contracta.* As the fluid exits, the velocity is restored but the pressure is only partially restored, creating a pressure drop as shown in Figure 2.25. Note that pressure recovery is how much pressure is restored from the vena contracta. As such, all else being equal, a high recovery valve has a low pressure drop and vice versa.

If the minimum pressure falls below the vapour pressure of the liquid, it partially vaporises. This is what causes cavitation and flashing, discussed in more detail later. At the vena contracta, as the pressure decreases, the density of the vapour phase, and mixture, decreases. Eventually, this decrease in density offsets any increase in the velocity of the flow so that no additional mass flow is realised, according to the continuity equation.

The issue becomes how choked flow is integrated into liquid valve sizing. Fisher Controls [9, 10] defines a pressure recovery coefficient as follows:

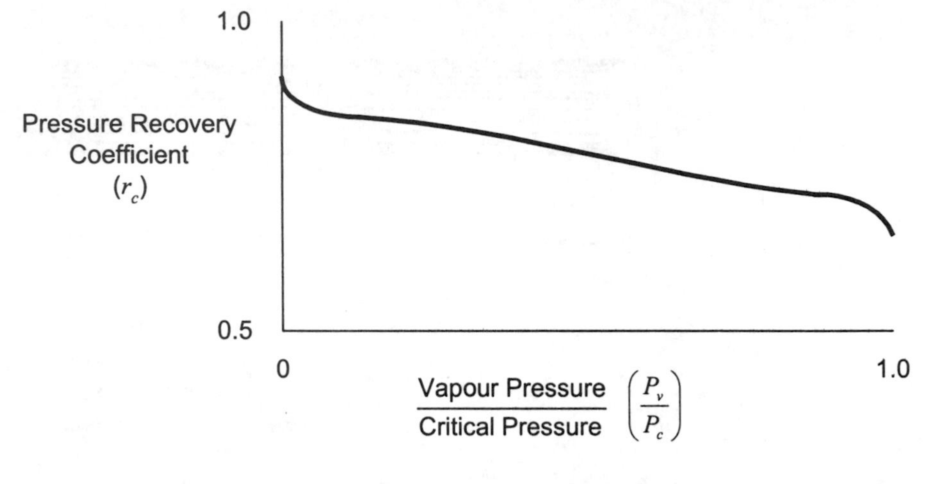

Figure 2.26 r_c versus P_v/P_c.

$$K_m = \frac{P_1 - P_2}{P_1 - P_{vc}} \tag{2.10}$$

where P_{vc} is the pressure at the vena contracta.

Experimentally, it has been found that $P_{vc} = r_c\, P_v$, where P_v is the vapour pressure. Typically, r_c is obtained from a graph, like Figure 2.26.

Note that water has a similar but different curve for obtaining r_c. Equation 2.10 can be rearranged as:

$$\Delta P_{allow} = (P_1 - P_2)_{allow} = K_m\,(P_1 - r_c\, P_v) \tag{2.11}$$

In Equation 2.11, K_m is constant for a particular valve. Other control valve vendors also have an expression for ΔP_{allow} that are functions of vapour and critical pressures. Overall, this results in the following liquid sizing procedure.

Liquid sizing procedure

1. Obtain conditions at minimum and maximum (flow, pressure drop) and properties SG, P_c and P_v.
2. Initially assume a valve size (one size smaller than piping) and valve type to obtain the pressure recovery coefficient, K_m, or equivalent.
3. Calculate $C_{v,\min}$ and $C_{v,\max}$ using the smaller of ΔP_{actual} and ΔP_{allow} in Equation 2.9 at both conditions.
4. If C_v falls in the range of 20% – 80% open, the valve is adequate, otherwise a larger valve is required. Note that the design C_v should be at about 50%–60% open.

The minimum flow often corresponds to maximum pressure drop and vice versa and these actual pressure drops are determined by system hydraulics (as discussed earlier). Also note that there are C_v corrections for viscous flow for cases where the valve Reynolds numbers are less than 5000. Finally note that the valve characteristic type and rangeability should be checked, as detailed earlier. Although sizing equations in this section use Fisher nomenclature, they are equivalent to ANSI/ISA versions if F_L^2 is substituted for K_m.

Gas, vapour and steam control valve sizing

The major differences between liquid service and gas, vapour or steam service are:

- that the fluid is compressible; and
- the phenomenon of critical flow.

When the ratio of $\Delta P/P_1$ exceeds 0.02, the gas is undergoing compression. Critical flow occurs when the flow is not a function of the square root of the pressure drop across the valve, but only of the upstream pressure. This phenomenon occurs when the fluid reaches sonic velocity at the vena contracta. Since gas cannot travel faster than sonic velocity, critical flow is a flow-limiting condition for gas. It has been found that critical flow occurs at different $\Delta P/P_1$ ratios depending on whether the valve is high or low recovery.

Fisher, as well as other vendors, has equations for gas, vapour and steam flow which have two parameters. One parameter represents flow capacity while the other parameter represents the type of valve and its effect on critical flow. The Fisher equations are as follows:

$$C_g = \frac{Q_{scfh}}{\sqrt{\frac{520}{SGT}}\,P_1 \sin\left[\left(\frac{C_a}{C_1}\right)\sqrt{\frac{\Delta P}{P_1}}\right]} \quad \text{(Ideal gas)} \quad (2.12)$$

$$C_g = \frac{W_{lb/hr}}{1.06\sqrt{\rho_1 P_1}\,\sin\left[\left(\frac{C_a}{C_1}\right)\sqrt{\frac{\Delta P}{P_1}}\right]} \quad \text{(Non-ideal gas)} \quad (2.13)$$

$$C_g = \frac{W_{lb/hr}\,(1 + 0.00065T_{SH})}{P_1 \sin\left[\left(\frac{C_a}{C_1}\right)\sqrt{\frac{\Delta P}{P_1}}\right]} \quad \text{(Steam)} \quad (2.14)$$

The constant, C_a, in the denominator is 59.64 if the sine evaluation is in radians and is 3,417 if the sine evaluation is in degrees and $C_1 = C_g / C_v$.

For these equations, when $\Delta P / P_1 \leq 0.02$ and $\sin(x) \approx x$, the C_g equation reduces to a "gas version" of the basic equation for liquids because the pressure drop is far below the critical value and the compressibility is negligible.

$$C_v = \frac{Q_{scfh}}{\sqrt{\frac{520}{SG\,T}P_1}\,C_a\sqrt{\frac{\Delta P}{P_1}}} \tag{2.15}$$

At the critical pressure drop, $\sin(x) \approx 1$ and C_g is only a function of P_1:

$$C_g = \frac{Q_{scfh}}{\sqrt{\frac{520}{SGT}P_1}} \tag{2.16}$$

Overall, this results in the following gas, vapour, or steam sizing procedure.

Gas, vapour or steam sizing procedure

1. Obtain conditions at minimum and maximum (flow, pressure drop) and ρ_1, if necessary.
2. Initially assume a valve size (one size smaller than piping) and valve type to obtain C_1.
3. Calculate $C_{g,\min}$ and $C_{g,\max}$ or $C_{s,\min}$ and $C_{s,\max}$.
4. If C_g or C_s fall in the range of 20% – 80% open, the valve is adequate, otherwise a larger valve is required. Note that the design C_g or C_s should be at about 50% to 60% open.

The ANSI/ISA valve sizing equation for gas is:

$$C_v = \frac{Q}{N\,P_1 Y\sqrt{\frac{\Delta P/P_1}{GTZ}}} \tag{2.17}$$

In Equation 2.17, Y is an expansion factor (ratio of flow coefficient for a gas to that for a liquid) that plays a similar role to C_1. Although the form of this equation seems much different than the Fisher one, the results are equivalent. Again, the valve characteristic type and rangeability should be checked.

Cavitation and flashing

As stated previously, if the pressure in the vena contracta falls below the vapour pressure of a liquid, it will partially vaporise. If the pressure recovers above the vapour pressure, the gas bubbles collapse on the metal and tend to break it away in small pieces. This is known as cavitation. Because the pressure drop across the valve varies as the opening varies, cavitation may not occur across the entire range of valve opening. If the pressure does not recover above the vapour pressure, flashing occurs which can erode the valve plug and seat.

Fisher uses a similar equation to Equation 2.18 to describe cavitation pressure drop:

$$(P_1 - P_2)_{cav} = \Delta P_{cav} = K_c (P_1 - r_c P_v) \tag{2.18}$$

The values for K_c are constant for a particular type of valve. Fisher has related K_c to K_m, with a few examples given in Table 2.2.

Other control valve vendors use a similar equation.

Control valves can be designed to prevent cavitation or at least minimise the damage from flashing. For example, cavitation control type valve trims use the concept of reducing the pressure in several small increments through several stages, instead of one larger pressure drop in a single stage. This avoids the pressure in the vena contracta dropping below the liquid vapour pressure. Flashing is determined by the system and not the control valve because the outlet pressure is below the vapour pressure of the liquid. However, flashing damage can be minimised by using specially designed valve trims.

Table 2.2 K_c/K_m values for some valve types

Valve type	K_c/K_m
Globe valve (cavitation control trim)	1.00
Globe valve (standard trim)	0.85
Ball valve	0.67
Butterfly valve	0.50

Figure 2.27 Typical appearance of flashing damage (courtesy of Fisher Controls International, Inc.).

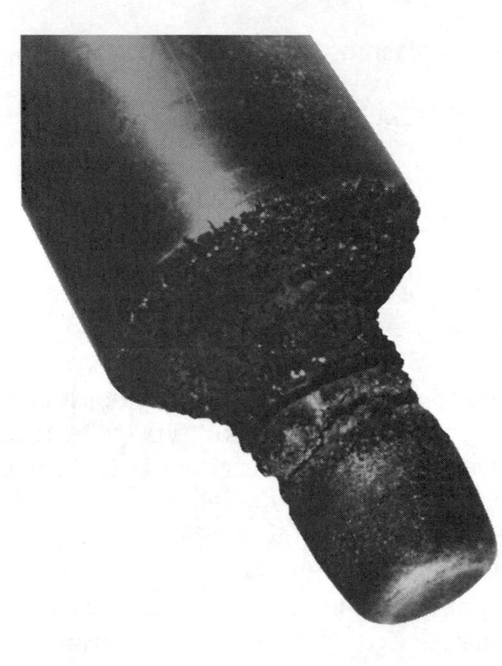

Figure 2.28 Typical appearance of cavitation damage (courtesy of Fisher Controls International, Inc.).

Valve positioners

Valve positioners are used to assist positioning control valves under difficult service applications where the control valve may otherwise be out of balance. Their operation employs the negative feedback principle. The position of the valve stem is balanced via cam and beam with the signal from the controller. The out of balance motion is detected by a nozzle, which increases the air to the top of the valve via a relay until equilibrium is obtained. Figure 2.29 shows a modern control valve utilising a digital valve positioner, while Figure 2.30 illustrates the function of a positioner for a diaphragm actuator.

Valve positioners should be used when any of the following conditions apply:

- single ported valves with high pressure drops that require large stem thrusts;
- viscous liquids, sludges, and slurries;
- large distances between the controller and control valve.;
- three-way control valves;
- unusually tight packing required because of corrosive fluids, low emissions, or high temperatures;
- large valves that use high volumes of control air; or
- split range operation, which is when two or more valves are operated by one controller.

In addition, with increasing emphasis on economic performance, valve manufacturers currently recommend that positioners be considered for valve applications where process variability performance is important [16].

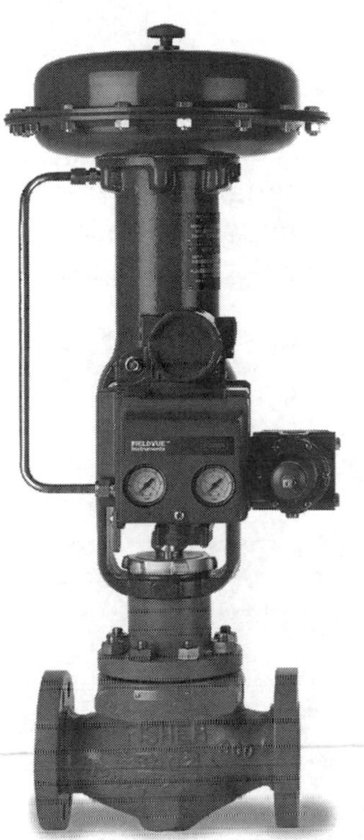

Figure 2.29 Modern control valve and digital valve positioner (courtesy of Fisher Controls International, Inc.).

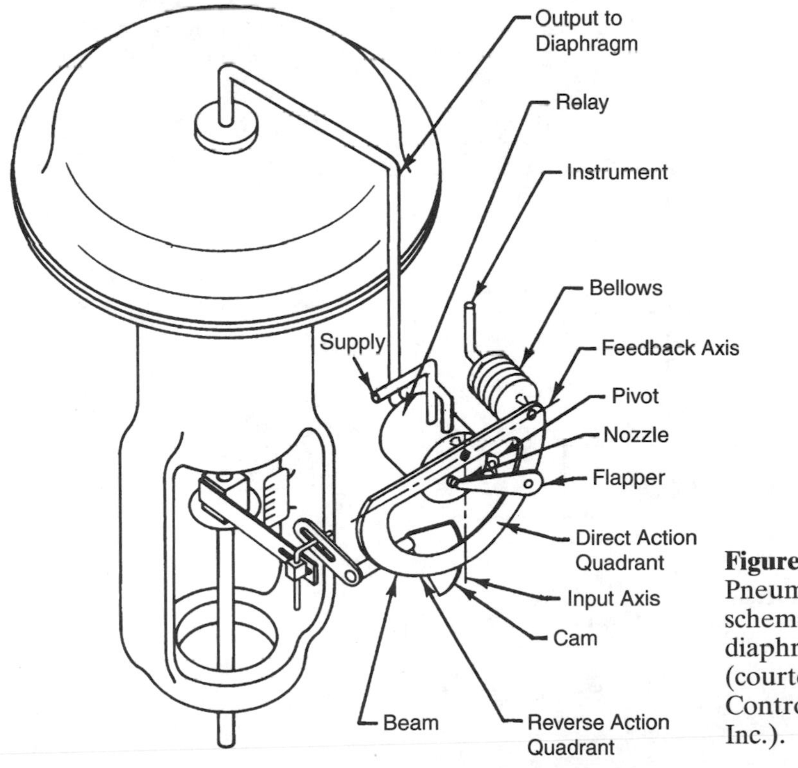

Figure 2.30 Pneumatic positioner schematic for diaphragm actuator (courtesy of Fisher Controls International, Inc.).

Fail-safe design

"Fail-safe" design means that if a plant has to close down because of instrument power supply failure, of either air or electricity, then the process is designed to shut down safely. This ensures safety for the environment, people, product and equipment. Several control valve designs are available that allow this purpose. The way a valve is classified is by the manner it closes under the action of the spring. There are fail-safe open and fail-safe close type designs. Figure 2.31 illustrates these designs.

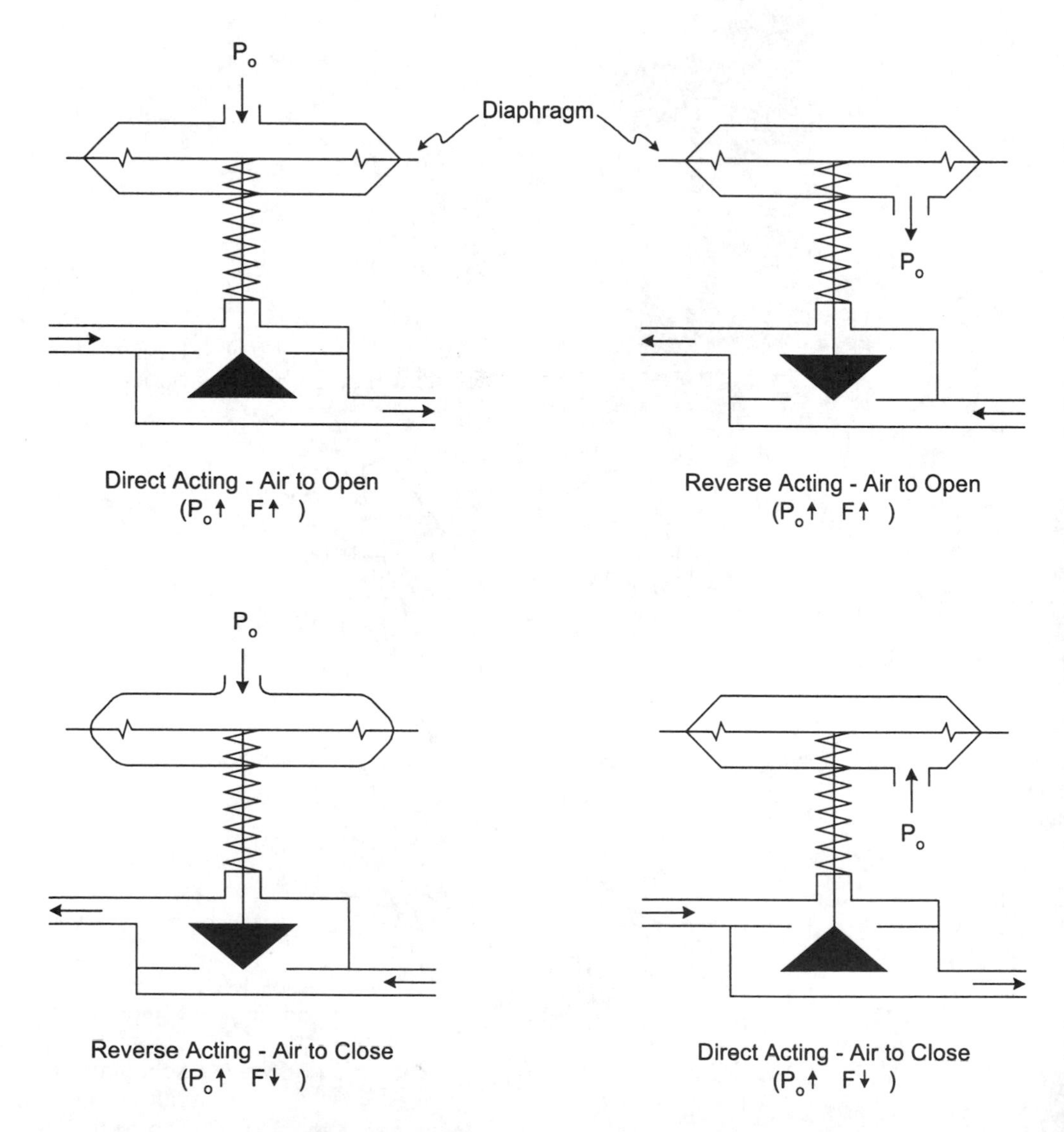

Figure 2.31 Fail-safe design of valves.

2.4 References

1. Giacobbe, J.B., Bounds, A.M., "Selecting and Working Bourdon Tube Materials". *Instrum. Mtg.*, 1952, **July–Aug**.
2. Goitein, K., "A Dimensional Analysis Approach to Bourdon Tube Design". *Instrum. Pract.*, 1952, **Sept**.: 748–55.
3. Considine, D.M. (Ed.) *Process Industrial Instruments and Controls Handbook*, 4th edition. McGraw-Hill, 1993.
4. Kompass, E.J. "SMART Transmitter Stores Calibration Digitally". *Control Eng.*, v30, 1983, **11**: 80–1.
5. Perry, R.H., Green, D.W. and Maloney, J. *Perry's Chemical Engineer's Handbook*. McGraw-Hill, 7th edition, 1997, pp. 8.49–8.50.
6. Johnson, F.L., "Temperature Measurement and Control Fundamentals". *Chemical Processing*, 1998, **June**: 89–99.
7. ISO Standards 5167, "Measurement of Fluid Flow by Means of Orifice Plates, Nozzles, and Venturi Tubes Inserted in Circular Cross-section Conduits Running Full" ISO 5167, 1980.
8. ANSI/ISA. "Control Valve Sizing Equations" Standard S75.01 (revised periodically), ISA, Research Triangle Park, NC, USA.
9. Control Valve Handbook, 3rd edition, Fisher Controls International Inc., USA, 1998.
10. Control Valve Sourcebook, Power A Severe Service, 2nd ed., Fisher Controls International, Inc., USA, 1990.
11. IEC Control Valve Standard, IEC 534–8–3, 1995.
12. Masoneilan. "Handbook for Control Valve Sizing". Bull. OZ-100, Masoneilan Division, McGraw-Edison Co., Norwood, Mass, 1981.
13. Purcell, M.K., "Easily Select and Size Control Valves", *CEP*, March 1999, pp. 45–50.
14. Bell, R.M., "Avoid Pitfalls When Specifying Control Valves", *Chem. Eng.*, 1996, **Dec.**: 75–7.
15. Connell, J.R., "Realistic Control Valve Pressure Drops", *Chem. Eng.*, 1987, **Sept. (28)**: 123–7.
16. Rheinhart, N.F., "Impact of Control Loop Performance on Process Profitability", Aspen World '97, Boston, Massachusetts, USA, October 15, 1997.

3 FUNDAMENTALS OF SINGLE INPUT/SINGLE OUTPUT SYSTEMS

In this chapter, we describe the basic components and concepts of single input/single output (SISO) control systems, along with some of the physical attributes commonly found in these systems. We will also explain the characterisation of system responses and provide an introduction to modelling various processes. After studying this chapter, the reader should understand:

- the basic components of a single input/single output process control loop;
- the difference between open-loop and closed-loop control;
- the concept of direct-acting and reverse-acting controls;
- what process capacitance is and what it contributes to process controllability;
- what process dead-time is and what challenges it poses to process controllability; and
- how to develop some of the basic equations that govern first-order system response with feedback control.

We recommend that the student review the fundamentals of differential equations and some of the more common numerical methods to aid his understanding of the mathematical development and solution of the various process models presented in this chapter. Some excellent sources for such a review are included in the references [1, 2, 3].

3.1 Open loop control

Most readers will be familiar with how the speed of an automobile is controlled. The basic "process" setup is quite standard, as illustrated in Figure 3.1. There is an air/fuel mixture feed, a throttle that regulates how much feed is introduced, and there is the engine itself that converts combustion energy into rotating mechanical energy that turns the wheels at a certain *rpm*. Consider a car on a straight, flat road on a still day. Move the throttle to just the right position and you will achieve the desired speed. Once set, there is no need to adjust it. After all, if nothing changes in the environment, the

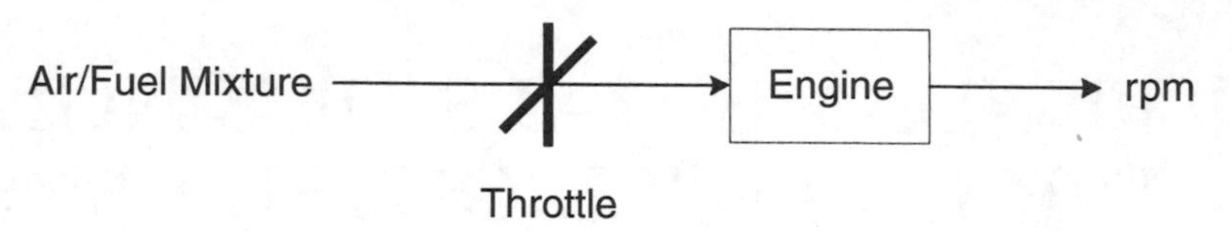

Figure 3.1 Illustration of car metaphor.

rpm of the engine should stay right where it is. This is a familiar example of *open loop control.*

Figure 3.2 illustrates a more generic process than our automobile example, but the basic elements are the same. Here, instead of air/fuel mixture feed, we use the term *Mass and Energy* feed. Instead of a throttle, we call it a *Final Control Element*, or *FCE*. Instead of an engine, we have a *Process*. And instead of an output rpm, we call the measured output of the process the *Controlled Variable/Process Variable.*

By definition, *open loop control* places the FCE in a fixed position, or a prescribed series of positions, with the expectation that nothing will change (i.e. there will be no disturbances) to cause the desired state of the system (set point value) to drift. Other examples of open loop control are traffic lights and a washing machine cycle. Once the control action is initiated, it will proceed through the prescribed steps, or remain fixed without any knowledge of the actual status of the process. Sometimes this actually works. Much of the time, however, it does not. Consider our automobile example, if the road suddenly rises steeply, or a strong head wind is encountered.

A more realistic view of a process or plant is shown in Figure 3.3.

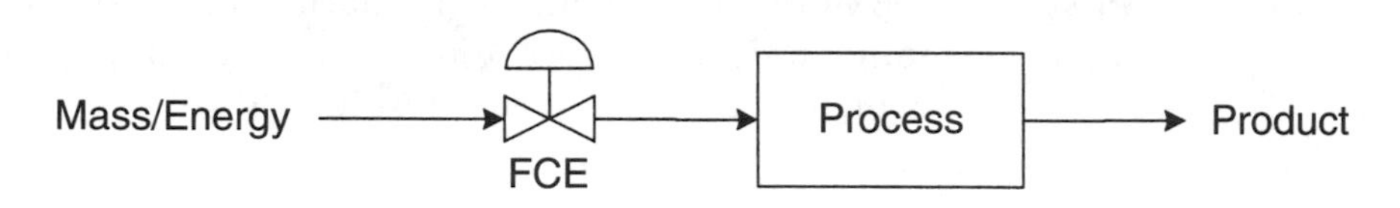

Figure 3.2 A simplified process view.

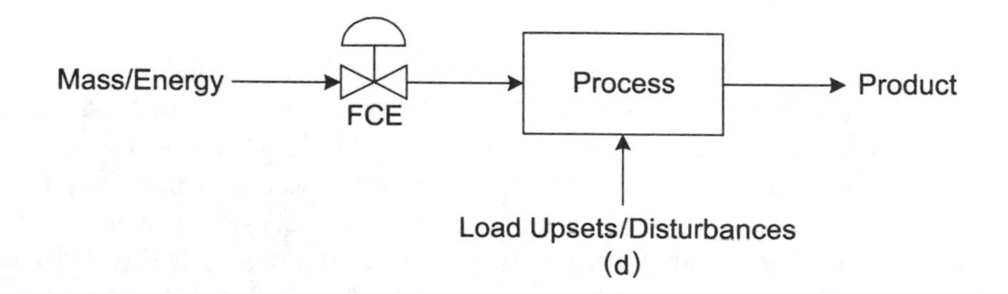

Figure 3.3 A realistic process view.

3.2 Disturbances

The process shown in Figure 3.3 adds the more realistic dimension of upsets or disturbances, *d*. Upsets and disturbances typically come in three types: input disturbances, load disturbances and set point disturbances. An *input disturbance* is a change in the mass or energy of the supply, or input, to the process that may cause the condition of the process variable to drift from its set point value, *SP*. A *load disturbance* is any other upset, except for an input mass or energy change, which may alter the quality of the process variable from the desired set point value. A *set point disturbance* occurs when the desired state of the process variable (*PV*) changes, and the process must adjust to a new state. The biggest difference between input disturbances and load disturbances – and the reason we distinguish between them – is that load disturbances cannot typically be anticipated, and they are often not measured directly. The only way we find out about them is by observing the effect they have on the product conditions or quality. While input disturbances may also be difficult to anticipate, they are often measured, and corrective action can more readily be taken. For this reason, we will primarily focus on load disturbances for the remainder of this chapter.

Returning to our automobile example and adding the more realistic dimension of disturbances, we see that in order to ensure we keep a steady rpm, we need to be able to constantly adjust the throttle position in order to keep a constant speed. This is essentially the function that cruise control carries out, and is an example of automatic feedback control. Simply put, *automatic feedback control* provides an automatic adjustment to the FCE in an attempt to maintain the conditions of the process variable at the desired set point value, *SP*, in the presence of disturbances, *d*.

3.3 Feedback control – overview

The simplest and most widely used method of process control is the feedback control loop shown in Figure 3.4. Note that we take a measurement of the process variable (indicated by *CV*, or controlled variable, in Figure 3.4) we care about (*Transmitter* in Figure 3.4 means "measuring device") and this value is compared to a set point (*SP*) to create an error, or departure from aim. This error is used to drive the corrective action of the FCE via the controller. Note that the output of the controller "manipulates" the mass and energy into the system via the FCE. Thus the property that the controller manipulates is referred to as the *manipulated value*, or *mv*. The action of the controller may be aggressive or sluggish; it depends on the internal equations of the controller (sometimes called the control algorithm or control law) and the tuning that is used. We will discuss controller types in Chapter

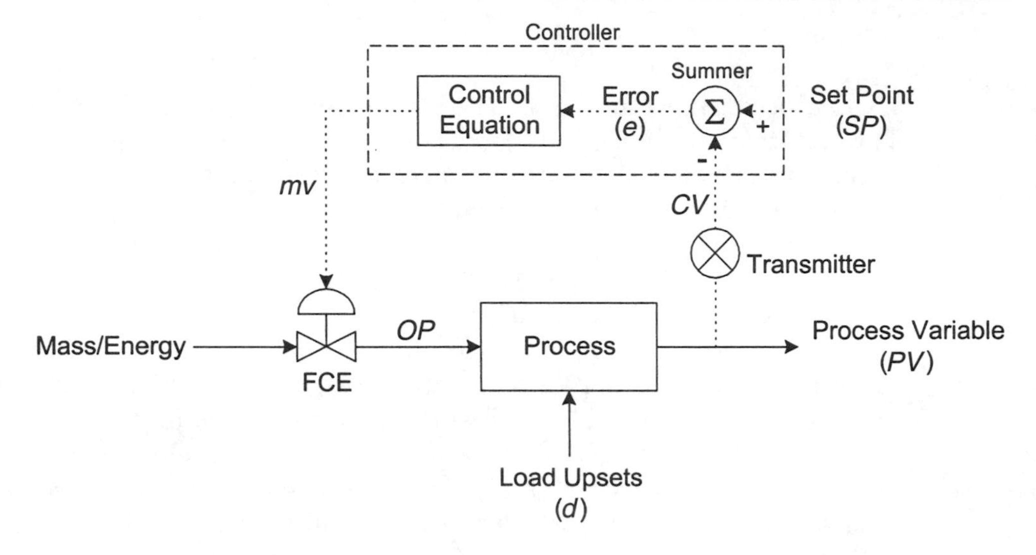

Figure 3.4 Single input / single output feedback control loop.

4 and tuning in Chapter 5. In order to successfully control a process, it is important to select both the right process variable and the right manipulated value. The process variable is typically a temperature, pressure, flow, composition or level, and is typically a variable that (a) is important to the product quality or the process operation, and (b) is responsive to changes in the selected manipulated variable.

It is interesting to note that if an automatic feedback controller succeeds in keeping the *PV* at the desired *SP* in the presence of load disturbances then, by necessity, there will be changes in the *mv* dictated by the controller. So in effect, *process control takes variability from one place and moves it to another*. Thus, the trick to process control is understanding where variability can be tolerated and where it cannot, and designing schemes that manage variability to acceptable levels.

The heat exchanger shown in Figure 3.5 [4] illustrates this transference of variability. The temperature of the "Hot Feed to Downstream Unit" stream is important (its temperature is the *PV* for this controller). The "Hot Utility" stream flow is manipulated (it's the *mv* for this controller) in order to keep the *PV* at its set point in the presence of load disturbances introduced in the "Feed from Upstream Unit" stream. With no control, these disturbances would make their way into the "Hot Feed to Downstream Unit" stream. With control, the *mv* absorbs these disturbances while keeping the *PV* at or near the set point. Thus, in effect, this controller transfers disturbances to the *mv* that would otherwise pass to the *PV*. As the control algorithm and/or tuning changes, so too does the amount of variability transferred.

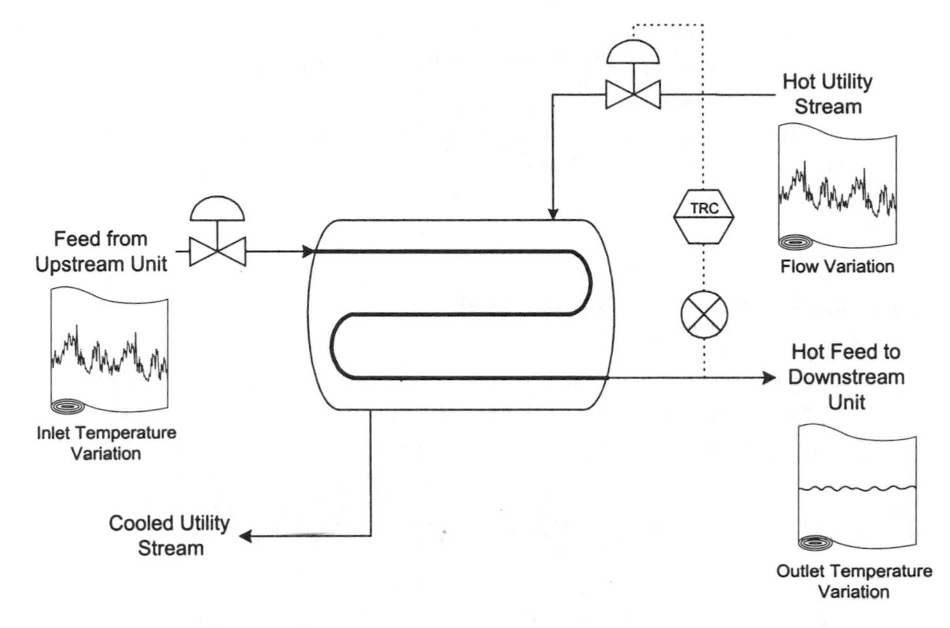

Figure 3.5 Transformation of variation from temperature to flow (courtesy of CACHE Corporation) [4].

Let us summarise our discussion so far:

- Open loop control suffices when no disturbances are present to threaten the desired state of the product.
- "Real" processes must operate in the presence of disturbances, and therefore require some sort of control. Automatic feedback control is the most common form of control.
- The basic elements of a feedback controller are:
  1. The process variable, *PV*, which represents the variable that is important to maintain under control;
  2. The set point, *SP*, which represents the desired value of the *PV*;
  3. The error, *e*, which represents the magnitude of the difference between the *PV* and the *SP*;
  4. The controller, whose "control law" and tuning drive the corrective action and influence the response of the SISO system;
  5. The final control element, FCE (typically a valve) to which the controller output is attached and through which the controller exercises its influence on the *PV*; and
  6. The manipulated variable, *mv*, which represents the variable in the process to which the *PV* is sensitive, and to which the FCE is attached.
- A feedback controller works by measuring the *PV* and comparing it to the *SP* to generate an error. The error, conditioned by the controller type

and tuning, drives appropriate changes in the FCE (and thus the *mv*) such that the *PV* is driven back in the direction of the *SP*.

Having provided an overview for the need for and basic operation of feedback control, we will now take a closer look at how such control loops are configured.

3.4 Feedback control – a closer look

Mathematically, the error drives the action of the controller. The sign of this error is an important consideration, and requires more development than one might expect. Let us begin with the notion of positive and negative feedback.

3.4.1 POSITIVE AND NEGATIVE FEEDBACK

Positive feedback represents a controller contribution that reinforces the error, and therefore precludes stability. Consider the audio feedback that occurs when a microphone is placed too close to the speaker that amplifies the microphone's output. Sound from the microphone is amplified through the speaker. If this sound re-enters the microphone, it adds to itself, and so on until the speaker saturates with a deafening tone. This is an example of positive feedback. Since positive feedback has no useful purpose for automatic control, we will consider it no further.

Negative feedback represents a controller contribution that diminishes the error, and therefore tends to add to stability. The cruise control in our automobile example works with negative feedback. If the speed is too high, the controller cuts back on the flow of the air/fuel mixture, thereby reducing the error. The opposite happens when the speed is too low.

As you can see, only negative feedback presents a viable control loop capable of maintaining stability. However, there are many different elements in a typical control loop, each one with a potentially reinforcing or subtracting contribution. Thus we need to understand the "action" of each component in the loop in order to determine whether, in the aggregate, the control loop will provide negative feedback. By action, we typically mean the sign relationship between an element's input and output. The next section will explain this.

3.4.2 CONTROL ELEMENTS

Let's first look at the action of the process element of the controller. Consider a furnace that heats your home in the winter. When the energy that drives the furnace increases, the temperature in the surrounding rooms increases as well. This is known as an Increase/Increase (I/I) relationship, or a direct acting element [5]. *Direct action* refers to a control loop element that, for an increase in its input, also experiences an increase in its output as well.

Now consider an air conditioner that cools your home in the summer. When the energy that drives the air conditioner increases, the temperature in the surrounding rooms decreases. This is known as an Increase/Decrease (I/D) relationship, or a reverse acting element. *Reverse action* refers to a control loop element that, for an increase in its input, experiences a decrease in its output [5].

Consider a general component with I/I action as shown in Figure 3.6. Ignoring the relative amplitudes between input and output, if there is an increasing or decreasing input there will be a corresponding increasing or decreasing output.

Consider a general I/D component as shown in Figure 3.7. For this element, if there is an increasing or decreasing signal, a resulting decreasing or increasing output signal will result.

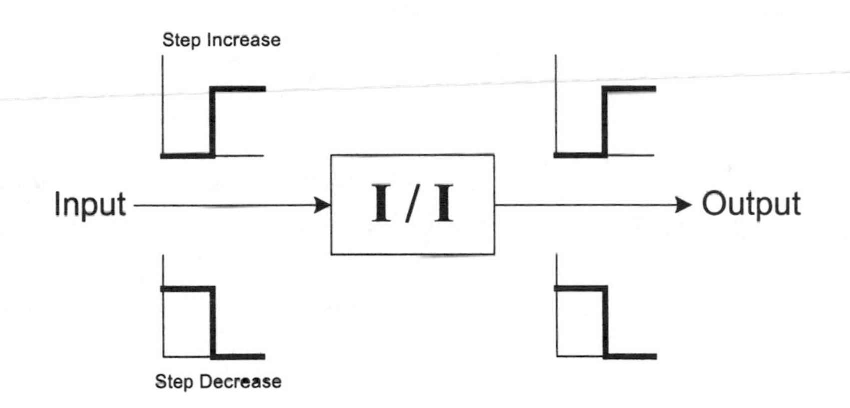

Figure 3.6 Increase/Increase component action.

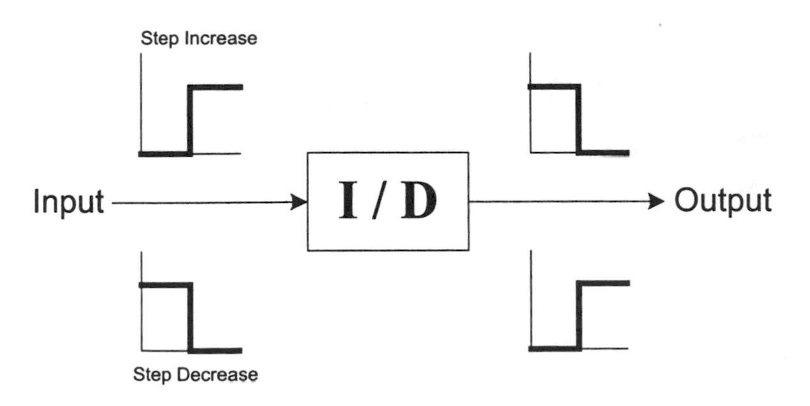

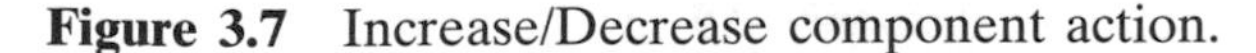

Figure 3.7 Increase/Decrease component action.

Connecting several I/I components in a series, as shown in Figure 3.8, will result in an overall I/I action.

As seen in Figure 3.9, if a single I/D component is placed anywhere in the sequence the overall action is I/D.

Figure 3.10 shows that when two I/D blocks are in a series there is an overall I/I action. It can further be shown that whenever there are an *even* number of I/D blocks in a series the overall effect is I/I, and whenever there is an *odd* number of I/D blocks in a series, the overall action is I/D.

Every component in the typical loop (shown in Figure 3.11) including the sensor/transmitter, the controller, the FCE and the process is either direct or reverse acting. Recall that only negative feedback presents a viable control loop capable of maintaining stability. Thus in the aggregate, the overall action of the control loop must be I/D or reverse acting. D/D or direct action generates, by definition, positive feedback.

Since the overall action of the control loop is determined by the action of each of the individual components, let's take a look at each of the typical control loop elements in turn.

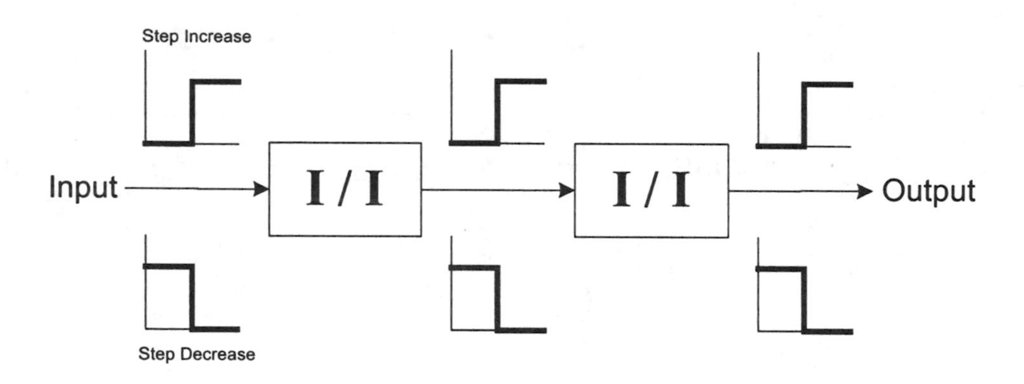

Figure 3.8 Increase/Increase components in series.

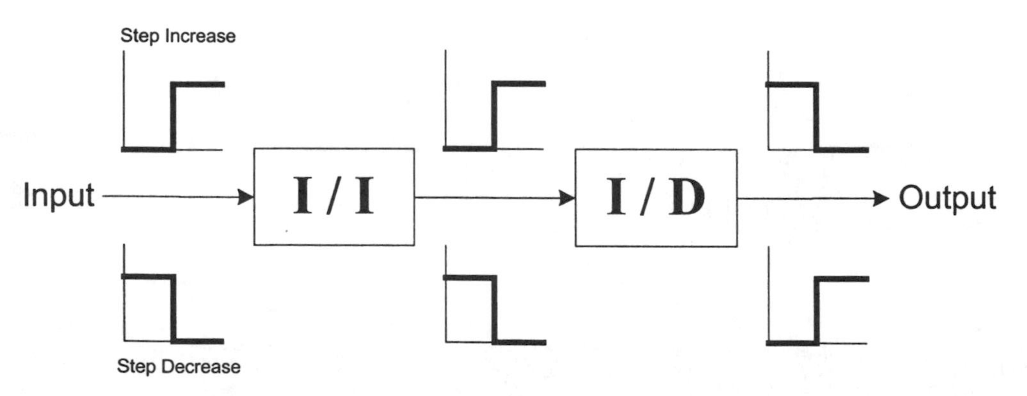

Figure 3.9 Combined components in a series.

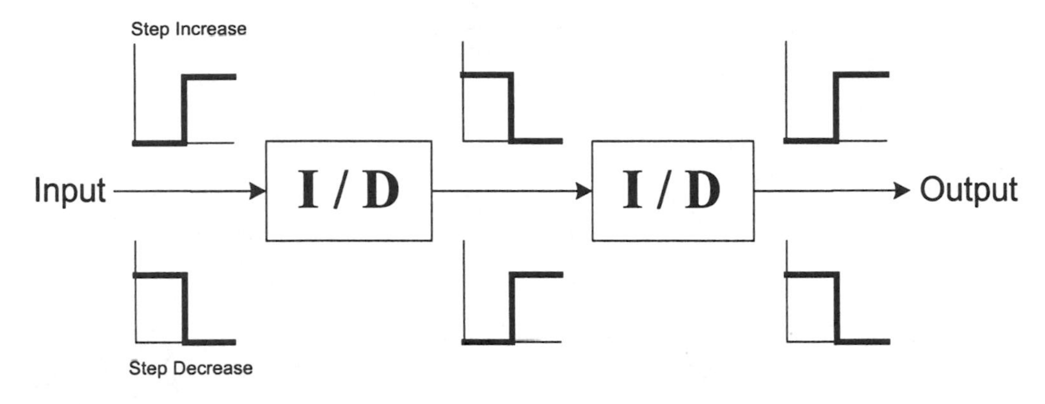

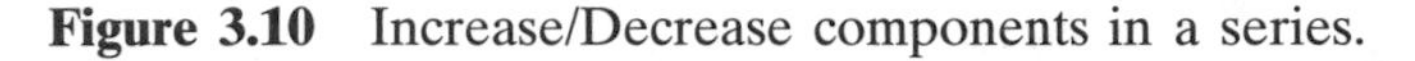
Figure 3.10 Increase/Decrease components in a series.

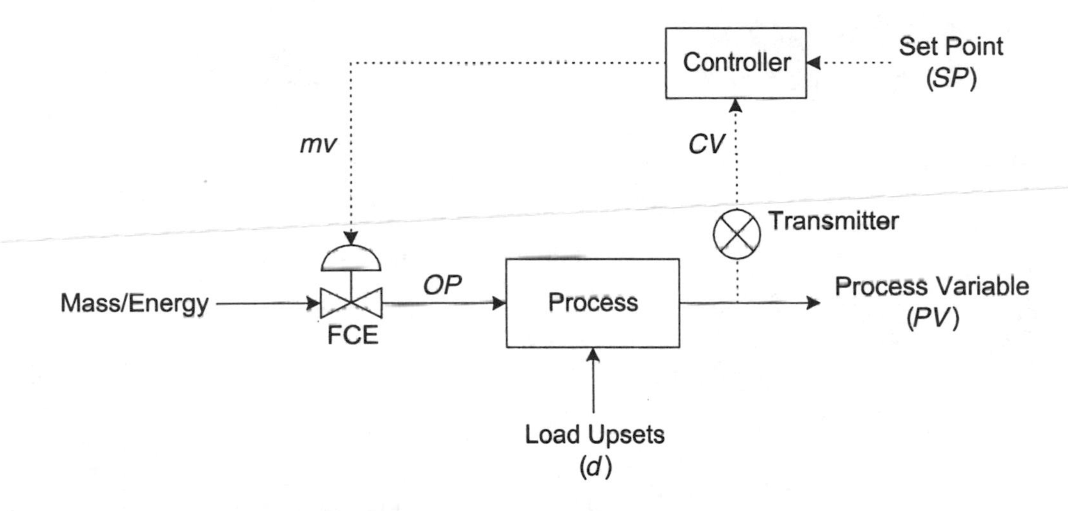

Figure 3.11 Typical single input/single output (SISO) loop.

3.4.3 SENSOR/TRANSMITTER

For the majority of applications, the sensor/transmitter produces an increasing output for an increasing input; therefore the sensor/transmitter is typically direct acting. There are some special cases where a sensor/transmitter may be reverse acting. However, this is generally not the case, and even if it were, as will be shown later, this poses no problem.

3.4.4 PROCESSES

Most processes are direct acting; however they can also be reverse acting. Let us examine several major types of processes and determine the relationships between the sign of their inputs and outputs.

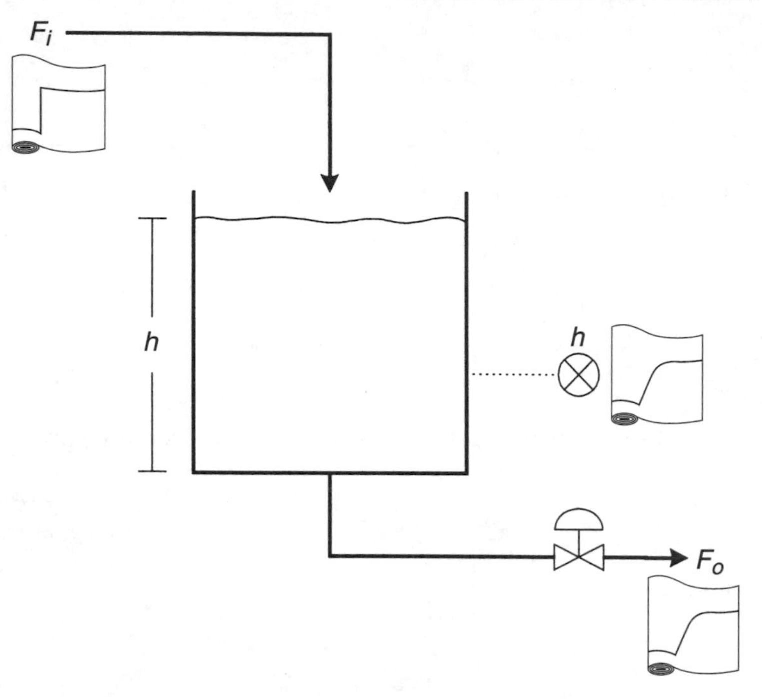

Figure 3.12 Mass flow process.

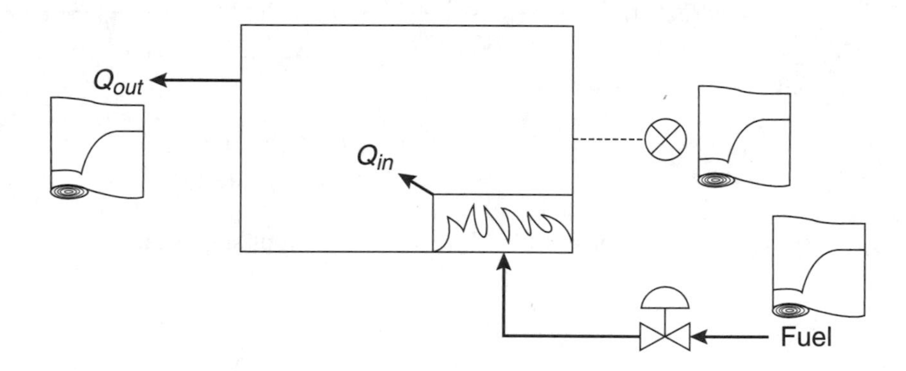

Figure 3.13 Energy flow process.

The first process is a single tank shown in Figure 3.12. For this process an increase in the input, F_i, causes an increase in the level, h, for a fixed valve position. Hence this is a direct acting process.

Now consider an energy flow or heating process illustrated in Figure 3.13. In this case increasing the fuel flow results in an increase in temperature. Hence, this is also an I/I process, or direct acting.

Finally consider the case of the reactor shown in Figure 3.14. We assume that the feed and catalyst are mixed and the resulting chemical reaction generates

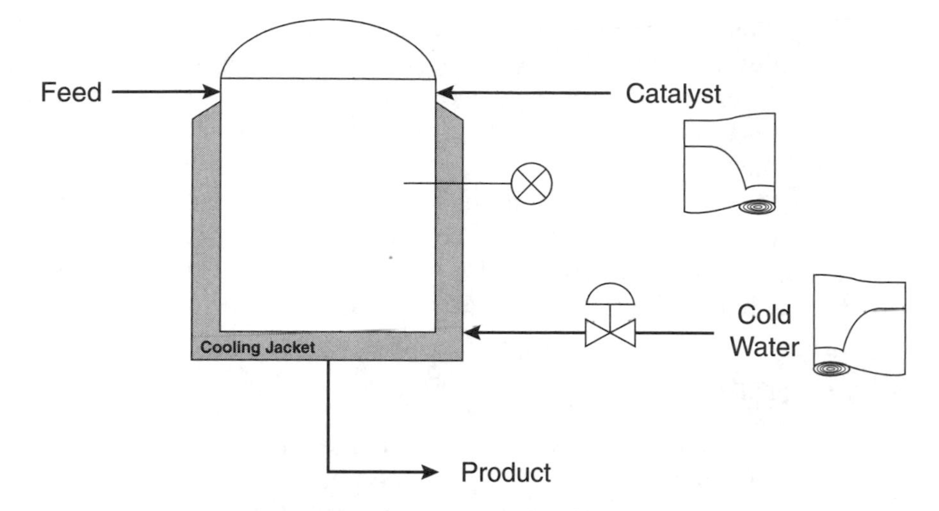

Figure 3.14 Exothermic reactor.

heat, in other words the reaction is exothermic. The rising temperature from this generated heat is the process output, and the cold water flow to the reactor jacket is the process input. The result is a reverse acting process, since an increase in cold water flow will result in a decrease in reaction temperature.

3.4.5 FINAL CONTROL ELEMENT

A final control element, FCE, can be almost anything that controls the flow of mass or energy into or out of a process. It may be a motor speed control on the fan blades of an air-cooled heat exchanger, a star valve on a bin containing solids, etc. However, in the fluid processing industries about 90% of all final control elements are valves. Hence, it is necessary to understand the action of a control valve. For a manual control valve, as the stem position of the valve is moved upwards or open, the flow through the valve also increases, resulting in direct action. Most valve actuators in process control applications are pneumatically activated. In the case of an air-to-open actuator, an increasing air signal causes the actuator to stroke open and therefore the flow through the valve increases. This is direct acting. For an air-to-close actuator, an increasing air signal to the actuator closes the valve and flow decreases, resulting in a reverse acting valve.

Therefore, in the case of the FCE being a valve, it may have either direct or reverse action depending on the actuator chosen for the valve. The desired action is chosen so that fail-safe operation is achieved. For fail-safe operation, the engineer must consider whether a "fail-open" or "fail-closed" valve would provide the best safety in the event of a failure. A "fail-close" valve simply means that if the energy supply to the valve was to fail then the valve would close, allowing no flow. Conversely, a "fail-open" valve opens when the energy supply fails. Cooling water to a reactor is best by an air-to-close valve. Loss of instrument air would fail the valve in the open position (because it takes air to close it), ensuring that there is sufficient cooling and preventing damage to the reactor. Conversely, the valve controlling the steam flow to a reboiler should be an air-to-open valve. Loss of instrument air here would fail the valve closed (since it takes air to open it), ensuring that the column will not overheat during the failure. See the Fail-safe design section in Chapter 2 for an illustration of air-to-open and air-to-close conventions.

3.4.6 CONTROLLER

All controllers, whether implemented as standalone or as part of a distributed control system (DCS) application, have a switch which will allow either direct or reverse action. In general, the action of the controller is the last to be specified, since there is typically little choice in the action of the other elements in the loop. Once the other elements' actions are known, the controller action may be set such that the overall loop action is reverse acting, or I/D.

For the components shown in Figure 3.15, assume the action shown in Figure 3.16, and also assume an air-to-open actuator (I/I) for the valve.

Note that the valve, process and sensor/transmitter are all direct acting. Therefore, in order to get the desired negative feedback action (I/D overall loop action) the controller must be set to reverse action.

Next consider the situation for an air-to-close (I/D) actuator, shown in Figure 3.17. In this situation, the controller must be set to direct action in order to achieve the overall negative feedback required for the loop.

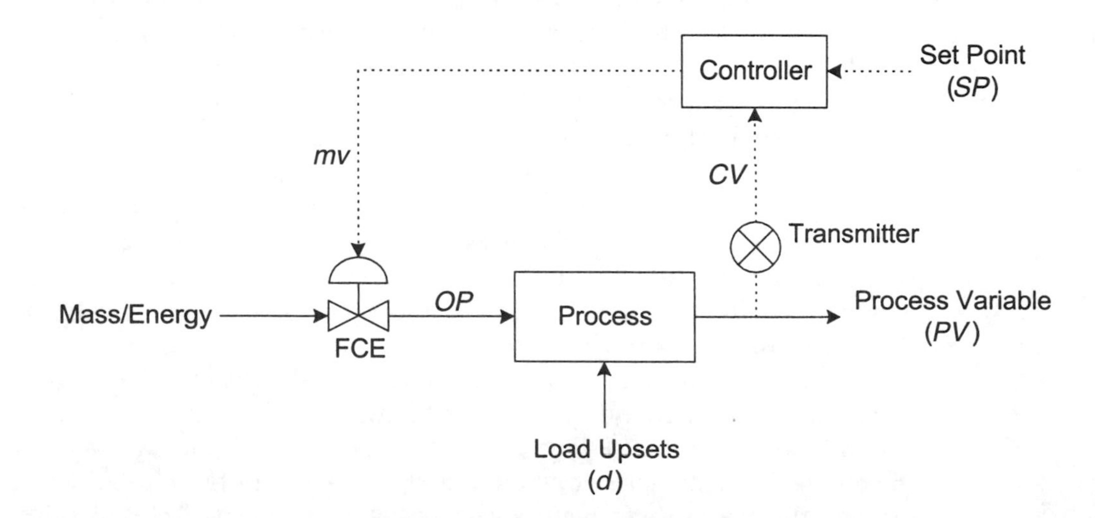

Figure 3.15 SISO feedback control loop.

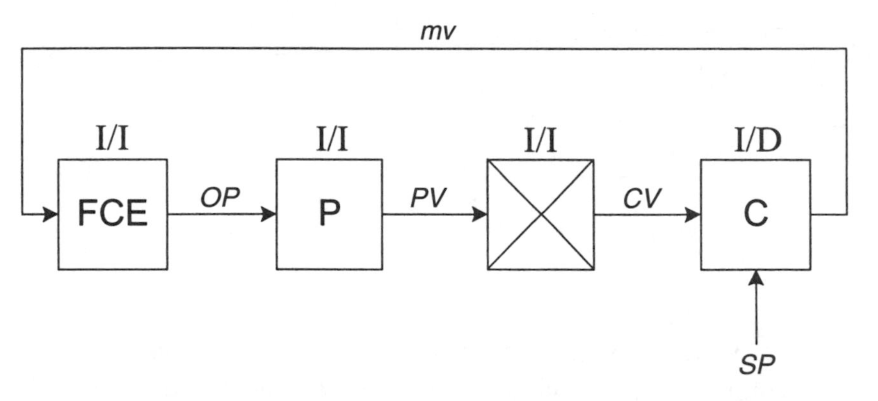

Figure 3.16 Component input/output for air-to-open actuator.

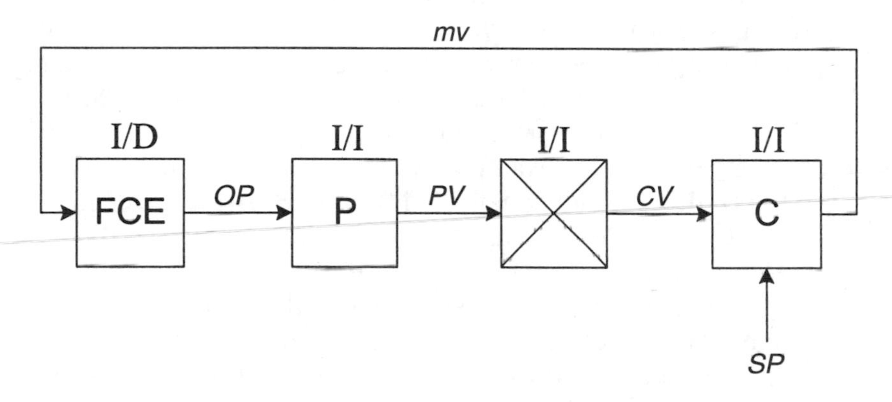

Figure 3.17 Component input/output for air-to-close actuator.

3.5 Process attributes – capacitance and dead time

As we will see later in this chapter and elsewhere, the equations that govern the dynamics of some of the unit operations in typical process plants can be quite complex. Despite this complexity, many processes behave as if they were first-order systems; exhibiting transport delay or dead time. Because of this, it is important to understand two fundamental process dynamic characteristics: capacitance and dead time.

3.5.1 CAPACITANCE

Simply stated, *capacitance* represents a system's ability to absorb or store mass or energy. Capacitance may also be defined as the resistance of a system to the change of mass or energy stored in it, i.e. inertia. A common example of a capacity-dominant process is one that stores energy, Figure 3.18.

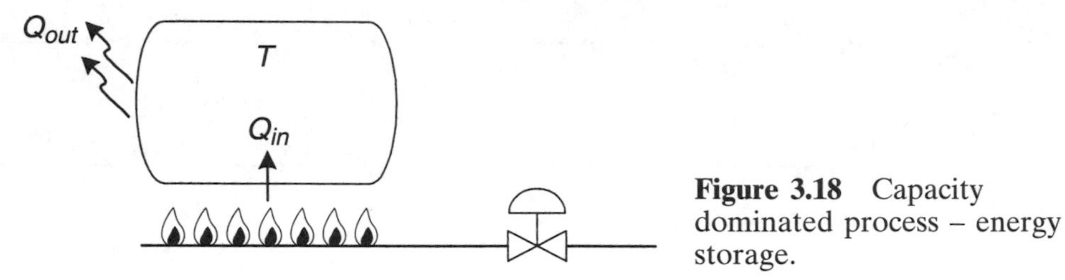

Figure 3.18 Capacity dominated process – energy storage.

In this example the process consists of an oven which is storing heat to maintain a particular temperature, T. The gas flow creates a flow of energy in, Q_{in}. Q_{out} is the flow of energy to the ambient or, in other words, the heat loss to the ambient. For an increase or decrease in the valve position changing the gas flow in, the temperature would correspondingly increase or decrease. It would not, however, change instantaneously with a change in valve position. This behaviour is due to the system's capacitance.

Consider the classical capacity dominant system shown in Figure 3.19: the surge tank. In this example the tank has a volume in which a mass of liquid is stored. Consider what would happen to the level in the tank, H, if the inflow, F_i, were increased. One would certainly expect the level to rise. However, if F_i was increased by 10%, the level would not also increase by 10% instantaneously. It would eventually reach a level 10% higher, but the capacitance of the tank limits the rate of change in level, thus it takes some time to reach a 10% increase in level. In other words, the tank has inertia and self-regulation. *Self-regulation* occurs when a process, in this case tank level, eventually lines out to a steady-state value for each input step change, rather than ramping off indefinitely.

The rate of change of volume in the tank can be written as a lumped parameter model, where all the resistance to flow is assumed to be associated with the valve, and all the capacitance of the process is assumed to be associated with the tank. This model is shown in Equations 3.1 and 3.2. The basis of Equation 3.1 is the principle of conservation, mass balance in this case, (i.e. what goes in must come out or get accumulated in the system).

$$\text{In} - \text{Out} = \text{Accumulation} \tag{3.1}$$

$$\rho Q_{in} - \rho_{out} Q_{out} = \frac{d(\rho_{out} V)}{dt} = \frac{d(\rho_{out} AH)}{dt} \tag{3.2}$$

In Equation 3.2, Q is the volumetric flow rate of water, V is the volume of the tank, A is the cross-sectional area of the tank, ρ is the density, and H

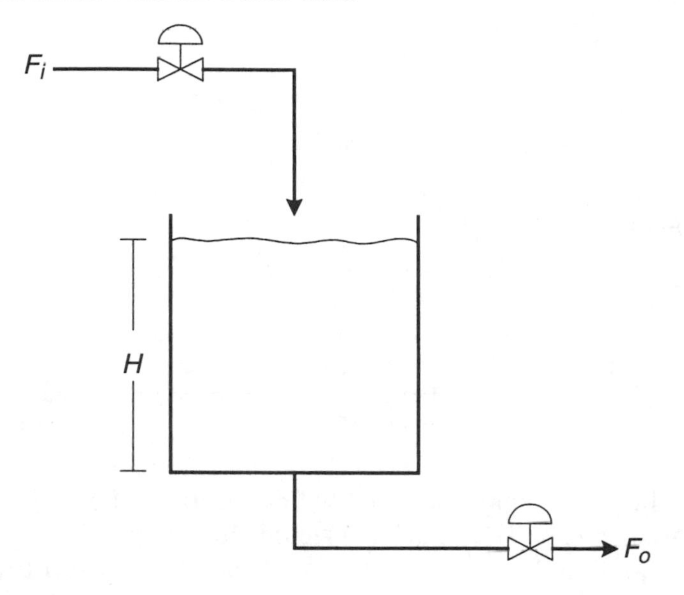

Figure 3.19 Capacity dominated process – surge tank.

is the water level. Assuming the density and cross-sectional area are constant results in Equation 3.3,

$$Q_{in} - Q_{out} = \frac{A\,dH}{dt} \tag{3.3}$$

The flow, Q_{out}, is determined by the valve characteristic $V(X_p)$, with X_p being the valve plug-stem expressed in percentage opening, the valve flow constant C_V, and the square root of the pressure drop across the valve as given in Equation 3.4:

$$Q_{out} = V(Xp)C_V\sqrt{\Delta p} = V(Xp)C_V\sqrt{\rho g H} \tag{3.4}$$

Note that g is the acceleration of gravity. Substituting for Q_{out} into Equation 3.3, which is a first order differential equation, would result in a nonlinear first order differential which unfortunately has no analytical solution. The response of head (level), H, to changes in Q_{in} or valve position could only be determined by numerical methods. However, if Q_{out} is linearised using a Taylor series expansion about a desired operating level the first order differential equation can be solved analytically for a disturbance in flow into the tank. For a single variable the Taylor series can be written as shown in Equation 3.5.

$$F(H) = F(H_o) + \left(\frac{\partial F}{\partial H}\right)_{Ho}(H - H_o) + HOT \tag{3.5}$$

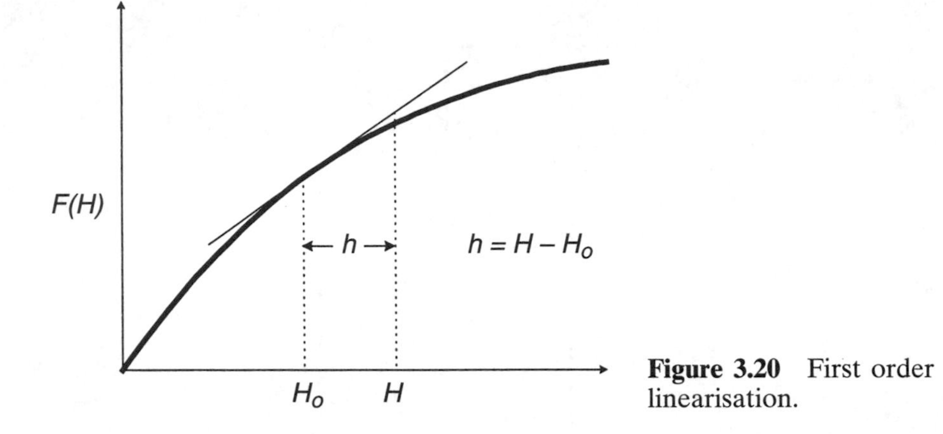

Figure 3.20 First order linearisation.

In performing a first order linearisation, shown in Figure 3.20, the higher-order terms (HOT) are neglected since $h = (H - H_o)$ is small. Setting $F(H)$, (Equation 3.6) as a function of head, H, and substituting into Equation 3.5 results in the linear form shown in Equation 3.10.

Let :
$$F(H) = V(Xp)C_V\sqrt{\rho g}\sqrt{H} = K_V\sqrt{H} \tag{3.6}$$

$$F(H) = F(H_o) + \left(\frac{\partial F}{\partial H}\right)_{H_o}(H - H_o) \tag{3.7}$$

$$\left(\frac{\partial F}{\partial H}\right)_{H_o} = \frac{1}{2}\left(\frac{K_V}{\sqrt{H_o}}\right) \tag{3.8}$$

Or:

$$F(H) = F(H_o) + \left(\frac{K_V}{2\sqrt{H_o}}\right)(H - H_o) \tag{3.9}$$

Or:

$$F(H) - F(H_o) = \left(\frac{K_V}{2\sqrt{H_o}}\right)(H - H_o) \tag{3.10}$$

Let us now complete the derivation of the linear differential equation that describes the response (also called time behaviour or personality) of the level to feed flow disturbances, starting with Equation 3.11.

$$A\frac{dH}{dt} = Q_{in} - F(H) \tag{3.11}$$

At the initial or steady-state condition, Equation 3.11 can be written as Equation 3.12:

$$A\frac{dH_o}{dt} = Q_{in_o} - F(H_o) \tag{3.12}$$

Subtracting the above two equations results in Equation 3.13.

$$A\frac{d(H_o - H)}{dt} = Q_{in_o} - Q_{in} - [F(H_o) - F(H)] \tag{3.13}$$

Equation 3.21 can be rewritten in terms of deviation or variation variable, $h = (H_o - H)$ and $q_{in} = (Q_{in_o} - Q_{in})$, as is shown in Equation 3.14 and in a slightly more simplified form as Equation 3.15:

$$A\frac{dh}{dt} = q_{in} - \frac{K_V}{2\sqrt{H_o}}h \tag{3.14}$$

$$A\frac{dH}{dt} = q_{in} - \frac{h}{R} \tag{3.15}$$

In Equation 3.15 R is the resistance to flow in units of time/length^2 (min/ft^2). Equation 3.15 can be written in the standard form for a first order linear differential equation (LDE), Equation 3.16.

$$RA\frac{dH}{dt} + h = Rq_{in} \tag{3.16}$$

where $RA = \tau =$ time constant (units of time).

Using the classical mathematical approach to solutions of a first order LDE one can proceed as follows:

$$\frac{dC}{dt} + P(t)C = Q(t) \tag{3.17}$$

Equation 3.17 has a solution of the form shown in Equation 3.18 [6].

$$C = e^{-\int P(t)\,dt} \int Q(t)e^{+\int P(t)dt}\,dt + C_1 e^{-\int P(t)\,dt} \tag{3.18}$$

C_1 is a constant of integration evaluated from the initial conditions. Writing the LDE for the tank level in general form gives Equation 3.19

$$\frac{dH}{dt} + \frac{h}{\tau} = Kq_{in} \tag{3.19}$$

where $K = 1/A$.

When there is a step input of size q_{in}, a solution only exists for times greater than zero and is shown as Equation 3.20. C_1 is evaluated at initial conditions yielding $C_1 = -Rq_{in}$.

$$h(t) = Rq_{in}\,(1 - e^{-t/\tau}) \tag{3.20}$$

The time constant, τ, characterises the response of the first order system and is discussed in greater detail in the next section. All higher order systems can be broken down into sets of first order systems, and the time constants of these LDEs can be used to ascertain the relative importance of each from a dynamic response perspective. That is, the dominant, or largest, time constant will determine the speed of the response. The commonly used rule of thumb is that any subsystem with a time constant an order of magnitude (10 times) less than the dominant time constant can be described by steady-state or algebraic equations.

Some practical perspectives on capacitance

While the workshop associated with this chapter will illustrate capacitance with simulation, it is worthwhile examining the practical characteristics of capacitance because, as you'll find out, capacitance can be a control engineer's best friend. A capacity-dominated system is described by Equation 3.21.

$$h(t) = Kq_{in}\,(1 - e^{-t/\tau}) \tag{3.21}$$

Consider a step change in q_{in}. Mathematically, the change in $h(t)$ begins immediately, even though the full impact of the change in q_{in} will take some time. Now, consider a controller whose aim is to keep $h(t)$ at some set point. Although we have not yet introduced controller algorithms and tuning, consider the most simple of control functions:

$$\Delta mv = K_c(SP - PV) = K_c e \tag{3.22}$$

where mv is the manipulated variable, K_c is the controller gain, PV is the process variable, SP is the set point and e is the error. In short, corrective action carried out by the mv is simply a constant multiplier of the error.

Returning to our capacity-dominated process, the instant the PV (in this case $h(t)$) deviates from the set point an error, e, will be generated, and the *mv* will make some adjustment. The larger K_c is, the greater the corrective action. The fact that changes in the input (*mv* or q_{in}) have an immediate effect on the output (*PV* or $h(t)$) helps immensely, since the corrective action required to drive the error to zero is a straight algebraic function of the error. In the limit, as K_c approaches infinity, the error will be driven to zero and perfect control is achieved. Unfortunately, real life is not perfect, and controller gains never function at infinity. In practical terms, there is nothing that shows an absolutely "immediate" response either, i.e. there is no 100% pure capacitance system. However, this line of argument illustrates an important point: *for capacity-dominated systems, effective control can often be achieved using simple control and large gains*. As we will see in the next chapter, "simple control" can be as simple as proportional-only control.

With such a simple approach to controlling capacity-dominated systems, it's easy to see why capacitance is often regarded as the control system engineer's ally. As with most things, too much is not good either. Recognise that large capacities typically increase capital cost. In addition, although capacitance acts as a buffer to upsets, if too large a volume of "upset" material is allowed to accumulate, it can take a long time to work out of the system. Thus, the wise process designer will typically use dynamic simulations to balance the trade-offs between the capital-cost optimum and the dynamic operability and control optimum.

3.5.2 DEAD TIME

We sometimes say that capacitance is our friend because it has the tendency to dampen out disturbances and lends itself to simple controls and straightforward tuning. Dead time, on the other hand, is typically regarded as the arch enemy of the process control engineer. Let's find out why.

Everyone has probably had the frustrating experience of showering in a very old building. The distance between the hot and cold water taps and the shower head must be quite substantial for, when you turn the tap, it takes several seconds for you to feel the effect. Assuming you've been lucky enough to get the water to a comfortable temperature, there's always the inconsiderate or unaware house guest who flushes the toilet mid-shower. Quickly, you race to cut back on the scalding hot water source. There is no immediate effect and seconds seem like an eternity so you crank the valve some more. The effect of your manipulation starts to make itself known, but within a few seconds, you realise that you overdid it, and the water is suddenly freezing cold! Now your manual adjustments to the tap begin all over again, but may not settle down until a few more freeze/scald cycles play out. This example illustrates the menace of dead time. Although it may not be as entertaining, let's look at dead time mathematically and try to understand a little more about this dynamic process characteristic.

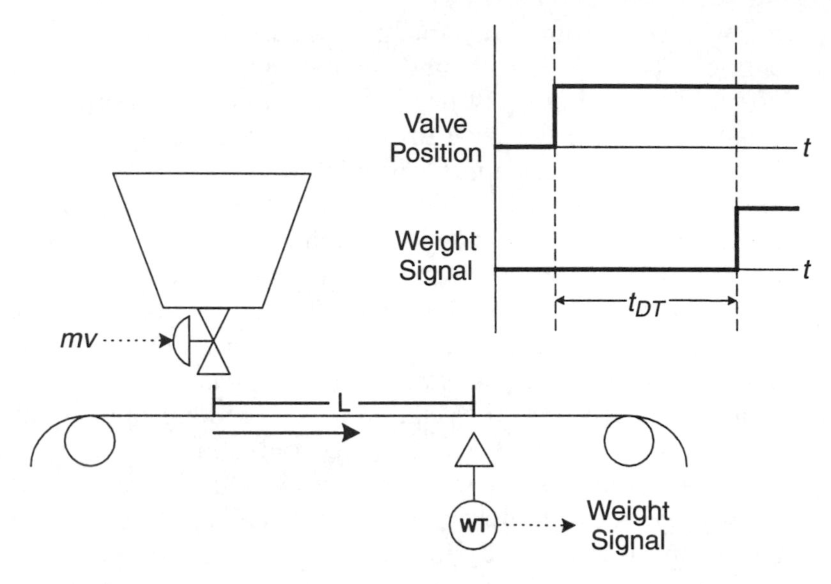

Figure 3.21 Continuous weighing system.

Dead time is a characteristic of a physical system that causes an input disturbance to be delayed in time, but unaffected in form. Whereas capacitance changes the form of the input disturbance (i.e. a step is filtered into a typical first order curve), dead time is a *pure* delay of the input disturbance. Dead time is also referred to as transport lag or distance-velocity lag. A typical example process is the continuous weighing system shown in Figure 3.21.

For instance, assume there is a conveyor belt, L metres long, moving at some velocity, v. The dead time, t_{DT}, is calculated as shown in Equation 3.23:

$$t_{DT} = \frac{L}{v}\left[=\right]\left[\frac{\mathrm{m}}{\mathrm{m/min}}\right][=][\mathrm{min}] \tag{3.23}$$

If the valve is opened by some amount, increasing the material on the belt, there will be a delay equal to t_{DT} minutes before the increased weight is sensed at the weight sensor/transmitter.

Another classic example is a liquid flowing in a pipe. If the liquid is flowing at a velocity v, through a pipe length l, an analogous situation exists to the weighing system. If a slug of liquid were followed through the pipe at the instant the valve is opened, it would take an amount of time t_{DT} for the slug to go from one end of the pipe to the other. The delay times in these two cases would not necessarily be the same, but the delay effect is similar.

For a pure dead time element, assume that a step input of magnitude A occurs. The magnitude of the output step would also be A, except displaced

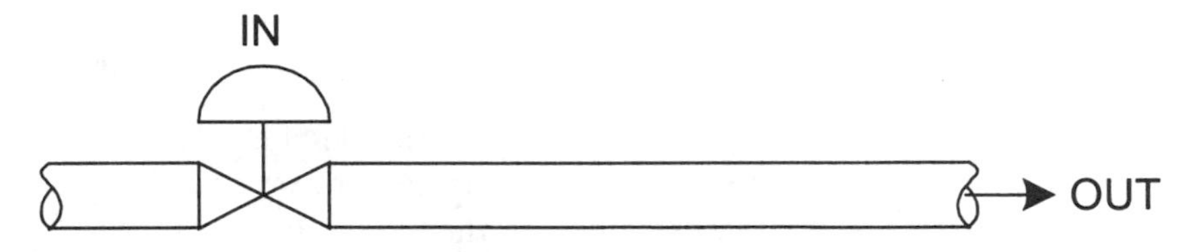

Figure 3.22 Valve/pipe flow system.

in time by the dead time amount. The static gain, K_{ss}, would therefore by definition be dimensionless and equal to one, as in Equation 3.24.

$$K_{ss} = \frac{A}{A} = 1 \tag{3.24}$$

Using the weighing system example of dead time, a 10 kilogram increase in material on the conveyor belt would result in a 10 kilogram increase at the weight sensor or a static gain of one. Similarly, it can also be shown that the above analysis holds for the pipe flow situation. In each of these cases a pure dead time exists since $K_{ss} = 1$. However, consider the following scenario shown in Figure 3.22.

In this scenario, the input to the valve is the input to the system and the output from the system is the flow through the pipe. If the opening of the valve is increased by some percent, A, an increase in flow of B m^3/sec is delayed by an amount t_{DT}, where t_{DT} is the time it takes to see an increase in flow at the exit or measurement point in the pipe.

$$K_{ss} = \frac{\Delta \text{out}}{\Delta \text{in}} = \frac{B(m^3/\text{s})}{A(\%)} \tag{3.25}$$

For the case shown in Equation 3.25, not only are there units but also the ratio of B to A is not necessarily one. It should be kept in mind that for pure dead time, i.e. the pipe alone, $K_{ss} = 1$. However, for the case where another component is involved in the dead time, the component serves to supply units to the overall gain.

Some practical perspectives on dead time

The workshop associated with this chapter will illustrate dead time with simulation, and will show just how its presence makes it difficult to control a process. Let's look at why.

A system with capacitance and dead time (actually, quite a common combination) is described by Equation 3.26.

$$h(t) = Kq_{in}(t - DT)(1 - e^{-(t-DT)/\tau}) \tag{3.26}$$

Consider a step change in q_{in}. Mathematically, the change in $h(t)$ will not be seen until DT time elapses. From that point on, the response in $h(t)$ will be exactly as that illustrated in the capacity-dominated system. Consider again, a controller whose aim is to keep $h(t)$ at some set point. Also consider again the most simple of control algorithms:

$$\Delta mv = K_c(SP - PV) = K_c e \tag{3.27}$$

where mv is the manipulated variable, K_c is the controller gain, PV is the process variable, SP is the set point and e is the error. Note again that the corrective action carried out by the mv is simply a constant multiplier of the error. Unlike in the capacity-dominated system, the PV (in this case $h(t)$) will not react immediately to the change in q_{in}. For DT time units, $h(t)$ will go unaffected. Only after DT time will $h(t)$ begin to change. At that time mv will, as before, act to drive the error to zero. However, because there is no longer an instantaneous response of $h(t)$ to the mv, the error can no longer be driven to zero by a large gain. In fact, the larger K_c becomes, the more the controller is apt to over react. Recall our shower example! Thus, dead time, by "hiding" the disturbances that lurk in the system, makes the job of rejecting disturbances extremely difficult. *The larger the dead time, in proportion to the amount of capacitance, the more difficult control becomes*. It is largely the presence of dead time (along with process interactions and nonlinear process response) that keeps control engineers earning a decent living. In general, the more dead time can be designed or engineered out of a system the better. Also, for any given amount of dead time, the more capacitance the better. Can you think why? Hint: two minutes of dead time in a chemical process with long response times (large capacitance) will not cause too much trouble. What about two minutes of dead time in the control loops of a jet airplane?

Let's summarise what we've just covered:

- two types of feedback exist: *positive* and *negative* feedback. Only negative feedback produces stable control;
- each element in a control loop has a particular action, or sign relationship between its input and output response. This action is important to understand, so that the action of the overall control loop produces negative feedback;
- elements whose output increases with an increase in their input are said to be *direct acting* (I/I) elements. Elements whose output decreases with an increase in their input are said to be *reverse acting* (I/D) elements;

- two important dynamic process response characteristics are *capacitance* and *dead time*
- capacitance acts to absorb and store mass and energy and as such, tends to be a natural buffer to disturbances. Thus this aids in process control ... to an extent. Too much can create other problems such as high capital cost and overly sluggish recovery from upsets. In general, capacity-dominated systems can be controlled with simple controls and large controller gains;
- dead time imposes a pure delay on disturbances, effectively hiding the disturbance from the process, the measurements and the controls until it's well into the system. Dead time deteriorates controllability, especially if it is large relative to the amount of capacitance in the system with which it is associated. Dead time should generally be minimised as far as possible.

3.6 Process dynamic response

By this point, the reader should have an understanding of the need for and function of feedback control, should understand the elements of the feedback loop, and understand some of the qualitative features of the process dynamic response. While standard feedback control does not require extensive understanding of the process being controlled, some process understanding is important. In fact, the more we understand about the process, the easier our overall control system design may become. We'll touch on this more in Chapter 10 on plant-wide control. For now, let's simply take a look at the typical process dynamic responses seen in process plants. Process response determines how easily a process can be controlled, and also impacts the tuning required to achieve acceptable performance.

Up to now, we have looked primarily at control as a means of maintaining the *PV* at a fixed set point in the presence of load disturbances. Set point changes are also types of disturbances that a control loop must be able to handle. Production rate changes, for example, require that a flow rate set point be changed. We'll use the set point change disturbance and the load disturbance as a means of illustrating process dynamic response.

As illustrated in Figure 3.23, numerous real time or dynamic responses are possible in returning the *PV* to the set point. The response labelled C_1 in both cases shown above would be classed as over-damped, i.e. a slow, sluggish return to set point. C_2 presents the case for critical damping, i.e. the fastest return to set point without oscillation. C_3 is a case where there is oscillation, and C_5 shows the case where instability is occurring, i.e. showing a hard constraint.

It is possible to adjust the feedback control loop to give any of the above responses. The form of the response desired depends on the process being

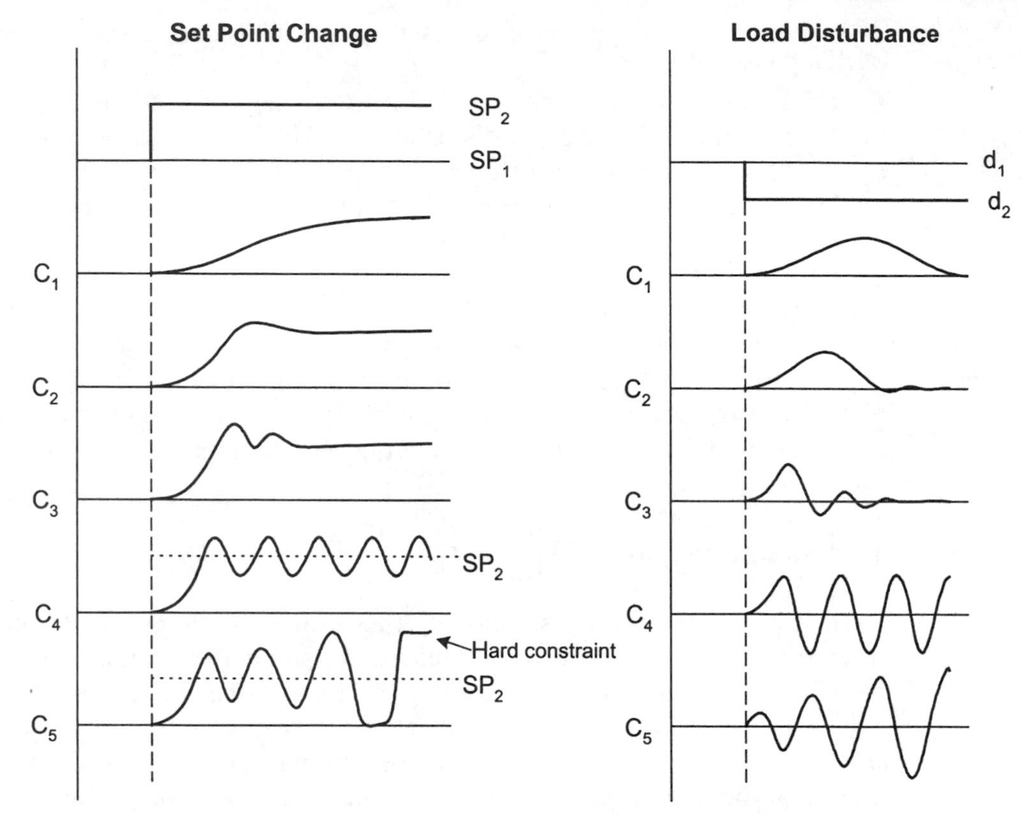

Figure 3.23 Typical *PV* response to set point and load disturbance upsets.

controlled. For the most part, C_1 to C_3 would give desired behaviour since each results in a return of the process variable to the desired set point. C_4 is useful in some cases for adjusting the controller, which is also referred to as tuning. Several methods developed for controller tuning depend on information gained from the uniformly oscillating loop show in C_4. C_5 results in instability and is not desirable for control.

3.7 Process modelling and simulation

Let's examine the response of SISO control systems in further detail. In order to examine a system and its response to disturbances, an understanding of the system equations is essential and a means by which to solve these model equations. The basic steps to examining a system dynamically are to determine the equations that describe the system, solve these equations for the desired solution, and then characterise the system response. The first two steps have already been done for the single tank scenario described by Figure

3.19. Now we will take this process one step further and examine the system response.

All process systems respond to various disturbances in different ways. Certain types of responses are characteristic of specific types of processes. Two of the most common personalities are those for first and second order systems. The single tank that was mathematically modelled in the previous section is an example of a first order type of system.

3.7.1 FIRST ORDER SYSTEMS

If a step input is applied to a capacity-dominated process such as a single tank, the output begins to change instantaneously but does not reach its steady-state value for a period of time. This is true of any process that is capacitive in nature. It takes approximately five time constants, 5τ, for the output of the capacity process to reach its final, steady-state value. A time constant, τ, is defined as the amount of time it takes the output of the system to reach 63.2% of its steady-state value. τ is a basic characteristic of capacity-dominated physical systems. The time constant, τ, can be defined in electrical terms as the product of the resistance times the capacitance (see Equation 3.28).

$$\tau = RC \tag{3.28}$$

where R is the resistance in the system and C is its capacitance with the units of each being appropriate for the system in question to make the time constant units be units of time, i.e. seconds, minutes, etc.

Figure 3.24 demonstrates the step response behaviour of the single tank example discussed previously. The equations describing the tank system were developed in the previous section and are clearly first order differential equations. Any single capacity system is typically a first-order system and will respond in the same manner illustrated in Figure 3.24.

3.7.2 SECOND ORDER AND HIGHER ORDER SYSTEMS

Higher order responses are the result of multi-capacitance processes that contain vessels in series, fluid or mechanical components of a process that are subjected to accelerations causing inertial effects to become important, or the addition of controllers to a system. In a chemical plant, higher order systems that result from a combination of capacities and controllers are very common. Typical examples are reactors in series, heat exchangers and distillation columns. In the case of distillation columns, when controllers are attached to the column, very high order, nonlinear differential equations result when the system is mathematically modelled. Mechanical component time constants and natural frequencies are very small relative to the process

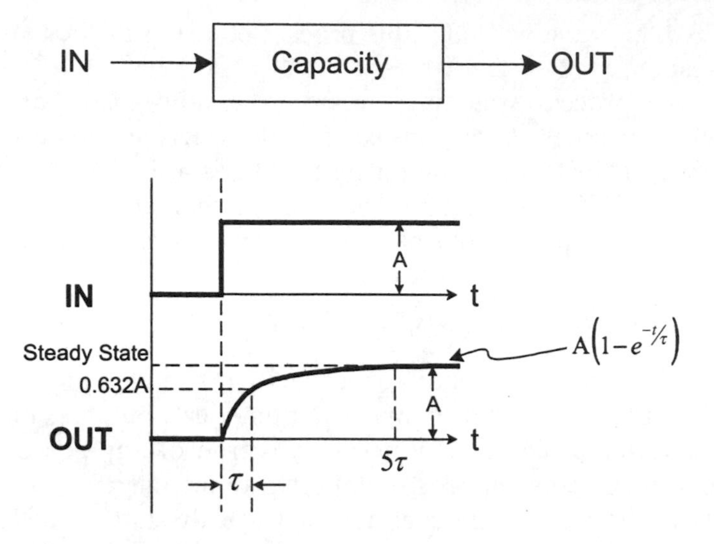

Figure 3.24 First order system response to a step input.

time constants and frequencies and, as such, the resultant effects are typically minor.

In order to get a feel for what a second order system looks like, we will first examine a familiar component from the single input / single output (SISO) system. An integrated part of the SISO system that results in a second order differential equation is the diaphragm operated control valve shown in Figure 3.25.

In order to derive the system equation we first apply Newton's second law, which states:

$$\sum F_i = M \cdot a \tag{3.29}$$

The spring force, viscous friction and acceleration terms are described as follows:

$$\text{spring force} = K \cdot X_v \tag{3.30}$$

$$\text{viscous force} = b \cdot \frac{dX_v}{dt} \tag{3.31}$$

$$\text{acceleration term} = W \cdot \frac{d^2X_v}{dt^2} \tag{3.32}$$

The force is a function of time equal to the pressure applied to the top of the valve, $P(t)$, times the area, A. This term is referred to as the forcing function.

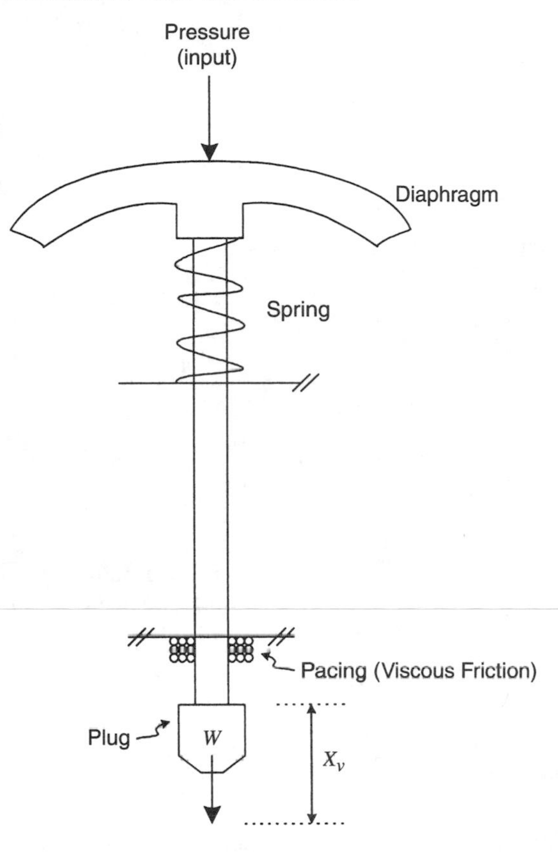

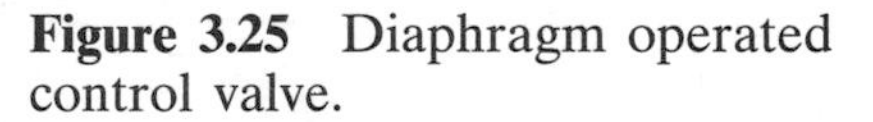
Figure 3.25 Diaphragm operated control valve.

$$W \cdot \frac{d^2X_v}{dt^2} + b \cdot \frac{dX_v}{dt} + K \cdot X_v = A \cdot P(t) \tag{3.33}$$

where: X_v = position of the plug (output)
P = pressure at the input
K = spring constant
b = coefficient of viscous friction
W = weight of the plug and stem

Equation 3.33 is obviously of the second order differential form and, hence, when simulated will give a typical second order response. To better understand what type of response these second order systems will display we will examine another common system and generalise its equations. The closed form solution is best illustrated by an example that is familiar, namely the spring, mass, and damper system presented in Figure 3.26. Note that this is really just a simplification of the control valve example just cited.

If we perform the same analysis on this system as on the previous one, the following equation is obtained describing the system where both the force and displacement are positive in the upward direction.

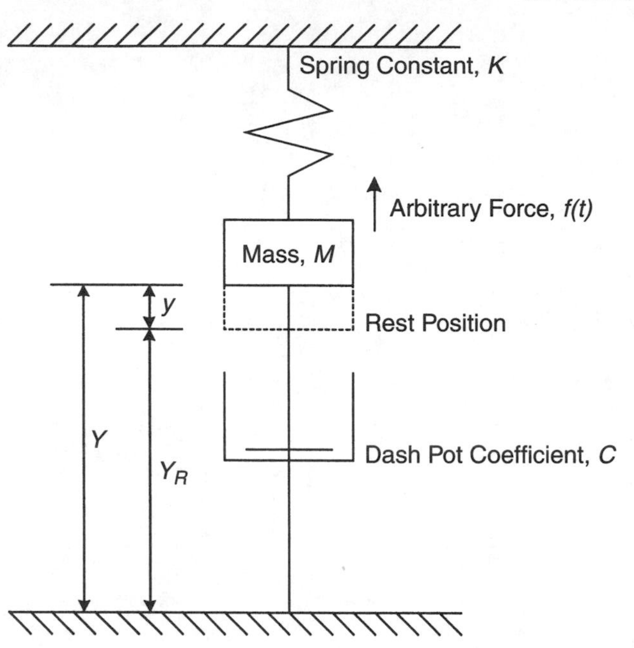

Figure 3.26 Spring, mass and dash pot (damper) system.

$$M\frac{d^2Y}{dt^2} + C\frac{dY}{dt} + K(Y - Y_R) = f(t) \tag{3.34}$$

Since the rest, or equilibrium, position Y_R is constant, Equation 3.34 can be rewritten in terms of the displacement from the rest position, y. In this manner, we will be looking at variations about the equilibrium position, i.e. steady-state. This is a common approach in system analysis because analysis of even nonlinear systems about a steady state results in a linear system, i.e. ordinary differential equations (ODE).

$$y = Y - Y_R \tag{3.35}$$

$$\frac{dy}{dt} = \frac{dY}{dt} \tag{3.36}$$

$$\frac{d^2y}{dt^2} = \frac{d^2Y}{dt^2} \tag{3.37}$$

Substituting Equations 3.35, 3.36, and 3.37 into Equation 3.34 results in the following:

$$M\frac{d^2y}{dt^2} + C\frac{dy}{dt} + Ky = f(t) \tag{3.38}$$

Another point to be made about the analysis of our system is that we used the lumped parameter simplification. All the mass, friction (dash pot) and self-regulation (spring) are considered to be lumped at one point. The use of a lumped parameter model instead of a distributed parameter simplifies the mathematics of the model by producing ordinary differential instead of partial differential equations.

In order to solve the second order equation which will give the position of the mass as a function of time we need the specific set of initial conditions. For a second order differential equation we use the following conditions:

$$Y(0) = \text{a negative constant} \tag{3.39}$$

$$\frac{d[Y(0)]}{dt} = 0 \tag{3.40}$$

In the simplest case the forcing function $f(t)$ can be set to zero. The resulting homogeneous differential equation can be solved by finding the roots of the characteristic equation given in Equation 3.41.

$$M \cdot r^2 + C \cdot r + K = 0 \tag{3.41}$$

For a second order algebraic, these roots are given by Equation 3.42 and are called the Eigenvalues.

$$r_{1,2} = \frac{-C \pm \sqrt{C^2 - 4KM}}{2M} \tag{3.42}$$

Provided that the roots are all real and unique, the solution is as follows:

$$y = C_1 e^{r_1 t} + C_2 e^{r_2 t} \tag{3.43}$$

C_1 and C_2 are constants evaluated using the two initial conditions. The resulting plot of y versus t will have one of the general responses shown in Figure 3.27. The system descriptive parameters, in this case C, K, and M govern the particular response or behaviour of the system.

Second order systems are very common in the chemical industry and have recently received much attention. The equation describing a second order system such as the spring, mass and damper example can be further generalised by dividing Equation 3.38 by M:

$$\frac{d^2y}{dt^2} + \left(\frac{C}{M}\right)\frac{dy}{dt} + \left(\frac{K}{M}\right)y = \frac{f(t)}{M} \tag{3.44}$$

Equation 3.44 can be further generalised by defining the following terms:

$$\omega_n = \sqrt{\frac{K}{M}} \tag{3.45}$$

$$\xi = \frac{C}{2\sqrt{MK}} \tag{3.46}$$

These generalised terms then characterise the response of the system. The first term, ω_n is called the undamped natural frequency, while ξ is known as the damping coefficient. Note that the natural frequency of the system is proportional to K (the tendency to self-regulate) and inversely proportional to M (the capacitance or inertia of the system). Note also that damping is directly proportional to C (system friction), but inversely proportional to M (capacitance or inertia) and K (self-regulation). Understanding how the frequency and damping in a system is affected by these fundamental process characteristics can be useful as the control scheme for a real chemical process is undertaken. Remember, sometimes some simple changes in the process itself can make the job of designing and tuning a regulatory control system much simpler!

The solution to Equation 3.44 can again be found by finding the roots of the characteristic equation, shown in Equation 3.47.

$$r^2 + 2\xi\omega_n r + \omega_n^2 = 0 \tag{3.47}$$

$$r = \frac{-2\xi\omega_n \pm \sqrt{4\xi^2\omega_n^2 - 4\omega_n^2}}{2} \tag{3.48}$$

which simplifies to:

$$r = -\xi\omega_n \pm \omega_n^2\sqrt{\xi^2 - 1} \tag{3.49}$$

or:

$$y(t) = e^{-\xi\omega_n t}\left[C_1 e^{(-\omega_n(\xi^2-1)^{1/2})t} + C_2 e^{(\omega_n(\xi^2-1)^{1/2})t}\right] \tag{3.50}$$

The response of the system will depend mainly on the damping coefficient, ξ. When $\xi < 1$, the system is underdamped and has an oscillatory response. The smaller the value of ξ, the greater the overshoot. If $\xi = 1$, the system is termed critically damped and has no oscillation. A critically damped system provides the fastest approach to the final value without the overshoot of an underdamped system. Finally, if $\xi > 1$, the system is overdamped. An overdamped system is similar to a critically damped system in that the response never overshoots the final value. However, the approach for an overdamped system is much slower and varies depending upon the value of ξ. These typical responses are illustrated in Figure 3.27.

Now let us examine the case of multiple capacities in a series. Consider two non-interacting tanks in a series, shown in Figure 3.28.

The mass balances for Tank 1 and Tank 2 are given by Equations 3.51 and 3.52, respectively.

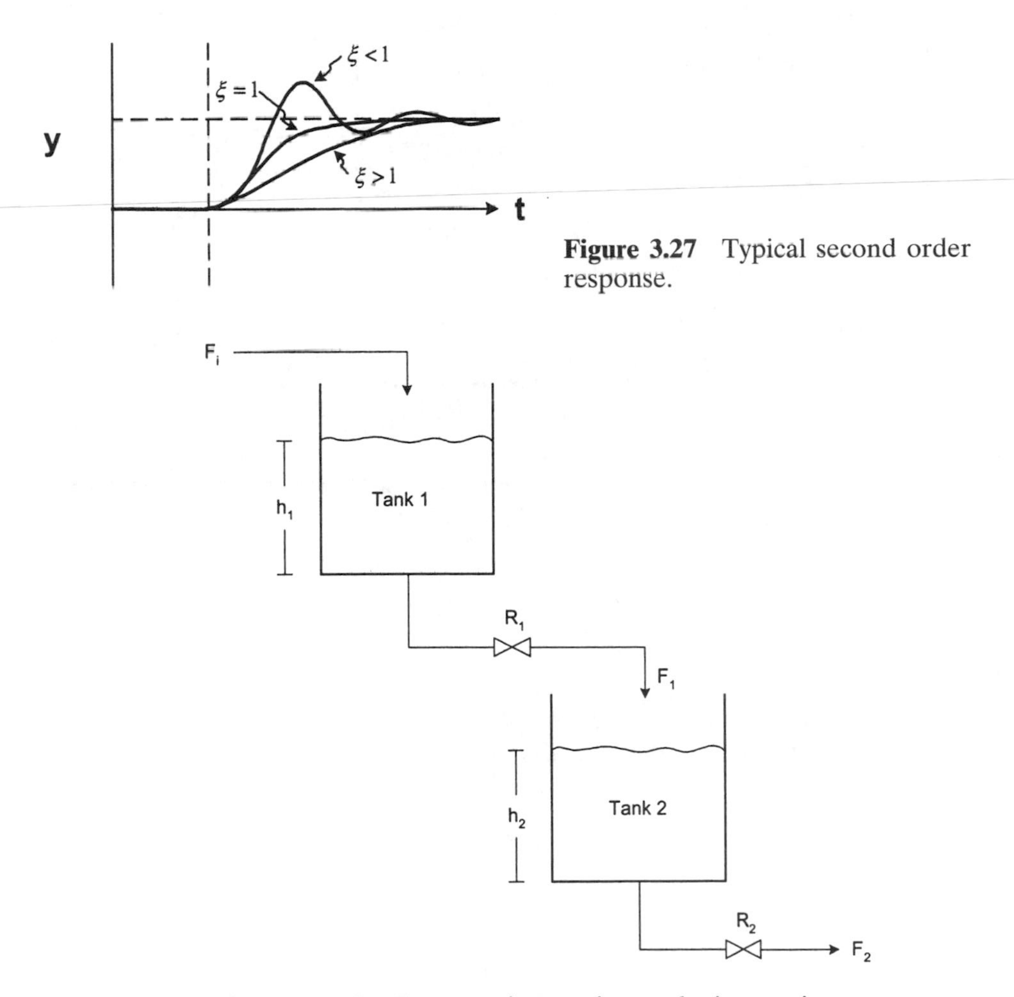

Figure 3.27 Typical second order response.

Figure 3.28 Two non-interacting tanks in a series.

$$A_1 \frac{dh_1}{dt} = F_i - F_1 \tag{3.51}$$

$$A_2 \frac{dh_2}{dt} = F_1 - F_2 \tag{3.52}$$

If linear resistance to flow is assumed for the valves in the system the following equations are obtained:

$$F_1 = \frac{h_1}{R_1} \tag{3.53}$$

$$F_2 = \frac{h_2}{R_2} \tag{3.54}$$

Now substitute Equations 3.53 and 3.54 into Equations 3.51 and 3.52 to give:

$$A_1 R_1 \frac{dh_1}{dt} + h_1 = F_i R_1 \tag{3.55}$$

$$A_2 R_2 \frac{dh_2}{dt} + h_2 = \frac{R_1}{R_2} h_1 \tag{3.56}$$

Differentiating the second equation with respect to time and rewriting h_1 in terms of h_2 gives a second order ODE:

$$\frac{d^2 h_2}{dt^2} + \left(\frac{1}{A_2 R_2} + \frac{1}{A_1 R_1}\right)\frac{dh_2}{dt} - \left(\frac{1}{A_1 A_2 R_1 R_2}\right) h_2 = \frac{F_i}{A_1 A_2 R_1} \tag{3.57}$$

We can then apply the same generalisation to Equation 3.57 as we did to Equation 3.44. This generalisation gives the following:

$$2\xi\omega_n = \frac{1}{A_2 R_2} + \frac{1}{A_1 R_2} \tag{3.58}$$

$$\omega_n^2 = \sqrt{\frac{1}{A_1 A_2 R_1 R_2}} \tag{3.59}$$

Now that the equations describing the system have been developed, the system can be simulated and its response to disturbances examined. Based on the equations developed for the single tank and the non-interacting tanks in A series, what type of response would the level in the first and second tank display?

3.7.3 SIMPLE SYSTEM ANALYSIS

Often questions arise in the design of a process concerning the controllability of the system, alternative control schemes and variations in the process design to achieve quality and/or throughput. In order to answer such questions, without building the plant, it is necessary to have available a rigorous mathematical model or modelling system. Once the system, which includes plant unit operations and controllers, is modelled and simulated, the effect of various parameters and control schemes can be predicted and evaluated.

The determination of the process mathematical model is often the most difficult and time-consuming step in control system analysis. This is a result of the dynamic nature of the process; in other words how the system reacts during upsets or disturbances. The problem is further complicated by process nonlinearities and time varying parameters. To illustrate the modelling procedure we will look at developing a model for a shell and tube heat exchanger with temperature control [7], shown in Figure 3.29.

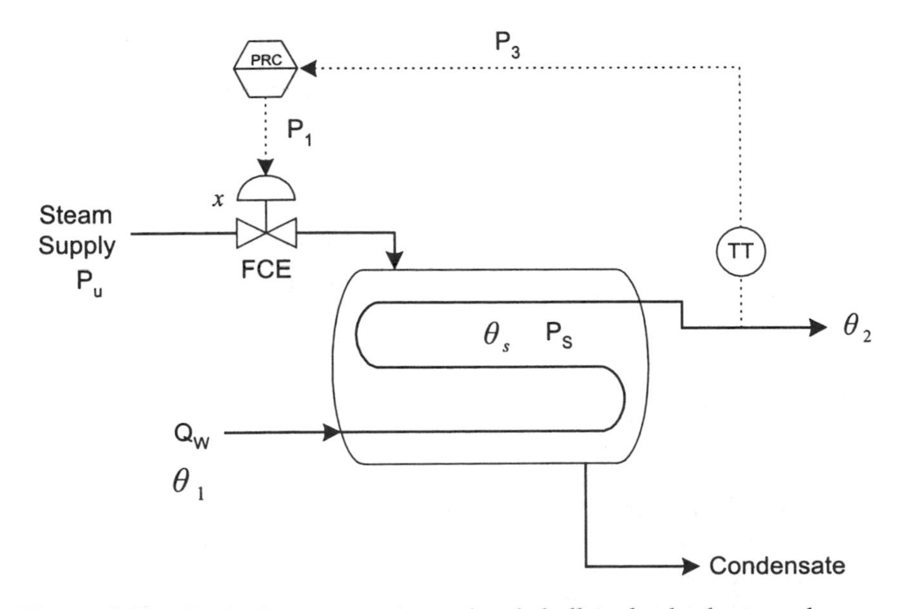

Figure 3.29 Typical process schematic of shell and tube heat exchanger.

Constructing a word block diagram

Before starting to analyse the process, it is helpful to construct a system block diagram, Figure 3.30. The purpose of the block diagram is to identify the components of the SISO and all load disturbances. Recall that a simple SISO feedback loop is comprised of four basic units:

1. Process.
2. Sensing element/transmitter.
3. Controller.
4. Final control element (typically valve and positioner).

In order to limit the complexity of the analysis all parameters will be assumed to be constant. The procedure will be to develop the ordinary differential equations for each of the components and combine these into a system model.

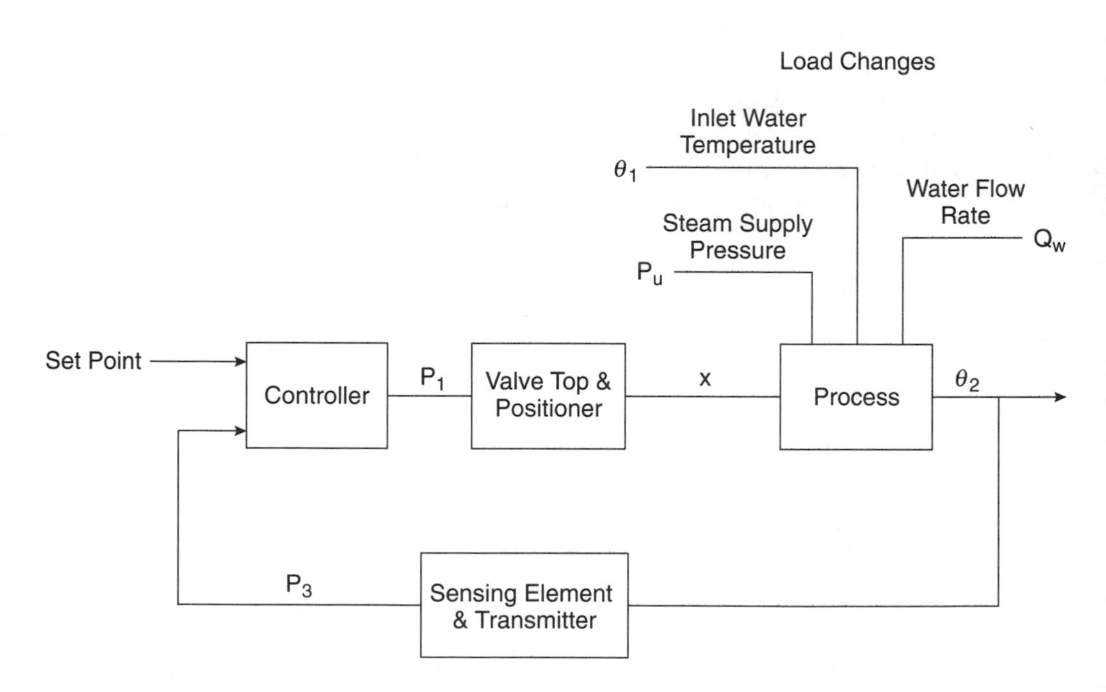

Figure 3.30 Word block diagram of shell and tube heat exchanger.

Valve top and positioner

One approach to modelling SISO loop components is through step testing. For the valve and valve positioner a first order ODE such as the following may be determined.

$$\tau \frac{d\Delta X}{dt} + \Delta X = K \, \Delta P_1 \qquad (3.60)$$

where X is the valve stem position, P_i is the valve top pressure, τ is the first order time constant, K is the steady-state gain, and t is the time. For this valve and valve position, the steady state gain is $K = 0.017$ mm/kPa and the time constant is $\tau = 0.033$ minutes. Therefore, we can write Equation 3.60 as follows:

$$0.033 \frac{d\Delta X}{dt} + \Delta X = 0.017 \, \Delta P_1 \qquad (3.61)$$

Equations 3.60 and 3.61 are written in terms of variation variables. Variation variables represent a change from or about a steady-state level of the variable. The gain was determined by dividing the valve stem travel by the pneumatic signal range, which is 14.3 mm divided by 80 kPa (the range in this case is from 20–100 kPa). The time constant of 0.033 is determined experimentally using a step input signal (discussed previously under First order systems).

Sensing element and transmitter

These components can also be represented by a first order ODE, as in Equation 3.62. In most cases this is an adequate representation for SISO loop analysis.

$$\tau \frac{d\Delta P_3}{dt} + \Delta P_3 = K \, \Delta \theta_2 \qquad (3.62)$$

where P_3 is the pressure signal from the temperature sensor/transmitter to the controller and θ_2 is the exit water temperature. The steady-state gain, K, is determined from the sensor/transmitter ranges, which are 0–100°C input and 20–100 kPa output. This results in a gain of 0.80 kPa/°C.

The time constant of the measuring element equals the thermal resistance times the capacitance. The thermal resistance can be modelled as a function of area and the heat transfer film coefficient,

$$R_t = \frac{1}{Ah} \qquad (3.63)$$

where: R_t = thermal resistance, °C/kW
A = surface area, m^2
h = film coefficient, kW/°C.m^2

The thermal capacitance is a function of mass and specific heat:

$$C_t = WC \tag{3.64}$$

where: C_t = thermal capacitance, kJ/°C
W = mass of sensing element, kg
C = specific heat, kJ/kg.°C

The time constant, which has units of hours as a result of the units of the film coefficient, can then be found using Equation 3.63 and Equation 3.64:

$$\tau = R_t C_t = \frac{WC}{Ah} \tag{3.65}$$

Once the sensing element has been selected, its area, weight, and specific heat are fixed. Therefore τ is only a function of h. If the manufacturer provides a time constant, τ, for a given set of conditions, other τ can be estimated based on the new conditions. If the system properties are about the same as those the manufacturer used during the step tests, h will primarily be a function of fluid velocity. Assuming that the bulb has a time constant over a limited velocity range as follows:

$$\tau = 1.15\left(\frac{1}{v^{0.58}}\right) \tag{3.66}$$

where: τ = time constant, seconds
v = fluid velocity, m/s

Thus, for a fluid velocity of 0.686 m/s, the time constant of the sensing element is 0.024 min.

Process model

The process model can be determined either from first principles (the mechanistic approach) or by "black boxing". Mechanistic approaches attempt to model transient energy and mass balance. Black box models are often even simpler and describe the input–output behaviour with no recourse to conservation principles. Using a linear system analysis approach, the heat exchanger is modelled as one lump (lumped parameter approach) in which a small change in the valve stem position, ΔX, and its effect on the outlet water temperature, θ_2, is predicted. A change in X results in more or less steam entering the shell, which changes the energy input to the heat exchanger. This change in energy input is accounted for by a change within the exchanger and a change in the energy leaving the exchanger. If the inlet water flow rate, Q_w, and inlet temperature, θ_1, are constant, any change in energy will show up as a change in water outlet temperature, θ_2.

The steam flow, Q_s, through the valve can be modelled as follows:

$$Q_s = f(X,P_s) = \left(\frac{3}{2}\right)C_v P_s = 0.00086XP_s \qquad (3.67)$$

where: Q_s = steam flow, kg/s
X = valve stem position or travel, mm
P_s = steam pressure, kPa

In terms of variation variables, the change in steam flow can be modelled as follows:

$$\Delta Q_s = \alpha\,\Delta P_s + \beta\,\Delta X \qquad (3.68)$$

where:

$$\alpha = \left(\frac{\partial Q_s}{\partial P_s}\right)_x = (0.00086)X_{op}$$

$$\beta = \left(\frac{\partial Q_s}{\partial X}\right)_{P_s} = (0.00086)P_{s_{op}}$$

If one assumes saturated steam in the shell, there is a unique relationship between changes in steam pressure, P_s, and changes in steam temperature, θ_s:

$$\Delta P_s = \gamma\,\Delta\theta_s = \left(\frac{\partial P_s}{\partial \theta_s}\right)\theta_s \qquad (3.69)$$

The coefficient γ can be evaluated from steam tables at the shell nominal operating pressure of 432 kPa, refer to Table 3.1. As can be seen from Figure 3.31, at an operating pressure of 432 kPa, γ can be linearised to give a value of 11.0 kPa/°C.

Table 3.1 Saturated steam temperature

P_s (kPa)	θ_s (°C)
300	133.5
350	138.9
400	143.6
450	147.9
500	151.8

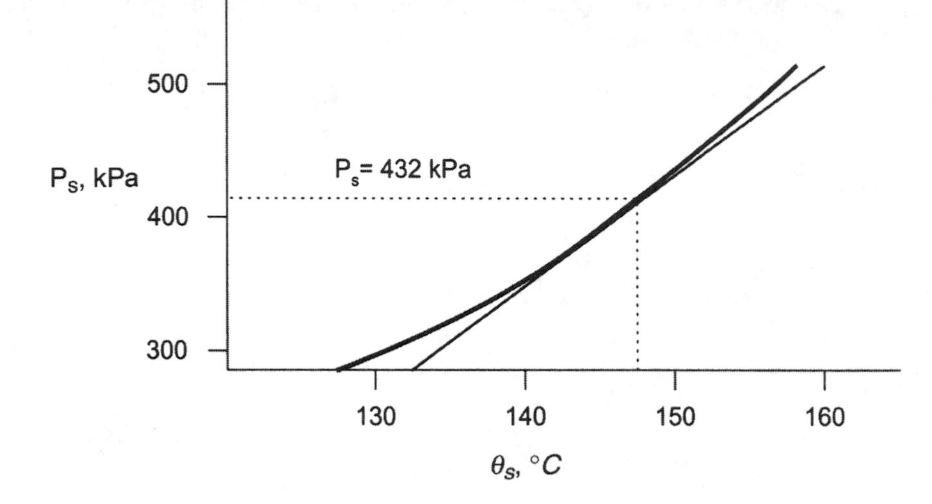

Figure 3.31 Steam pressure versus steam temperature.

An energy balance on the shell using the principle of energy conservation can be written as follows:

$$\Delta Q_s m = (W_t C_t + W_s C_s)\Delta\theta_s + h_t A_t\left(\theta_s - \frac{\Delta\theta_2}{2}\right) \tag{3.70}$$

where:
- m = latent heat of condensing steam, kJ/kg (2.129 kJ/kg for this example)
- W_t = tube weight, kg
- C_t = specific heat of tube material, kJ/kg.°C (0.5 kJ/kg.°C for this example)
- W_s = shell weight, kg (19.6 kg for this example)
- C_s = specific heat of shell material, kJ/kg.°C
- h_t = tube side film coefficient, kW/°C.m^2
- A_t = tube area, m^2 (h_t A_t = 0.376 kJ/s.°C)
- θ_s = steam temperature, °C
- θ_2 = outlet temperature, °C

The shell side energy balance contains a number of simplifying assumptions. The shell side heat transfer film coefficient is assumed to be negligible, thus the temperature of the tube and shell walls is equal to the condensing steam temperature. Shell steam capacity is also assumed to be negligible due to the small volume. Losses to the atmosphere are neglected, i.e. the shell is well insulated.

In the development of a mathematical model the validity of assumptions is always debatable and depends on the use of the model, required accuracy, equipment size and configuration. These need to be considered in light of where and how the model is to be used.

The water temperature was taken as the average between the inlet and outlet temperatures. This assumption is valid since the inlet temperature is assumed to be constant, hence the change in the average water temperature is half the change in the outlet water temperature. An energy balance for the tube side results in the following equation:

$$W_s C_s \frac{d(\Delta\theta_2/2)}{dt} + Q_w C_w \Delta\theta_2 = h_t A_t \left(\Delta\theta_s - \frac{\Delta\theta_2}{2}\right) \tag{3.71}$$

where: Q_w = water flow into tube, kg/s
C_w = specific heat of water, kJ/kg.°C (4.2 kJ/kg.°C for this example)
W_s = shell weight, kg
C_s = specific heat of shell material, kJ/kg.°C (W_s C_s = 3.96 kJ/°C)
h_t = tube side film coefficient, kW/°C.m^2
A_t = tube area, m^2
θ_s = steam temperature, °C
θ_2 = outlet temperature, °C

Controller

If we use a standard PI controller, it can be modelled using Equation 3.72. A PI controller takes remedial action proportional to the magnitude of both the error and the integral of the error, and is rigorously defined in Chapter 4.

$$\Delta P_1 = K_e \left[\Delta e + \frac{1}{T_i}\int_0^t \Delta e \, dt\right] \tag{3.72}$$

where K_c is the controller gain, an adjustable tuning parameter of the controller, T_i is the integral time, another adjustable tuning parameter of the controller and Δe is the error and is defined as the difference between the measured variable and the set point, which is 65°C in this example.

$$\Delta e = \Delta P_3^{SP} - \Delta P_3 = 65 - \Delta P_3 \tag{3.73}$$

Response

The time response of the outlet temperature to various load disturbances can be determined by integrating the set of ODEs, as developed previously. This can be accomplished by using one of the standard mathematical software packages, such as MATLAB™ [8]. For clarity these equations are repeated below with their original numbering and are in the order that they appear in the control loop, with the values of the constant parameters shown.

$$0.033\frac{d\Delta X}{dt} + \Delta X = 0.17\,\Delta P_1 \tag{3.61}$$

$$\Delta Q_s = (0.00086X_{op})\,\Delta P_s + (0.00086P_{s_{op}})\,\Delta X \tag{3.68}$$

$$\Delta P_s = (11.0)\,\Delta\theta_s \tag{3.69}$$

$$\Delta Q_s\,(2129) = (0.5W_t + (0.5)(19.6))\Delta\theta_s + (0.376)\left(\theta_s - \frac{\Delta\theta_2}{2}\right) \tag{3.70}$$

$$(3.96)\frac{d(\Delta\theta_2/2)}{dt} + Q_w(4.2)\Delta\theta_2 = (0.376)\left(\Delta\theta_s - \frac{\Delta\theta_2}{2}\right) \tag{3.71}$$

$$0.024\frac{d\Delta P_3}{dt} + \Delta P_3 = 0.80\,\Delta\theta_2 \tag{3.62}$$

$$\Delta P_1 = K_c\left[(65 - \Delta P_3) + \frac{1}{T_i}\int_0^t (65 - \Delta P_3)\,dt\right] \quad \text{(3.72 and 3.73)}$$

The resulting response for this system to a PI controller set point change of 10°C is shown in Figure 3.32.

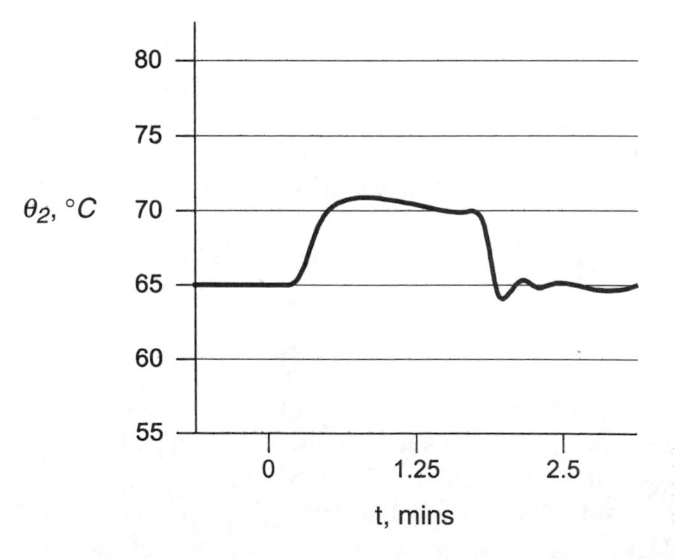

Figure 3.32 Typical response to a PI controller set point change.

The response in Figure 3.32 is based on a simple linear lumped parameter model, thus for large disturbances or set point changes that exceed the limits of the linear assumptions and other operating point, quite different responses will be obtained.

Mathematical models imbedded in today's simulation software provide a means to handle both variations in operating level and process nonlinearities. The important result is not the controller settings but the ability of the designer to manipulate process parameters to meet process specifications at a minimum cost. This can only be achieved by developing a deep understanding of the process and the system, which includes the process and the controllers.

3.8 References

1. Bequette, B.W., *Process Dynamics: Modeling, Analysis, and Simulation*. Prentice Hall, New Jersey, 1998.
2. Riggs, J.B., *An Introduction to Numerical Methods for Chemical Engineers*, 2nd edn. Texas Technical University Press, Texas, 1994.
3. Smith, C.A., Corripio, A.B., Principles and Practice of Automatic Process Control. John Wiley & Sons, New York, 1985.
4. Downs, J.J., Doss, J.E., "Present Status and Future Needs – A View From North American Industry". Proceedings of 4th International Conference on Chemical Process Control, Feb. 17–22, Padre Island, Texas, 1991, pp. 53–77.
5. Shinskey, F.G., "Process Control Systems: Application, Design and Tuning", 4th edn, McGraw-Hill, New York, NY, 1996, p. 4.
6. Johnson, R.E., Kiokemeister, F.L., "Calculus with Analytical Geometry". Allyon and Bacon, Boston, 1957, p. 604.
7. Wilson, H.S., Zoss, L.M., "A Practical Problem in Process Analysis". Control Theory Notebook, ISA Journal Reprints, 1962, pp. 25–8.
8. MATLAB™, Version 5.3, the MathWorks Inc., Natick, MA, USA, 1999.

4 BASIC CONTROL MODES

The previous chapter discussed basic feedback control concepts including the vital role of the controller. Again, the purpose of the controller in regulatory control is to maintain the controlled variable at a predetermined set point. This is achieved by a change in the manipulated variable using a pre-programmed controller algorithm. This chapter will describe the basic control modes or algorithms used in controllers in feedback control loops.

4.1 On-off control

The most rudimentary form of regulatory control is *on-off control.* This type of control is primarily intended for use with final control elements (FCE) that are non-throttling in nature, i.e. some type of switch as opposed to a valve. An excellent example of on-off control is a home heating system. Whenever the temperature goes above the set point, the heating plant shuts off, and whenever the temperature drops below the set point, the heating plant turns on. This behaviour is shown by Equation 4.1.

$$mv = 0\% \text{ for } PV > SP \text{ and } mv = 100\% \text{ for } PV < SP \tag{4.1}$$

The controller output, mv, is equal to 0% or off whenever PV exceeds the set point, SP. Whenever the process variable is below the set point, the controller output is equal to 100% or on.

The most useful type of process where on-off control can be successfully applied is a large capacitance process where tight level control is not important, i.e. for the case of flow smoothing. A good example of this type of process is a surge tank. A large capacitance is important due to the nature of the controller action and its effect on the operational life of the FCE. This leads us to one of the disadvantages of an on-off type of controller. Due to the continual opening and closing of the controller, the FCE quickly becomes worn and must be replaced. This type of control action is illustrated in Figure 4.1, which shows the typical behaviour of an on-off controller.

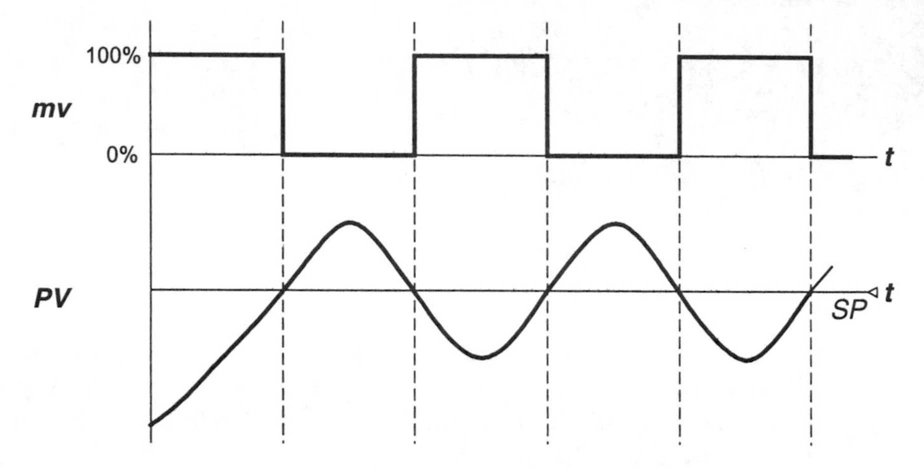

Figure 4.1 On-off controller response.

In this example, at time, $t = 0$, PV is less than SP, and mv is equal to 100%. When PV crosses the set point, mv becomes 0%. The temperature rises somewhat above the set point before the controller turns off because of dead time, the capacitance of the heating system, and heat transfer to the ambient. These factors are termed system dynamics. When the temperature drops below the set point, the controller opens the valve. However, again due to system dynamics, the temperature drops somewhat below the set point before PV begins to rise again. It is easy to see how the FCE would quickly become worn out when this action is continually occurring.

Since the controller cannot throttle the actuator, but only turn it on or off, the primary characteristic of on-off control is that the process variable is always cycling about the set point. The rate at which PV cycles and the deviation of PV from the set point are a function of the dead time and capacitance in the system or the system dynamics. The longer the lag time, the slower the cycling but the greater the deviation from the set point. This can better be illustrated by using an on-off controller with a differential gap or dead band as shown in Figure 4.2. Most on-off controllers are built with an adjustable differential gap or dead band, inside which no control action takes place. The intent of this differential gap is to minimise cycling of the controller output and extend the operational life of the FCE.

The controller switches off when the process variable exits the dead band on the high side and does not turn on again until PV is outside the dead band on the low side. The frequency of cycling is reduced, but the deviation from the set point is increased. If the dead band is reduced the frequency of cycling is increased but deviation from set point is decreased.

Typically, the dead band is adjusted as a percentage of the process variable span. Using the heating system example, suppose the temperature measure-

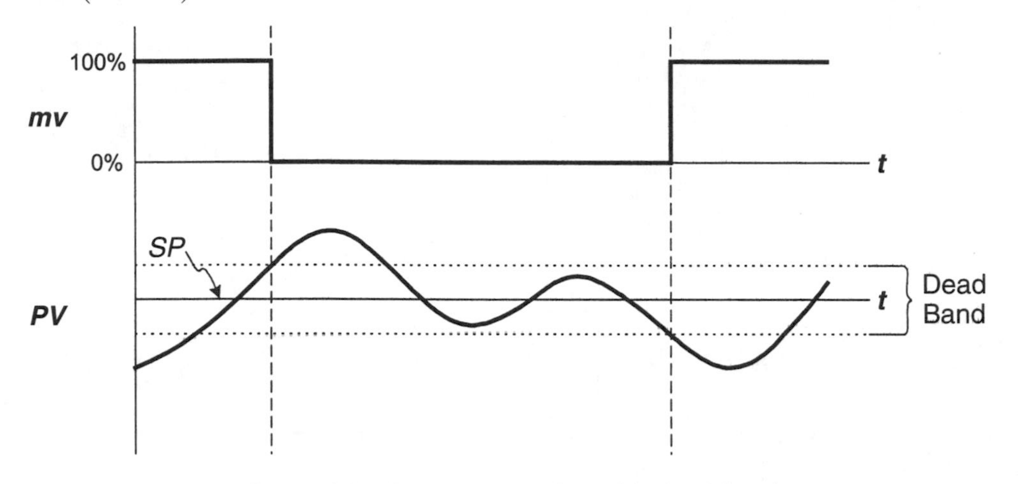

Figure 4.2 On-off controller with dead band.

ment range was from 20° to 120°. Setting the dead band equal to 10%, the dead band in degrees would be 10°. If the set point were 75°, then the upper edge of the dead band would be 75° + 5° = 80° and the lower edge of the dead band would be 75° − 5° = 70°, giving the dead band width of 80° − 70° = 10°.

With an on-off controller cycling cannot be eliminated. When a large lag is present in the process, the deviation from the set point may not be perceptible since the amount of time per cycle is longer. If this is acceptable, an on-off controller can be used. However, in order to eliminate cycling completely, another control mode would need to be implemented.

4.2 Proportional (P-only) control

Proportional control is the simplest continuous control mode that can damp out oscillations in the feedback control loop. This control mode normally stops the process variable, *PV*, from cycling but does not necessarily return it to the set point.

For example, consider the liquid level control situation given in Figure 4.3, in which the tank must not overflow or run dry. If the inflow, F_i, is equal to the outflow, F_o, then the level, as seen in the sight glass, remains constant. If F_o increases such that it is greater than F_i, the level will begin to drop. In order to stop the level from dropping F_i must be increased by opening the inflow valve until it is equal to F_o, and the level stops dropping. However, the tank level is no longer at the initial level; it has dropped to a new steady-state level. The amount the level drops depends on how much the inflow valve has had to move to make F_i equal to F_o. A similar situation would occur if F_o was less than F_i, only in this case the level would rise until the

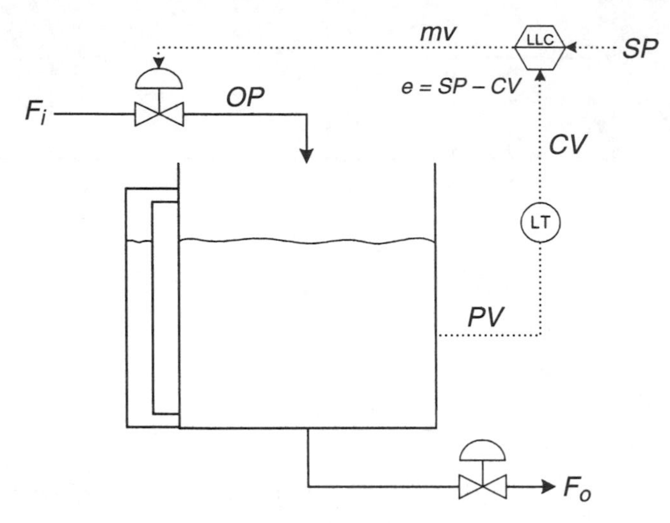

Figure 4.3 Liquid level control – proportional mode.

readjusted inflow equals the outflow. This scenario describes what a proportional controller would do if it were connected to the tank.

In equation form the output of a proportional controller is proportional to the error[1], Equation 4.2.

$$mv = K_c e \tag{4.2}$$

In Equation 4.2, K_c is the controller gain, e is the error, and mv is the manipulated variable. Remember:

$$e = SP - CV \quad \text{(for reverse acting)}$$

$$e = CV - SP \quad \text{(for direct acting)}$$

To allegorise proportional control we will use the liquid level loop shown in Figure 4.3. Initially, the proportional controller is placed in manual and the level in the tank is manually adjusted to equal the set point. With F_i equal to F_o, the level should stay at the set point. Also, set $F_o = 50\% = F_i$, $CV = SP = 50\%$, and $K_c = 2$. Now, if the controller is placed to auto what will happen to the output? At the instant the controller is placed to auto, the error will be zero since CV is equal to SP, and the controller output will also be zero:

$$mv = 2(50 - 50) = 0 \text{ (from Equation 4.2)}$$

[1] Error is the deviation of the measurement, CV, from the set point, SP

For a controller output of zero, what will the level do? The level will begin to drop. To stop this movement F_i and F_o must equal 50% again. If a linear relationship is assumed between inflow and controller output, then for F_i = 50% we will have mv = 50%.

Since	$mv = K_c e = 2(SP - CV) = 2\ (50\% - CV)$
For	$mv = 50\%,\ e = 25\%$
Therefore	$mv = 50\%$ when $CV = 25\%$

Thus, the controller output becomes 50% when the measurement, CV, drops by 25%, creating a 25% error. For this case, in order to stop the level from dropping, the proportional controller had to drop CV to create a large enough error so the controller could make $F_i = F_o$.

Suppose K_c was set equal to 4, giving $mv = 4e$. Now, the error would only need to be 12.5% for $mv = 4(12.5\%) = 50\%$. Logically, it would appear that the larger the controller gain, the smaller the error. In theory if K_c is set to infinity the error can be reduced to zero. The problem with this extrapolation is that the gain of the controller, K_c, is multiplied by all the gains of the other elements to give the loop gain, K_L. If K_c becomes large enough, the loop gain will be greater than one, thus causing the loop to become unstable. Because of this loop gain limit, there is a limit to how large the controller gain can be. However, there is another approach to reducing the error to zero. Suppose another term was added to the proportional controller equation, as shown in Equation 4.3:

$$mv = K_c e + b \tag{4.3}$$

This additional term is called the bias, b, and is simply defined as the output of the controller when the error is zero. Using the previous example again, let us set K_c equal to two. Also, manually adjust $CV = SP = 50\%$, $F_i = F_o = 50\%$, and set $b = 50\%$. Now, when the controller is set to auto, what will happen? Since CV is equal to SP, e is equal to zero and, hence, $K_c e = 2(0) = 0$. There is no proportional contribution to the output and the output, mv, is equal to the bias which is 50% (see Equation 4.3). Since F_o is equal to 50% and mv is also equal to 50%, the level will stay the same. In general, if the bias equals the load, mv, the error will always be zero.

Now, suppose F_o becomes 75%. In order to stop the level from dropping, mv must equal F_o, which in this case is 75%. From Equation 4.3, $mv = 2e + 50\% = 2(50\% - CV) + 50\%$, and CV must drop to 37.5% to make the output, mv, equal to 75%. When mv is equal to the outflow, the level will stop dropping. The level could also be prevented from dropping if the outflow, F_o, was decreased. Suppose F_o is equal to 25%. In this case, the level will stop rising when CV is equal to 62.5%, since that gives $mv = 2(50\% - CV) + 50\% = 25\%$.

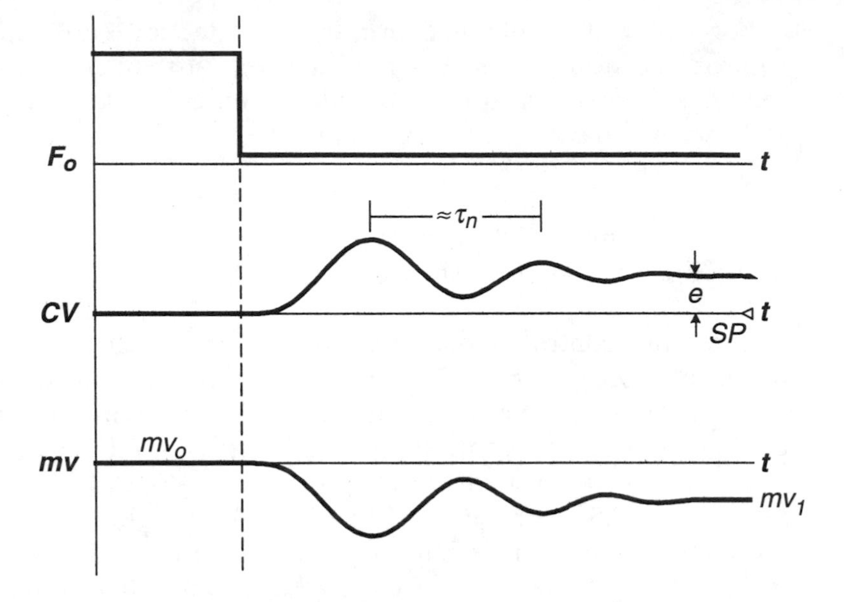

Figure 4.4 Typical proportional only controller response.

As previously mentioned, increasing K_c can decrease the error, but remember not to increase K_c such that it makes the loop unstable. There is a limit for each feedback control loop. If K_c has a value such that the loop gain, K_L, is equal to one, the loop will oscillate with a period that is a function of the natural characteristics of the process. This is called the natural period, τ_n. If K_c is adjusted such that the loop gain is equal to 0.5 and a change is made in F_o, the response shown in Figure 4.4 could be expected.

CV damps out with a quarter decay ratio[2] and a period approximately equal to the natural period. It then stabilises with an offset that is a function of both the controller gain and the bias. The offset is the sustained error, *e*, where *CV* does not return to the set point even when steady state is reached. This is a typical response for a loop under proportional only control.

Now let us look again at Equation 4.3, and recall that the gain of any loop element is defined by Equation 4.4.

$$K = \frac{\Delta\text{output}}{\Delta\text{input}} \tag{4.4}$$

The block diagram of a proportional controller can be represented as shown in Figure 4.5.

[2] Quarter decay ratio is discussed in greater detail in Chapter 5

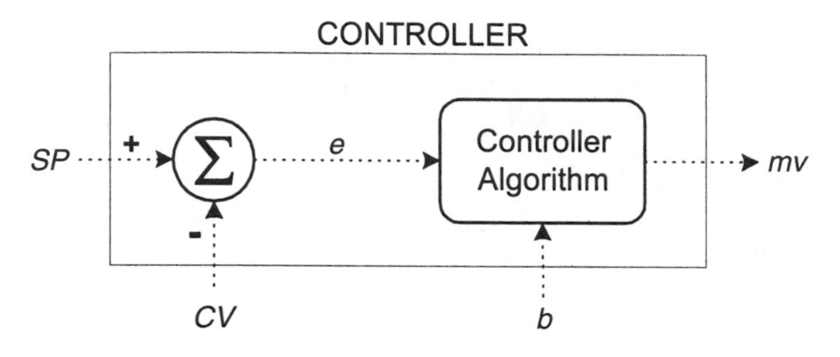

Figure 4.5 Block diagram of proportional only controller.

The controller gain is the ratio of the change in controller output to the change in error. Hence, the gain of the proportional controller, K_c, is given by Equation 4.5:

$$K_C = \frac{\Delta mv}{\Delta e} \tag{4.5}$$

Since there is a one-to-one relationship between CV and e the controller gain can be written as per Equation 4.6:

$$K_C = \frac{\Delta mv}{\Delta CV} \tag{4.6}$$

The controller gain can also be defined as a change in controller output for a change in the process variable, PV. This is true because the controlled variable, CV, is the transformed process variable from the transmitter to the controller. Therefore, CV is essentially PV, only in different units, i.e. % level instead of milliamperes.

Assuming that a linear relationship exists between CV and mv, as shown in Figure 4.6, Equation 4.7 may be written as follows:

$$K_C = \frac{\Delta mv}{\Delta CV} = \frac{100\%}{\Delta CV} \tag{4.7}$$

The controller gain, K_c, in Equation 4.7, is the amount that CV must change to make the controller output change by 100%. The gain of the transmitter is similar and is given by Equation 4.8.

$$K_T = \frac{\Delta out}{\Delta in} = \frac{100\%}{span} \tag{4.8}$$

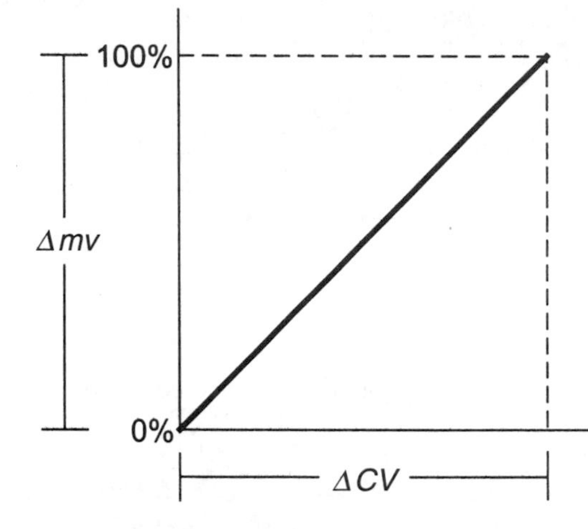

Figure 4.6 Controller input/output relationship.

In other words, the input of the transmitter changes the amount of the span to make the transmitter output change by 100%. The span is the difference between the upper and lower value of the range.

The case of the controller is analogous to that of the transmitter, but instead of calling ΔCV the span, it is called PB, the proportional band. The proportional band is defined as the change in CV that will cause the output of the controller to change by 100%. Using this definition of PB we can define the controller gain as shown in Equation 4.9:

$$K_c = \frac{\Delta mv}{\Delta CV} = \frac{100\%}{PB\%} \tag{4.9}$$

If the proportional band setting on the controller is set to 40%, the output of the transmitter, which is the input to the controller, changes over 40% of its output span. The output of the controller would change by 100%, or the controller gain, K_c, would be

$$K_c = \frac{100\%}{40\%} = 2.5$$

Virtually all modern controllers use a gain adjustment, however a few older controllers exist that still use a proportional band adjustment. Remember that $K_c = 100\%/PB\%$ or, as the PB gets larger, the gain gets smaller and vice versa. The equation for a proportional controller in terms of PB can be written as follows:

$$mv = \frac{100}{PB} e + b \tag{4.10}$$

Note that:

$$e = SP - CV \qquad \text{(for reverse acting)}$$

$$e = CV - SP \qquad \text{(for direct acting)}$$

In order to make the error equal to zero, one of the following two possibilities must occur:

1. Set $PB = 0$ ($K_c = \infty$)
2. Set $b = mv$

The first option, as previously discussed, is not plausible since as $PB \to 0$, $K_c \to \infty$ and the loop becomes unstable. Furthermore, it is not possible to set $PB = 0$, because on many controllers the minimum setting is usually 2%–5%. However, if PB was very small, i.e. 2% or $K_c = 50$, the error would certainly be minimised, provided the loop remained stable. This case can be illustrated using Figure 4.7.

If in Figure 4.7, $K_v \times K_p \times K_T < 1/50$, then the loop would be stable since the loop gain, $K_L < 1$ (Equation 3.2). If the process had a lower gain, K_P, then a higher controller gain or smaller PB in the P-only controller could be used to minimise the error. One type of process where this is the case is a very large capacitance process, i.e. a large surge tank. Due to the low process gain, a P-only controller is often used for level control.

The second option to make the error zero is to set the bias equal to the controller output, mv. Some controllers have an adjustable bias and hence make this option viable, as in Equation 4.11.

$$e = \frac{1}{K_c}(mv - b) \tag{4.11}$$

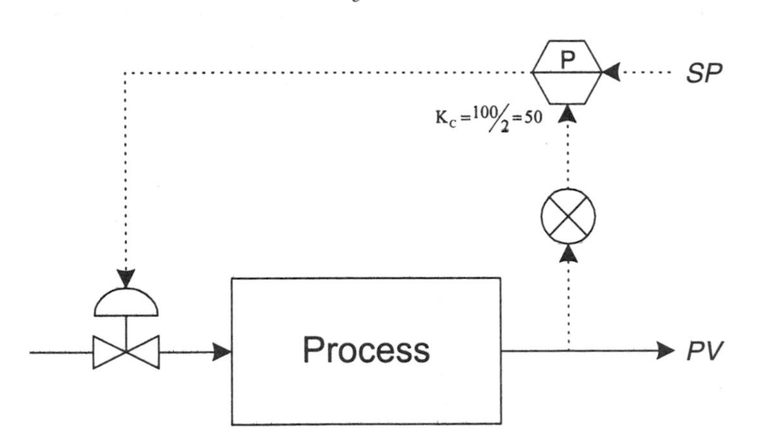

Figure 4.7 SISO feedback control loop.

However, this approach is only an option for processes that experience few load upsets, since a manual readjustment of the bias is required each time there is a load upset. There would be no error as long as the bias was equal to the load. Hence, if the process had infrequent load upsets, the operators could readjust the bias to give zero error, and it would be possible to use a P-only controller.

In general, a proportional controller provides a fast response compared to other controllers but a sustained error occurs where the *PV* does not return to the set point even when steady-state is reached. This sustained error is called offset and is undesirable in most cases. Therefore, it is necessary to eliminate offset by combining proportional control with one of the other basic control modes.

4.3 Integral (I-only) control

The action of *integral control* is to remove any error that may exist. As long as there is an error present, the output of this control mode continues to move the FCE in a direction to eliminate the error. The equation for integral control is given in Equation 4.12.

$$mv = \frac{1}{T_i} \int e \, dt + mv_o \tag{4.12}$$

mv_o is defined as either the controller output before integration, the initial condition at time zero, or the condition when the controller is switched into automatic. The block diagram for an integral only controller is given in Figure 4.8.

The action or response of the integral control algorithm for a given error is shown in Figure 4.9, assuming Increase/Decrease action.

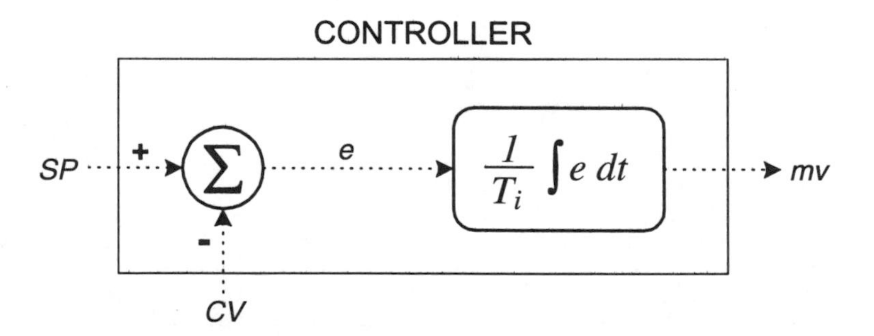

Figure 4.8 Block diagram of integral only controller.

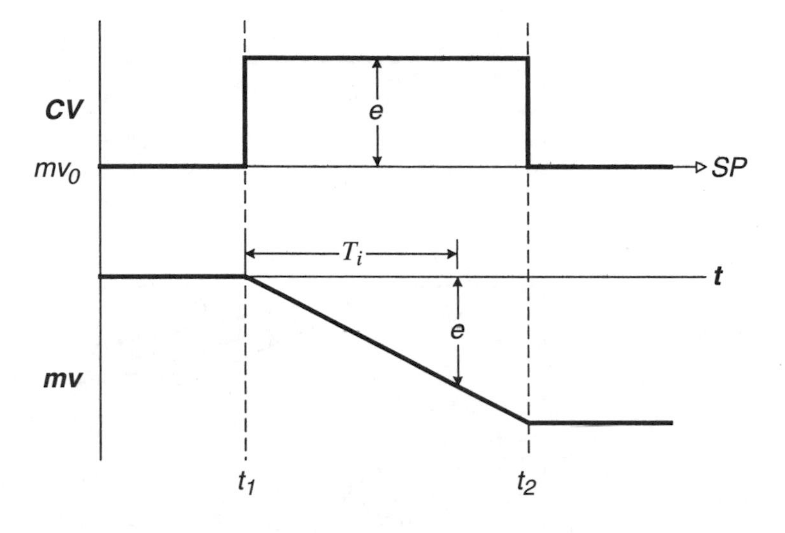

Figure 4.9 Integral controller response to square wave input.

If the measurement, CV, was increased in a step-wise fashion at time t_1 and then returned to the set point at t_2, the output would ramp up over the interval $t_1 < t < t_2$ since the controller is in effect integrating the step input. When the measurement is returned to the set point at $t = t_2$, the output would hold the value that the controller had integrated to, since the controller would think this was the correct value or the set point, i.e. $e = 0$.

The rate at which the controller output ramps is a function of two parameters: the integral time, T_i, and the magnitude of the error. Obviously, the controller output, *mv*, would ramp in the opposite direction if CV had been moved below the set point.

The integral time, T_i, is defined as the amount of time it takes the controller output to change by an amount equal to the error. In other words, it is the amount of time required to duplicate the error. Thus T_i is measured in minutes per repeat. Because of the form of Equation 4.12 some manufacturers measure the reciprocal of T_i or repeats per minute in a controller, Equation 4.13.

$$\frac{1}{T_i}\left[-\right]\left[\frac{1}{\text{mins/repeat}}\right]\left[-\right]\text{[repeats/min]} \tag{4.13}$$

As a result of this reciprocal relationship, if the controller is adjustable in min/rep, then increasing the adjustment gives less integral action, whereas in rep/min, increasing the number produces greater integral action. Therefore, it is important to be aware of how an individual controller adjusts T_i. The rate of change of *mv* also depends on the magnitude of *e* as shown in Figure 4.10, in which T_i is fixed.

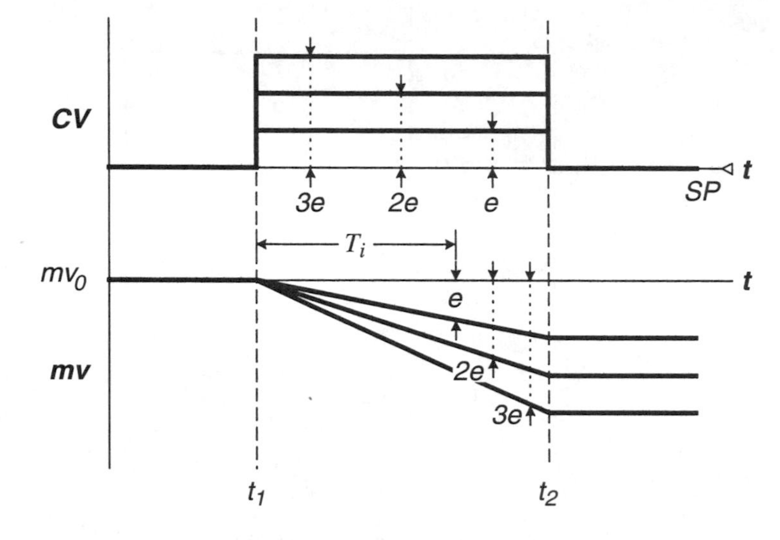

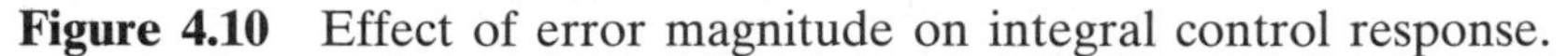

Figure 4.10 Effect of error magnitude on integral control response.

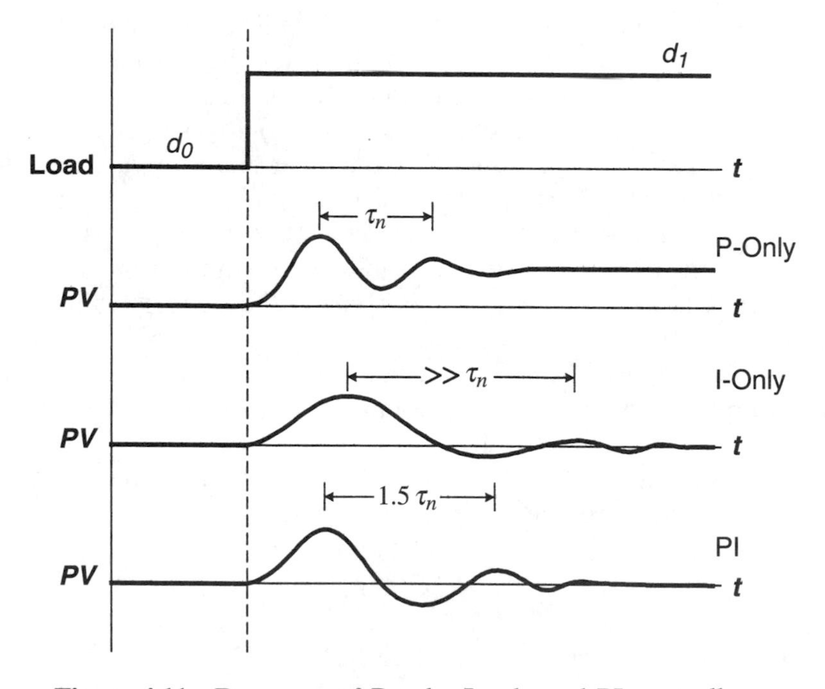

Figure 4.11 Response of P-only, I-only and PI controllers.

Figure 4.11 illustrates the responses of P-only, I-only and PI controllers to a step input. Although an integral only controller provides the advantage of eliminating offset, there is a significant difference in its response time when compared to a proportional only controller. As mentioned earlier, the output of the proportional only controller changes as quickly as the measurement changes; in other words, the controller tracks the error. So, if the measurement changes as a step, the controller output also changes as a step by an amount depending on the controller gain. For a step input to an integral controller, the output does not change instantaneously but rather by a rate that is affected by T_i and e.

Hence, integral only control, due to the additional lag introduced by this mode, has an overall response that is much slower than that for proportional only control. The period of response for the *PV* under integral only control can be up to ten times that for proportional only; so a trade-off is made when using an I-only controller. If no offset is required then a slower period of response must be tolerated. If the requirement is a return to the set point with no offset, and a faster response time is necessary, then the controller must be composed of both proportional and integral action.

4.4 Proportional plus integral (PI) control

A *proportional plus integral controller* will give a response period that is longer than a P-only controller but much shorter than an I-only controller. Typically, the response period of the process variable, *PV*, under PI control is approximately 50% longer than for the P-only ($1.5\tau_n$, Figure 4.11). Since this response is much faster than I-only and P-only control, the majority (> 90%) of controllers found in plants are PI controllers. The equation for a PI controller is given in Equation 4.14.

$$mv = K_c\left(e + \frac{1}{T_i}\int e\,dt\right) = K_c\,e + K_c \cdot \frac{1}{T_i}\int e\,dt \tag{4.14}$$

The PI controller gain has an effect not only on the error, but also on the integral action. When we compare the equation for a PI controller (Equation 4.14) to that for a P-only controller, (Equation 4.11) we see that the bias term in the P-only controller has been replaced by the integral term in the PI controller. Thus, the bias term for PI control is given by Equation 4.15:

$$b = K_c \cdot \frac{1}{T_i}\int e\,dt \tag{4.15}$$

Therefore, the integral action provides a bias that is automatically adjusted to eliminate any error. The PI controller is faster in response than the I-only controller because of the addition of the proportional action, as illustrated in Figure 4.12.

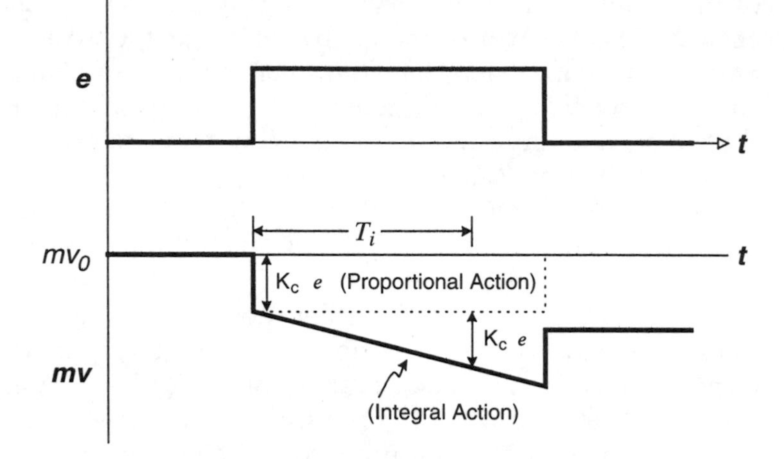

Figure 4.12 Proportional plus integral controller response to square wave input.

As shown in Figure 4.9, it takes T_i minutes for the output of the I-only controller to duplicate the error. With the addition of proportional action there is an immediate proportional step followed by integral action. The integral time in this case is defined as the amount of time it takes for the integral portion of the controller to replicate the proportional action. When the measurement is returned to the set point, the proportional action is lost since $e = 0$, and the controller output is determined solely by integral action.

As can be seen from Equation 4.16, the gain of the PI controller, K_{PI}, is the sum of the two component gains. These component gains are proportional action, K_P, and integral action, K_I.

$$K_{PI} = K_P + K_I = K_c + \frac{K_c}{T_i} \tag{4.16}$$

The PB and T_i are used to adjust the PI controller gain to give the loop a desired response. Suppose $T_i = \infty$, which would result in $K_I = 0$, regardless of the value of PB. In effect, the response would be that of a P-only controller with a period equal to τ_n and a sustained error. While $T_i = \infty$ is not realisable, it can be set to a very large number in min/rep to minimise the integral action.

Now, suppose T_i were set to a very small value. In this case, the PI controller gain would approach that of an integral only controller, since $K_I >> K_P$. The control action in the loop would now be that of an I-only controller with a return to the set point but a long response period.

In general, starting with only proportional action, as more integral action is added, the *PV* begins to return to the set point. We only want enough integral gain to return to the set point, as a greater amount will only serve to lengthen the response period. As more integral action is added by reducing T_i, we must compensate for the increased integral gain by reducing the proportional gain. Adjusting T_i will have an effect on K_I and thus affects K_{PI} and ϕ_{PI}, which in turn affects both the damping and the response period. Adjusting K_c affects both K_I and K_P equally, thus K_c only has an effect on K_{PI}, affecting the damping and not the response period. These interacting effects will also be considered in more detail under controller tuning in Chapter 5.

Although the response period of a loop under PI control is only 50% longer than that for a loop under P-only control, this may in fact be far too long if τ_n is as large as three or four hours. In order to increase the speed of the response it may be necessary to add an additional control mode.

4.5 Derivative action

The purpose of *derivative action* is to provide lead to overcome lags in the loop. In other words, it anticipates where the process is going by looking at the rate of change of error, *de*/*dt*. For derivative action, the output equals the derivative time, T_d, multiplied by the derivative of the input, which is the rate of change of error (see Equation 4.17).

$$output = T_d \frac{de}{dt} \tag{4.17}$$

Figure 4.13 shows how the output from a derivative block would vary for different inputs given a fixed value of T_d.

As the rate of change of the input gets larger, the output gets larger. Since the slope of each of these input signals is constant, the output for each of these rate inputs will also be constant. However, what happens as the slope approaches infinity as in the case of a step change, (4) in Figure 4.13? Theoretically the output should be a pulse that is of infinite amplitude and zero time long. This output is unrealisable since a perfect step with zero rise time is physically impossible, but signals that have short rise and fall times do occur. These types of signals are referred to as noise. Thus, the output from the derivative block would be a series of positive and negative pulses, which would try to drive the FCE either full open or full close. This would result in accelerated wear on the FCE and no useful control.

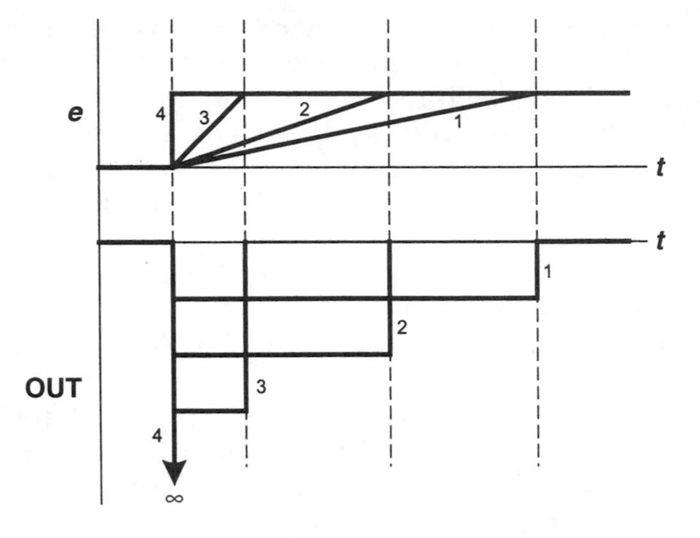

Figure 4.13 Effect of error on derivative mode output.

Consider a temperature measurement with a small amplitude and high frequency noise. One might assume that since the noise is of such small amplitude in comparison to the average temperature signal that a controller would not even notice it. This is only the case if the controller does not have derivative action. If the controller contains derivative action, the temperature signal would be completely masked by the noise into the derivative mode of the controller, and the controller output would be a series of large amplitude pulses, entirely masking any output contributed by the other control modes. Fortunately, in a case such as this the noise is either easily filtered out or is eliminated by modifying the installation of the primary sensor.

However, there are cases where noise is inherent in the measurement of *PV* and the rise and fall time of the noise is of the same magnitude as that of the measurement itself. In such a case, noise filtering would only serve to degrade the accuracy of the measurement of *PV*. A good example of a situation like this is a flow control loop. Flow measurement by its very nature is noisy; and therefore, derivative action cannot be successfully applied.

It is important to note that derivative control would never be the sole control mode used in a controller. The derivative action does not know what the set point actually is and hence cannot control to a desired set point. Derivative action only knows that the error is changing.

4.6 Proportional plus derivative (PD) controller

The minimum controller configuration containing derivative action is a combination of *proportional plus derivative action* shown in Equation 4.18. This combination is not used very often and is primarily applied in batch pH control loops. However, it will help in the definition of derivative time, T_d.

$$mv = K_c\left(e + T_d\left(\frac{de}{dt}\right)\right) + b \tag{4.18}$$

In Equation 4.18, the PD controller equation contains a bias term. A bias term will normally appear in any controller algorithm that does not contain integral action. This bias term does not appear when integral action is present since integral action is in effect an automatic adjustment of bias. As with the PI controller, the proportional controller gain acts on the error as well as the derivative time, T_d. Figure 4.14 shows the controller output (mv), for a typical input (e) test signal for the proportional and derivative portions of a PD controller.

In Figure 4.14, mv_P is the proportional portion of the output and mv_D is the derivative portion. In the example, the measurement changes at a fixed rate of change and, therefore, the derivative portion of the output is constant and depends on the rate of change, the derivative time, T_d, and proportional gain (K_c). This dependency is evident from Equation 4.18. The proportional output is a ramp whose slope is a function of the proportional controller gain, K_c.

Now, let us superimpose mv_P and mv_D to get the actual output for a PD controller:

For a ramp input it takes a period of time for the proportional action to reach the same level as the derivative action. This period of time is called

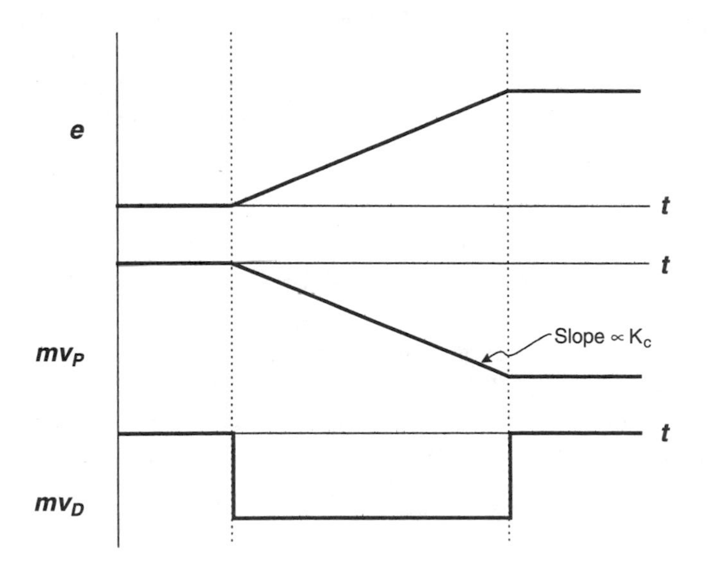

Figure 4.14 Responses for P-only and D-only portions of a PD controller.

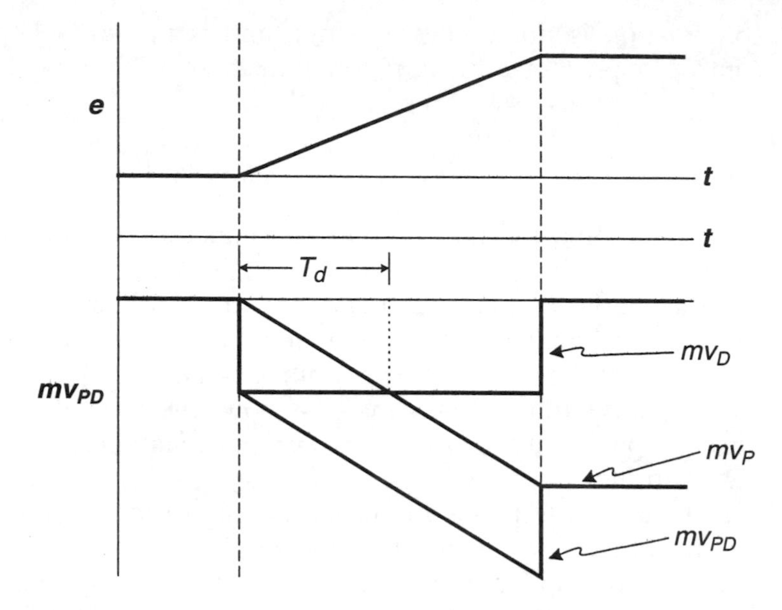

Figure 4.15 Combined response of a PD controller.

the derivative time, T_d, and is measured in minutes. Increasing the derivative time, increases mv_D or the contribution of the derivative action to the movement of the final control element.

In Equation 4.18, for the PD controller the derivative action acts on the error. Since $e = SP - CV$ for I/D action, de/dt is a function of both the derivative of the set point, dSP/dt, and the derivative of the controlled variable, dCV/dt.

$$\frac{de}{dt} = \frac{dSP}{dt} - \frac{dCV}{dt} \tag{4.19}$$

If there is a load upset to the process, the process variable, PV, will change at some rate, dCV/dt which will result in the error also changing at the same rate ($de/dt = -dCV/dt$, assuming there is no set point change. Now, if a set point change of even a few percent is made and if the set point is changed quickly, then dSP/dt can become very large. This would cause a large pulse to be generated at the output of the controller. To overcome this potential problem, the controller can be made so that the derivative mode simply ignores set point changes as shown in Equations 4.20–22.

$$\frac{de}{dt} = \frac{dSP}{dt} - \frac{dCV}{dt} \tag{4.20}$$

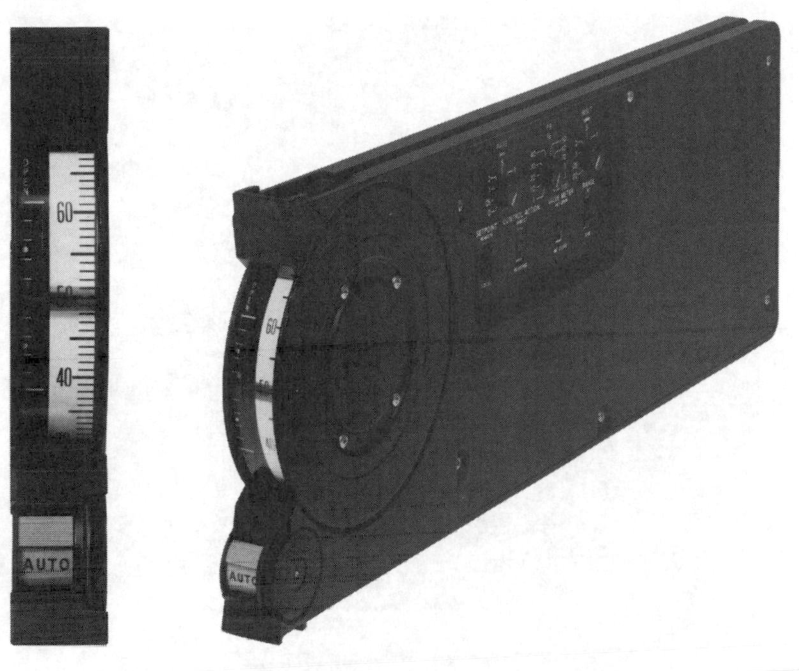

Figure 4.19 Electronic analog controller (courtesy of Fisher Controls International, Inc.).

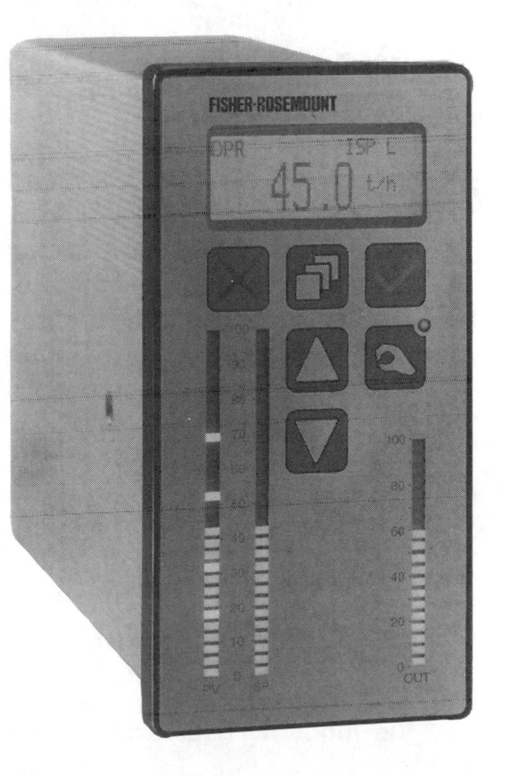

Figure 4.20 Electronic digital controller – Fisher DPR series (courtesy of Fisher Controls International, Inc.).

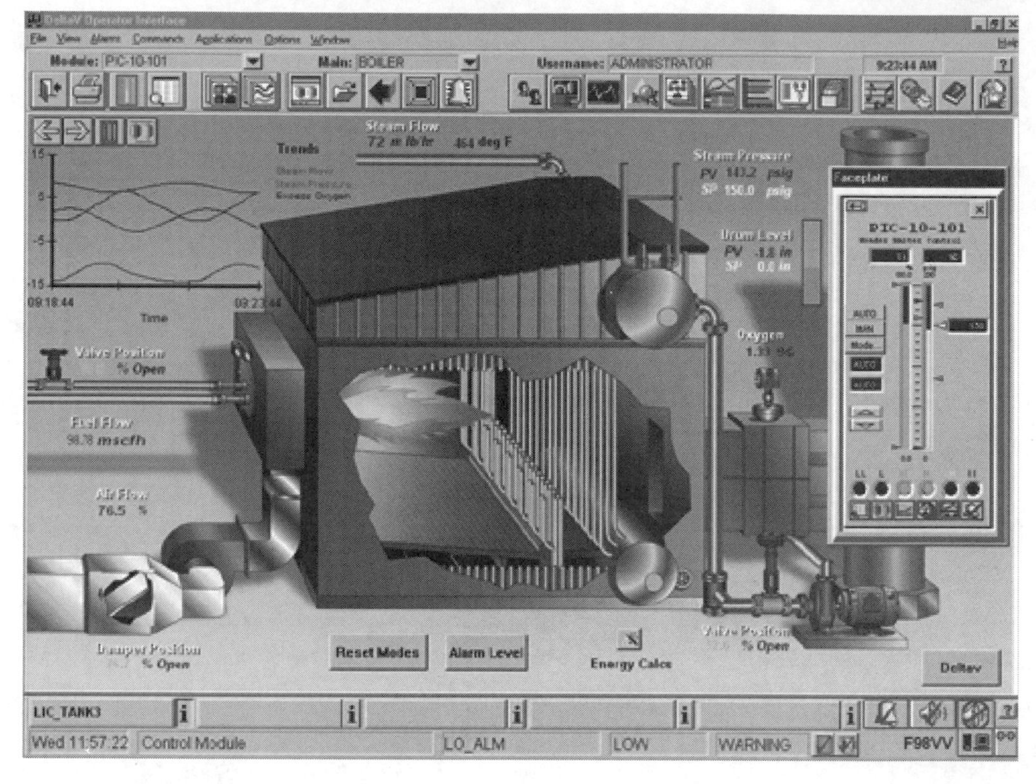

Figure 4.21 Screenshot from a DeltaV distributed control system (courtesy of Fisher-Rosemount Systems, Inc.).

Figure 4.21 shows a screenshot from a modern distributed control system (DCS). Simply put, a DCS is an electronic digital control system where computers spread functionality over multiple processes in large scale plants. The advantage of a DCS is that it allows operators to monitor and control entire plants from a central control room.

Distributed control systems were introduced in the mid-1970s with the advent of the microcomputer. DCSs enabled more flexible and complex control, monitoring, alarming and historic data trending than local, single-loop control or the centralised control previously possible with mini-computers.

Modern DCSs feature the use of digital, multi-drop communications that can interconnect sensors, actuators and the control room. Control can be allocated to the digital devices that can communicate directly to each other to fully exploit each other's capabilities with remote diagnostics and supervisory control and data acquisition (SCADA). This control technology is known as Fieldbus, [1,2].

The personal computer explosion of the 1980s and 1990s has also impacted on modern DCSs with the advent of PC-based control systems [3,4] which feature object linking and embedding (OLE) software for process control (OPC) [5].

4.10 References

1. Hodson, W.R., "A Fieldbus Primer". *Oil Patch Magazine*, 1998, **Nov./Dec.:** 7–9.
2. **http://www.fieldbus.org/**
3. DeltaV™ Fisher Rosemount Systems, Inc., USA, 1996.
4. Santori, M., "The Emergence of PC Technology". *Chemical Engineering*, 1997, **Dec.**: 70–8.
5. **http://www.opcfoundation.org/**

5 TUNING FEEDBACK CONTROLLERS

There is no absolute right way or, for that matter, wrong way to tune a controller. Controller settings depend on what the engineer/operator deems to be good performance in terms of the desired response to process upsets. The type of process, the process gain, the time constant and dead time all play a role in determining the controller settings.

The settings also depend on the anticipated type of disturbances that the process will encounter. A controller would be tuned differently for stability due to set point changes (servo control) than for load disturbances (regulatory control). In process control systems load disturbances are most frequently encountered and, hence, most systems are optimised for regulatory control.

This chapter will discuss control quality and optimisation, including the performance criteria that need to be considered when tuning a controller. A number of methods that can be used to determine controller settings in order to achieve the desired control are also described.

5.1 Quality of control and optimisation

Controller tuning can be defined as an optimisation process that involves a performance criterion related to the form of controller response and to the error between the process variable and the set point. When tuning a controller some of the questions that may be asked include:

- Can offset be tolerated?
- Is no overshoot desired?
- Is a certain decay ratio required?
- Is a fast rise time needed?

These questions address some of the performance criteria used in the tuning of a controller, including overshoot, decay ratio and error performance.

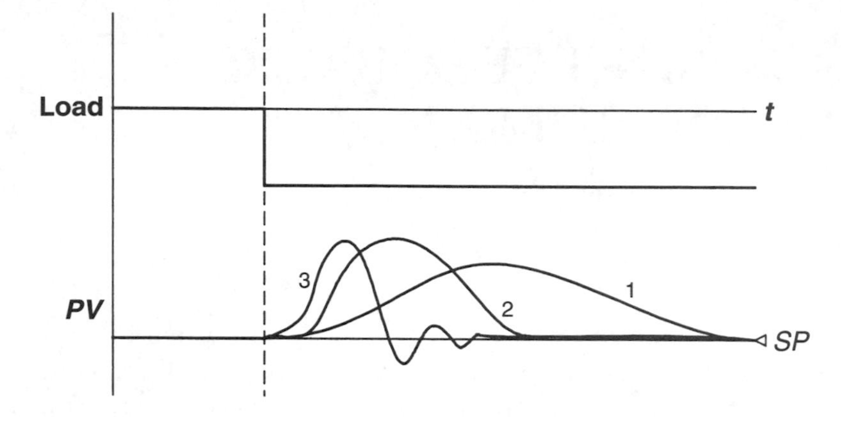

Figure 5.1 Typical responses to a load change.

5.1.1 CONTROLLER RESPONSE

Depending on the process to be controlled, the first consideration is to decide what type of response is optimal or at least acceptable. Typical process responses to a load change are illustrated in Figure 5.1.

The three possible general extremes of response that exist, as shown in the above figure, are:

1. overdamped – slow response with no oscillation
2. critically damped – fastest response without oscillation
3. underdamped – fast return to set point but with considerable oscillation

From these three general extremes, we can see that selection of good control is a trade off between the speed of response and deviation from the set point. A highly tuned controller may become unstable if large disturbances occur, whereas a sluggishly tuned controller provides poor performance but is very robust. What is typically required for most process control loops is a compromise between performance and robustness.

When examining the response, there are several common performance criteria that can be used for controller tuning, based on characteristics of the system's closed loop response. Some of the more common criteria include overshoot, offset, rise time and decay ratio. Of these simple performance criteria, control practitioners most often use decay ratio.

Cyclic radian frequency

The cyclic radian frequency, ω, is defined as:

$$\omega = 2\pi f \tag{5.1}$$

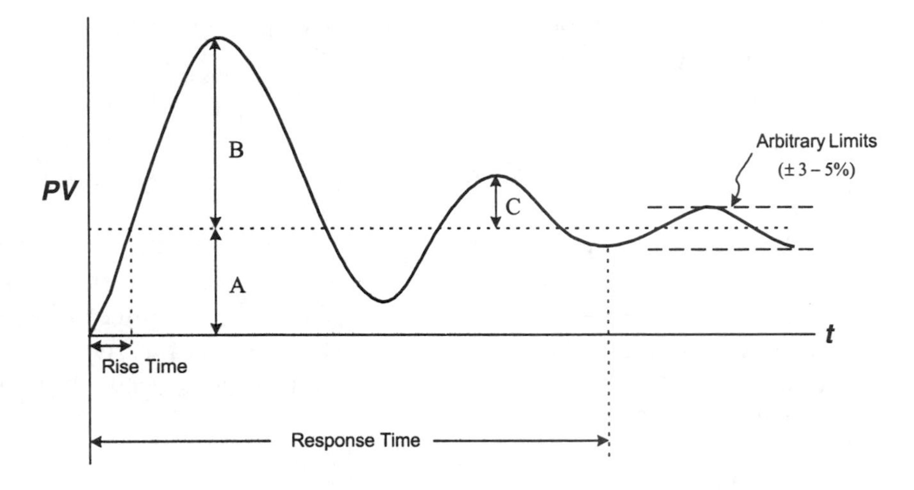

Figure 5.2 Second or higher order typical response to a set point change.

and

$$f = \frac{1}{\text{period}} \tag{5.2}$$

If Equation 5.2 is substituted into Equation 5.1, we obtain:

$$\omega = \frac{2\pi}{\text{period}} \tag{5.3}$$

The cyclic radian frequency can also be related to the undamped natural frequency, ω_n, and the damping coefficient, ξ:

$$\omega = \omega_n \sqrt{1 - \xi^2} \tag{5.4}$$

Overshoot

Overshoot is the amount by which the response exceeds the steady-state final value. Referring to Figure 5.2, the overshoot can be defined as:

$$\frac{B}{A} = e^{-\pi\xi(1-\xi 2)^{-1/2}} \tag{5.5}$$

Decay ratio

The decay ratio is the ratio of the amplitude of an oscillation to the amplitude of the preceding oscillation, C/B in Figure 5.2. More specifically, we can define the quarter decay ratio (QDR) which lies between critical damping and underdamping.

$$\text{QDR} = \frac{C}{B} = \frac{1}{4} \tag{5.6}$$

The decay ratio is often used to establish whether the controller is providing a satisfactory response. The quarter decay ratio has been shown through experience to provide a good trade off between minimum deviation from the set point after an upset and the fastest return to the set point. The penalty of quarter decay ratio is that some oscillation does occur. For a second order system it can be shown that:

$$\frac{C}{B} = e^{-2\pi\xi(1-\xi^2)^{-1/2}} \tag{5.7}$$

Rise time

The rise time is the time required by the transient response to reach the final steady-state value.

Response time

The response time is the time required for the response to settle within the specified arbitrary limits. These limits are typically set at ±3–5% of the *PV* steady-state value.

5.1.2 ERROR PERFORMANCE CRITERIA

The previously discussed simple performance criteria, i.e. decay ratio, overshoot, etc., use only a few points in the response and therefore are simple to use. On the other hand, error performance criteria are based on the entire response of the process and they are also more complicated.

Integrated error

The curve shown in Figure 5.3 represents the response of a loop due to a process upset. This graphical representation of the controlled variable's return to the set point is known as a response curve. The integrated error (IE) is the area under the response curve, and the idea of using this as an error criterion is to attempt to minimise this area.

In mathematical terms, with e representing the error as a function of time, we can write:

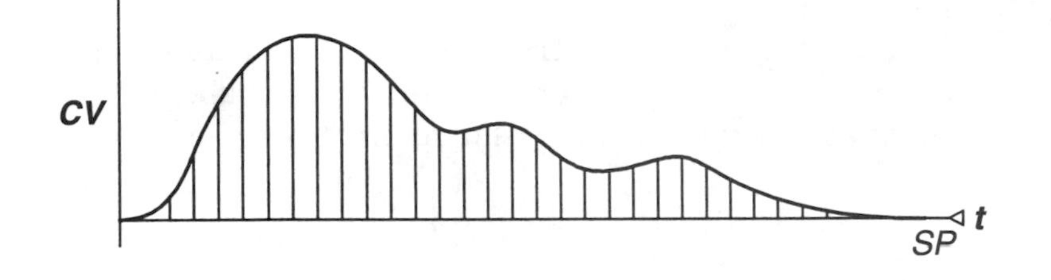

Figure 5.3 General response curve.

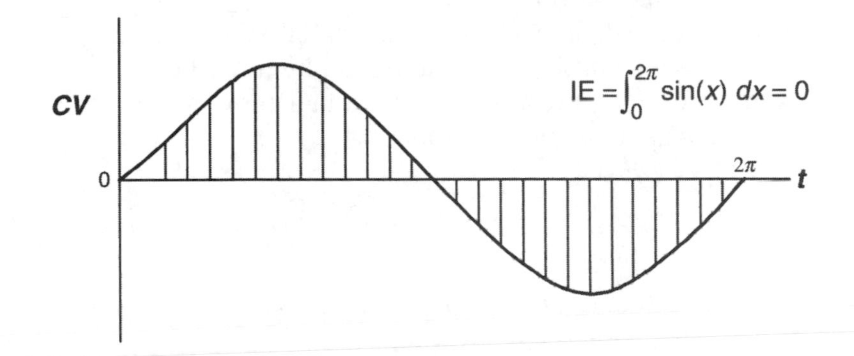

Figure 5.4 Sinusoidal error response.

$$IE = \int_0^\infty e\, dt \tag{5.8}$$

It may not always be possible to minimise IE without paying a penalty in some respect. For example, underdamping produces a minimum area under the response curve but has considerable oscillation.

The method of integrated error may not be 100% reliable if there is no averaging elsewhere in the process. For example, if there is a sinusoidal oscillation about the set point, the positive and negative areas tend to cancel each other out over time, which presents a misleading conclusion, as shown in Figure 5.4. However, barring this situation, integrated error is a perfectly adequate error criterion.

Integrated absolute error

Integrated absolute error (IAE) essentially takes the absolute value of the error. Negative areas are accounted for when IAE is used, thus dismissing the problem encountered with IE regarding sinusoidal responses.

$$IAE = \int_0^\infty |e|\, dt \tag{5.9}$$

Integrated squared error

The integrated squared error (ISE) criterion uses the square of the error, thereby penalising larger errors more than smaller errors. This gives a more conservative response, i.e. faster return to the set point.

$$ISE = \int_0^{\infty} e^2 \, dt \tag{5.10}$$

Integrated time absolute error

The integrated time absolute error (ITAE) criterion is the integral of the absolute value of the error multiplied by time. ITAE results in errors existing over time being penalised even though they may be small, which results in a more heavily damped response.

$$ITAE = \int_0^{\infty} t \mid e \mid dt \tag{5.11}$$

Figure 5.5 shows the various responses of a loop that is tuned to the above criteria, including IAE, ISE, and ITAE.

5.2 Tuning methods

The following presents a very brief description of some of the various accepted methods used for controller tuning. In each case the suggested controller settings are optimised for a particular error performance criterion, often quarter decay ratio. The first method described is based entirely on trial and error, while the rest are based upon some understanding of the physical nature of the process to be controlled. This understanding is generated from either open or closed loop process testing.

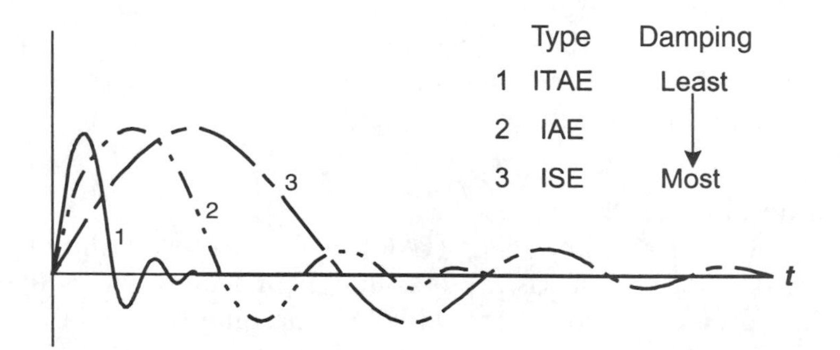

Figure 5.5 Response to various error criteria.

5.2.1 "TRIAL AND ERROR" METHOD

As the name suggests, tuning by trial and error is simply a guess and check type method. The following is a list of practical suggestions for tuning a controller by trial and error. These suggestions are also useful for fine-tuning controllers tuned by other methods.

1. Proportional action is the main control. Integral and derivative actions are used to trim the proportional response.
2. The starting point for "trial and error" tuning is always with the controller gain, integral action and derivative action all at a minimum.
3. Make adjustments in the controller gain by using a factor of two, i.e. 0.25, 0.5, 1.0, 2.0, 4.0, etc.
4. The optimal response is the quarter decay ratio (QDR).
5. When in trouble, decrease the integral and derivative actions to a minimum and adjust the controller gain for stability.

Rules of thumb

The following rules of thumb should not be taken as gospel or as a methodology but rather are intended to indicate typical values encountered. As such, these rules can be useful when tuning a controller using the trial and error method. However, it is important to remember that controller parameters are strongly dependent on the individual process and may not always abide by the rules outlined below.

Flow When dealing with a flow loop, P-only control can be used with a low controller gain. For accuracy, PI control is used with a low controller gain and high integral action. Derivative action is not used because flow loops typically have very fast dynamics and flow measurement is inherently noisy.

$K_c = 0.4 - 0.65$
$T_i = 0.1$ minutes (or 6 seconds)

Level Levels represent material inventory that can be used as surge capacity to dampen disturbances. Hence, loosely tuned P-only control is sometimes used. However, most operators do not like offset, so PI level controllers are typically used.

The following P-only settings ensure that the control valve will be 50% open when the level is at 50%, wide open when the level is at 75%, and shut when the level is at 25%.

$K_c = 2$
Bias term (b) = 50%
Set point (SP) = 50%

Typical PI controller settings are:

K_c = 2–20
T_i = 1–5 minutes

Pressure Pressure control loops show large variation in tuning depending on the dynamics of the pressure response. Typical ranges are as follows:

Vapour: K_c = 2–10
T_i = 2–10 minutes
Liquid: K_c = 0.5–2
T_i = 0.1–0.25 minutes

Temperature Temperature dynamic responses are usually fairly slow, so PID control is used. Typical parameter values are:

K_c = 2–10
T_i = 2–10 minutes
T_d = 0–5 minutes

5.2.2 PROCESS REACTION CURVE METHODS

In process reaction curve methods a process reaction curve is generated in response to a disturbance. This process curve is then used to calculate the controller gain, integral time and derivative time. These methods are performed in open loop so no control action occurs and the process response can be isolated.

To generate a process reaction curve, the process is allowed to reach steady state or as close to steady state as possible. Then, in open loop so there is no control action, a small disturbance is introduced and the reaction of the process variable is recorded. Figure 5.6 shows a typical process reaction curve generated using the above method for a generic self-regulating process. The term self-regulating refers to a process where the controlled variable eventually returns to a stable value or levels out without external intervention.

The process parameters that may be obtained from this process reaction curve are as follows:

L = lag time (min)
T = time constant estimate (min)
P = initial step disturbance (%)
ΔC_p = change in PV in response to step disturbance (%)
$N = \dfrac{\Delta C_p}{T}$ = reaction rate (%/min)

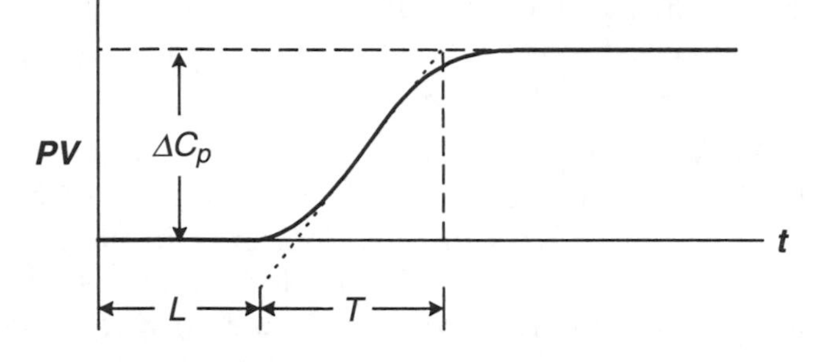

Figure 5.6 Process reaction curve.

$$R = \frac{N\,L}{\Delta C_p} = \text{lag ratio}$$

Methods of process analysis with forcing functions other than a step input are possible and include pulses, ramps, and sinusoids. However, step function analysis is the most common as it is the easiest to implement.

Ziegler-Nichols open loop rules

In 1942, Ziegler and Nichols [1] changed controller tuning from an art to a science by developing their open loop step function analysis technique. They also developed a closed loop technique, which is described in the next section on constant cycling methods.

The Ziegler-Nichols open loop recommended controller settings for quarter decay ratio are as follows:

P-only: $$K_c = \frac{P}{N\,L} \tag{5.12}$$

PI: $$K_c = 0.9\left(\frac{P}{N\,L}\right) \tag{5.13}$$

$$T_i = 3.33(L) \tag{5.14}$$

PID: $$K_C = 1.2\left(\frac{P}{N\,L}\right) \tag{5.15}$$

$$T_i = 2.0(L) \tag{5.16}$$

$$T_d = 0.5(L) \tag{5.17}$$

These settings should be taken as recommendations only and tested thoroughly in closed loop, fine-tuning the parameters to obtain QDR.

Cohen-Coon open loop rules

In 1953, Cohen and Coon [2] developed a set of controller tuning recommendations that correct for one deficiency in the Ziegler-Nichols open loop rules. This deficiency is the sluggish closed loop response given by the Ziegler-Nichols rules in the relatively rare occasion when process dead time is large relative to the dominant open loop time constant.

The Cohen-Coon recommended controller settings are as follows:

P-only: $$K_c = \frac{P}{NL}\left(1 + \frac{R}{3}\right) \tag{5.18}$$

PI: $$K_c = \frac{P}{NL}\left(0.9 + \frac{R}{12}\right) \tag{5.19}$$

$$T_i = L\left(\frac{30 + 3R}{9 + 20R}\right) \tag{5.20}$$

PID: $$K_c = \frac{P}{NL}\left(1.33 + \frac{R}{4}\right) \tag{5.21}$$

$$T_i = L\left(\frac{32 + 6R}{13 + 8R}\right) \tag{5.22}$$

$$T_d = L\left(\frac{4}{11 + 2R}\right) \tag{5.23}$$

As with the Ziegler-Nichols open loop method recommendations, the Cohen-Coon values should be implemented and tested in closed loop and adjusted accordingly to achieve QDR.

Simplified internal model control tuning rules

Many practitioners have found that the Ziegler-Nichols open loop and Cohen-Coon rules are too aggressive for most chemical industry applications since they give a large controller gain and short integral time. Rivera *et al.* [3], developed the internal model control (IMC) tuning rules with robustness in mind. The tuning parameter from the IMC method (the closed loop speed of response) relates directly to the closed loop time constant and the robustness of the control loop. As a consequence the closed loop step load response exhibits no oscillation or overshoot.

Table 5.1 Simplified IMC rules

	$\frac{\tau}{L} > 3$	$\frac{\tau}{L} < 3$	$L < 0.5$
K_c	$\frac{P}{2(\text{NL})}$	$\frac{P}{2(\text{NL})}$	$\frac{P}{N}$
T_i	$5L$	τ	4
T_d	$\leq 0.5L$	$\leq 0.5L$	$\leq 0.5L$

Since the general IMC method is unnecessarily complicated for processes that are well approximated by first order dead time or integrator dead time models, the following simplified IMC rules were developed by Fruehauf *et al.* [4] for PID controller tuning (see Table 5.1).

Of course, these recommendations need to be tested in the closed loop situation and the final settings arrived at through the use of fine-tuning.

5.2.3 CONSTANT CYCLING METHODS

Ziegler-Nichols closed loop method

The closed-loop technique of Ziegler and Nichols [5] is a technique that is commonly used to determine the two important system constants, ultimate period and ultimate gain. It was one of the first tuning techniques to be widely adopted.

When tuning using Ziegler-Nichols closed loop method, values for proportional, integral and derivative controller parameters may be determined from the ultimate period and ultimate gain. These are determined by disturbing the closed-loop system and using the disturbance response to extract the values of these constants.

The following is a step-by-step approach to using the Ziegler-Nichols closed loop method for controller tuning:

1. Attach a proportional only controller with a low gain (no integral or derivative action).
2. Place the controller in automatic.
3. Increase the controller gain until a constant amplitude limit cycle occurs.
4. Determine the following parameters from the constant amplitude limit cycle (see Figure 5.7):

$$P_u \equiv \text{ultimate period} = \text{period taken from limit cycle}$$

$$K_u \equiv \text{ultimate gain} = \text{controller gain that produces the limit cycle}$$

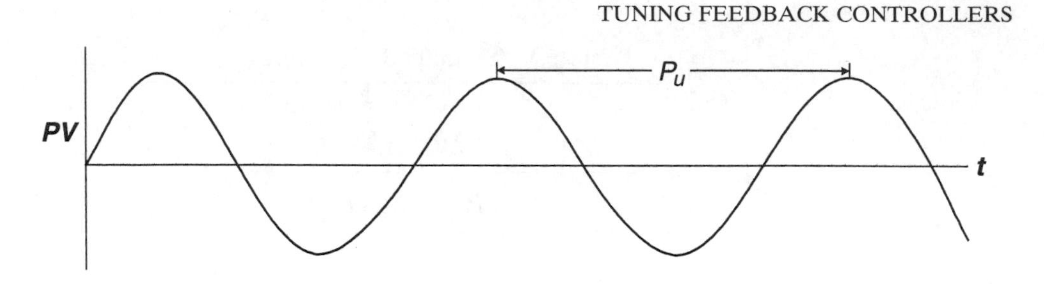

Figure 5.7 Constant amplitude limit cycle.

5. Calculate the tuning parameters using the following equations:

P-only:
$$K_c = \frac{K_u}{2} \tag{5.24}$$

PI:
$$K_c = \frac{K_u}{2.2} \tag{5.25}$$

$$T_i = \frac{P_u}{1.2} \tag{5.26}$$

PID:
$$K_c = \frac{K_u}{1.7} \tag{5.27}$$

$$T_i = \frac{P_u}{2} \tag{5.28}$$

$$T_d = \frac{P_u}{8} \tag{5.29}$$

6. Fine-tune by adjusting K_c, T_i, and T_d as required to find QDR.

Auto tune variation technique

The auto tune variation (ATV) technique of Åström and Hagglund [5] is another closed loop technique used to determine the two important system constants, the ultimate period and the ultimate gain. However, the ATV technique determines these system constants without unduly upsetting the process. Tuning values for proportional, integral and derivative controller parameters can be determined from these two constants. We recommend the use of Tyreus – Luyben [6] settings for tuning that is suitable for chemical process unit operations. All methods for determining the ultimate period and ultimate gain involve disturbing the system and using the disturbance response to extract the values of these constants.

In the case of the ATV technique, a small limit-cycle disturbance is set up between the manipulated variable (controller output) and the controlled variable (process variable). Figure 5.8 shows the instrument set up, and Figure 5.9 shows the typical ATV response plot with critical parameters defined. It is important to note that the ATV technique is applicable only to processes with significant dead time. The ultimate period will just equal the sampling period if the dead time is not significant.

General ATV tuning method for a PI controller:

1. Determine a *reasonable* value for valve change, h (typically 0.05, i.e. 5%). The value for h should be small enough that the process is not unnecessarily upset but large enough that the amplitude, a, can be measured.
2. Move the valve $+h$ units.
3. Wait until the process variable starts to move, then move the valve $-2h$ units.
4. When the process variable (PV) crosses the set point, move the valve $+2h$ units.
5. Repeat until a limit-cycle is established, as illustrated in Figure 5.9.
6. Record the value of a by picking it off the response graph.
7. Perform the following calculations to determine the ultimate period, ultimate gain, and the controller gain and integral time.

$$P_u \equiv \text{ultimate period} = \text{period taken from limit cycle}$$

$$K_u \equiv \text{ultimate gain} = \frac{4h}{3.14a}$$

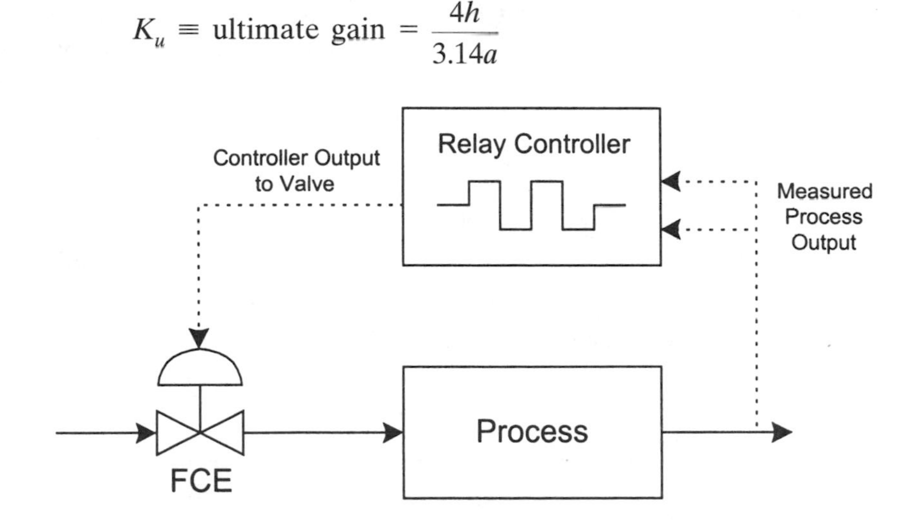

Figure 5.8 ATV tuning instrument set up.

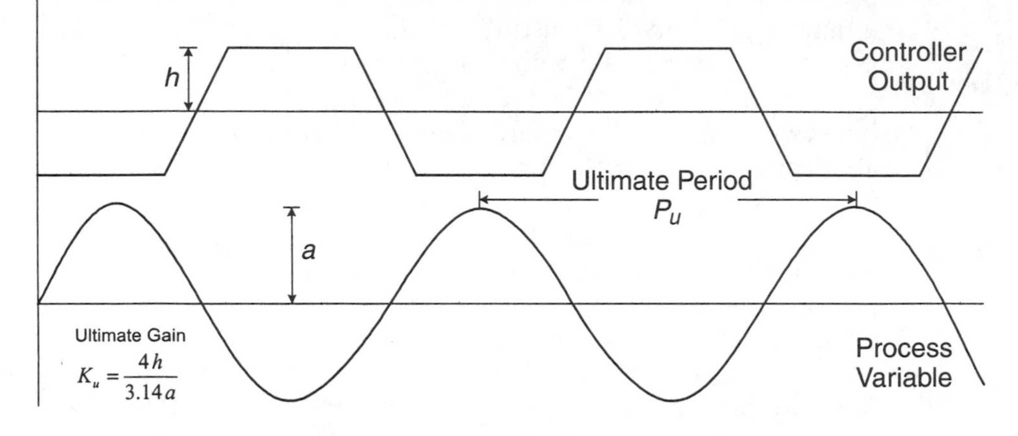

Figure 5.9 ATV critical parameters.

$$K_c = \frac{K_u}{3.2} \tag{5.23}$$

$$T_i = 2.2(P_u) \tag{5.24}$$

Comparison of ATV and Ziegler-Nichols closed loop tuning techniques

Table 5.2 compares the tuning constants between the ATV and Ziegler-Nichols Closed Loop (Z-N CL) tuning techniques. Notice that the Ziegler-Nichols tuning is more aggressive with a larger controller gain and shorter integral time. This technique was originally developed for electro-mechanical systems control and is based on the more aggressive QDR criterion. ATV tuning was developed for fluid and thermal processes and emphasises minimising overshoot. ATV is therefore often the preferred technique for process control.

Table 5.2 Tuning comparison

	Z-N CL	ATV	(Z-N/ATV) ratio
Controller gain (K_c)	$\frac{K_u}{2.2}$	$\frac{K_u}{3.2}$	1.45
Integral time (T_i)	$\frac{P_u}{1.2}$	$2.2(P_u)$	0.38

5.3 References

1. Ziegler, J.G. and Nichols, N.B., "Optimum Settings for Automatic Controllers". *Trans. ASME*, 1942, **64**: 759.
2. Cohen, G.H. and Coon, G.A., "Theoretical Consideration of Retarded Control". *Trans. ASME*, 1953, **75**: 827.
3. Rivera, D.E., Morari, M. and Skogestad, S., "Internal Model Control, 4. PID Controller Design". *Ind. Eng. Chem. Proc. Des. Dev.*, 1986, **25**: 252.
4. Fruehauf, P.S., Chien, I-L. and Lauritsen, M.D., "Simplified IMC-PID Tuning Rules". *ISA*, Paper# 93–414, p. 1745, 1993.
5. Åström, K.J. and Hägglund, T., "Automatic Tuning of Simple Regulators with Specifications on Phase and Amplitude Margins". *Automatica*, 1984, **20**: 645.
6. Tyreus, B.D. and Luyben, W.L., "Tuning of PI Controllers for Integrator/Deadtime Processes", *Ind. Eng. Chem. Res.*, 1992, **31**: 2625.

6 ADVANCED TOPICS IN CLASSICAL AUTOMATIC CONTROL

Up to this point discussion has been restricted to feedback control loops, the most common control method used in process industries. However, there are a number of more complex control techniques that should also be considered [1], which can provide improved and economic process control. Some of the control schemes that are discussed in this chapter include cascade, feedforward, ratio and override control. These schemes are classified as "advanced classical" [1] topics and are widely used in industry.

6.1 Cascade control

Cascade control is a common control technique that uses two controllers with one feedback loop nested inside the other [2,3,4]. The output of the primary controller acts as the set point for the secondary controller. The secondary controller controls the final control element (FCE). A typical cascade control loop is illustrated in Figure 6.1.

To better understand cascade control, we will examine a typical feedback control scheme and consider how it may be improved through the use of cascade control. Let us consider the following feedback control loop for a heat exchanger, shown in Figure 6.2.

In this example of a heat exchanger, the energy is provided by steam flow, F_s, and is used to heat an entering fluid, F_w, from an inlet temperature, T_1, to an outlet temperature, T_2. If the controller was properly tuned and there is a change in F_w, the change in T_2 will be sensed and the temperature controller will then change its output by repositioning the valve to bring the outlet temperature back to the set point. In other words, the controller moves the variance from one stream to another and thus mitigates changes in the process variable, PV.

Let us now consider another possible type of upset: a supply side upset. If the steam, F_s, comes from a supply header which is also servicing other users, there is a possibility that as the other users' needs vary, pressure upsets will occur causing changes in the steam supply F_s. Suppose that another user demanded steam quantities that caused a pressure drop in the header, thus resulting in a drop in steam flow to the exchanger. The only way that this

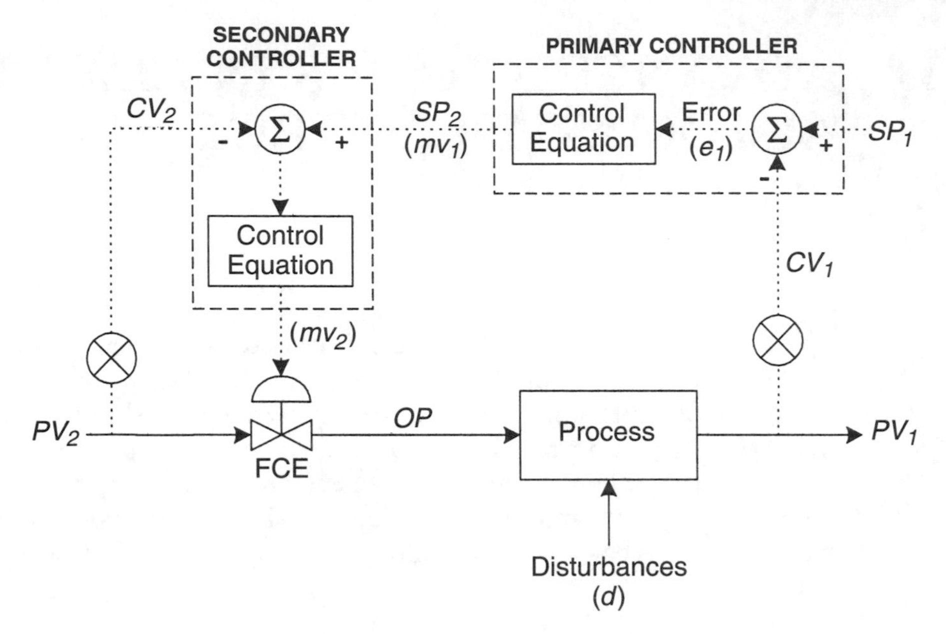

Figure 6.1 Cascade control loop.

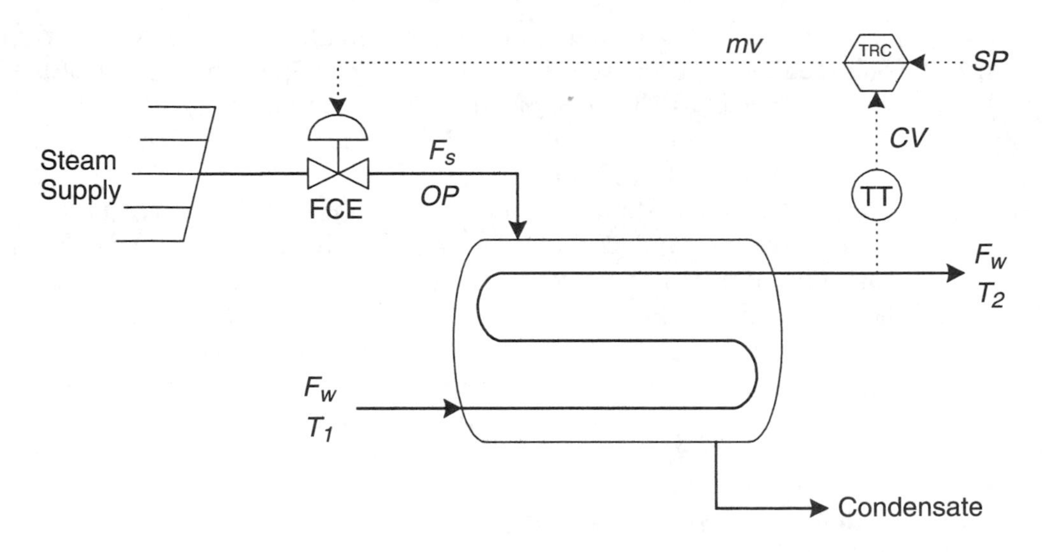

Figure 6.2 Feedback control loop for a heat exchanger (steam supply loop).

drop in F_s could be measured would be as a drop in the outlet temperature, T_2. This deviation from set point would be sensed and compensated for in the same manner as for the stream load upset, F_w. The response of the process variable, T_2, to the two situations is shown in Figure 6.3:

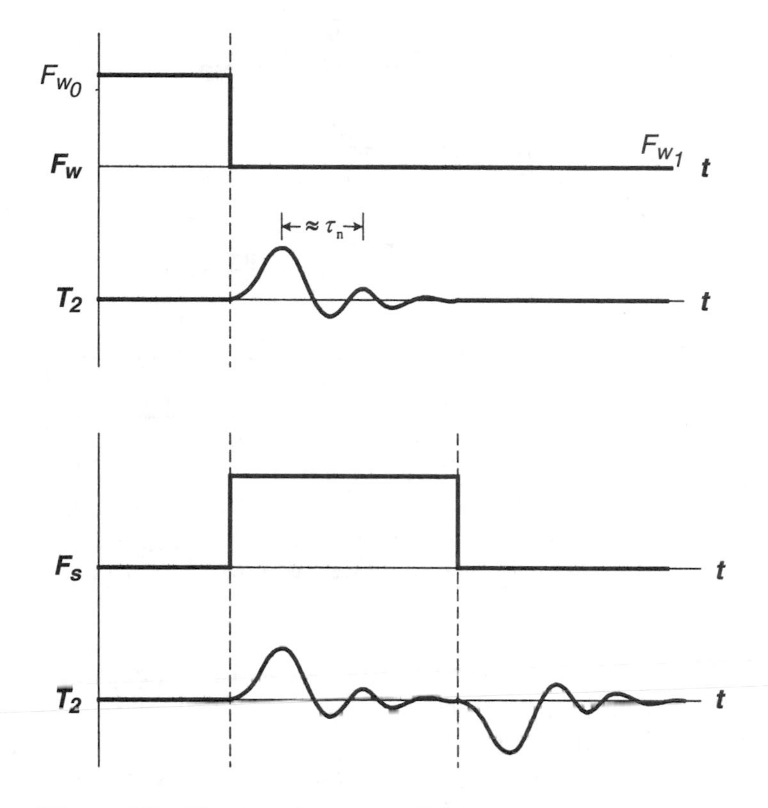

Figure 6.3 Heat exchanger outlet temperature response.

In both cases the process variable T_2 would dampen out in the period τ_n. However, for the supply upset situation the feedback control may cause the temperature to be in a constant state of flux. For instance, if the valve (when open to 50%) supplies the needed amount of steam, the outlet temperature will be at the set point. However, the flow through the valve is a function of the pressure drop across it. Therefore, if there is a decrease in inlet pressure the flow will decrease, even though the valve is 50% open. This disturbance will propagate through the dead time and capacitive lag of the heat exchanger before it emerges as an outlet temperature deviation, at which time the necessary control action will be taken to bring the temperature back to the set point. The response will damp out with a period τ_n back to the set point. If τ_n were a minute or two or even longer, the process might be in a constant state of flux and never settle to the desired set point.

The problem is that the controller output sets a valve opening rather than setting an energy supply requirement. With a constant pressure drop across the valve, the relationship between valve position and steam flow is constant, but if the pressure drop changes this relationship changes. Thus, it is better to set the steam requirement rather than valve position. As long as the valve

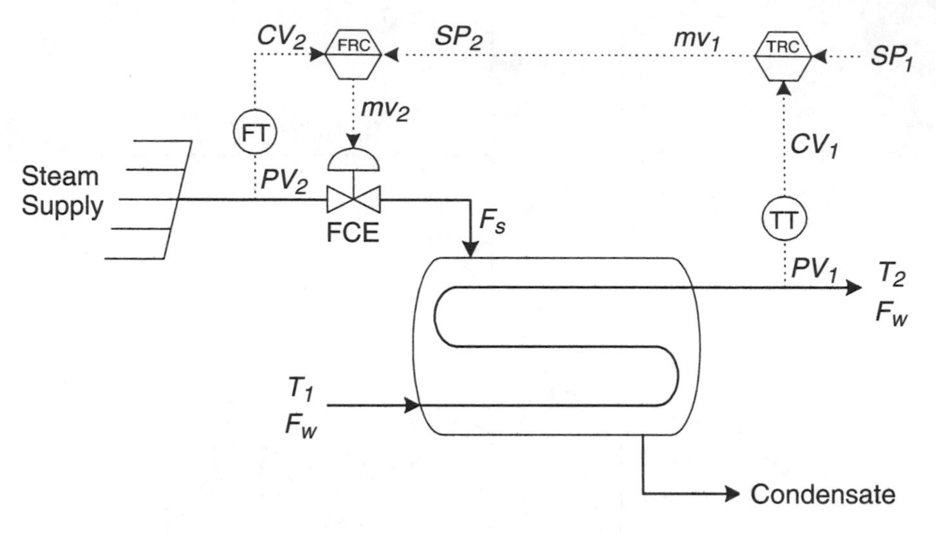

Figure 6.4 Typical cascade control loop for a heat exchanger.

can supply the energy requirement, it does not matter how wide open the valve is, only how much energy is being delivered. Hence, for this process better control can be achieved by using cascade control.

Now we will apply cascade control to the same heat exchanger to control supply side upsets. In Figure 6.4, the temperature loop (primary loop) is used to adjust the set point of the flow control loop (secondary loop).

With cascade control in place, if a set point change is made in the temperature controller or if a load upset occurs that changes F_w, the output of the temperature controller will change the steam flow controller set point. The flow loop operates so much faster than the temperature loop that the temperature controller does not in fact know whether its output is going directly to a valve or as a set point to another controller. In general, the control loop closest to the controlled variable is called the primary, outer or master loop. The control loop closest to the supply to the process is called the secondary, inner or slave loop. In the heat exchanger example given above, the temperature loop is the primary loop and the flow loop is the secondary loop.

Both the primary and secondary loops have their own response period, independent of whether they are in a cascade configuration or not. The response period of the primary loop is τ_1 and that of the secondary loop is τ_2. In order that cascade control works effectively, $\tau_1 > 4\,\tau_2$. What this rule of thumb implies is that the primary loop should never know that there is a secondary loop. The secondary loop should be able to respond as quickly as the FCE. If this rule is followed then there will be little interaction between the two loops and the control scheme will function normally.

6.1.1 STARTING UP A CASCADE SYSTEM

To put a cascade system into operation:

1. Place the primary controller in manual or the secondary controller to local set point. This will break the cascade and allow the secondary controller to be tuned.
2. Tune the secondary controller as if it were the only control loop present.
3. Return the secondary controller to remote set point and/or place the primary controller in auto.
4. Now tune the primary loop normally. If the system begins to oscillate when the primary controller is placed in auto, reduce the primary controller gain.

   Note: When tuning the primary controller there should be no interaction between the primary and secondary loops. If there is, it means that the primary loop is not slow enough in comparison to the secondary.

One of the most common forms of cascade is the output of a primary controller acting as a set point to a valve positioner.

6.2 Feedforward control

One of the disadvantages of using feedback control is that a disturbance must propagate through the process before it is detected and action is taken to correct it. This type of control is sufficient for processes in which some deviation from the set point is acceptable. However, there are certain processes where this set point deviation must be minimised. Feedforward control can accomplish this because it corrects and/or minimises disturbances before they enter a process [3, 4]. A typical feedforward control system is shown in Figure 6.5.

In its simplest form, a feedforward controller merely proportions the corrective action to the size of the disturbance. In other words, the control equation is merely a gain based on steady-state, i.e. mass or energy balance at steady-state. This does not take into account any of the process dynamics of the system. If there is a difference, or lag, in the speed of the process response to the control action when compared to that of the disturbance, it may be necessary to introduce some dynamic compensation into the control equation. The dynamic compensation correctly times the control action and response thus giving increased accuracy in the feedforward control.

In general, the feedforward dynamic elements will not be physically realisable. In other words, they cannot be implemented exactly. For instance, if

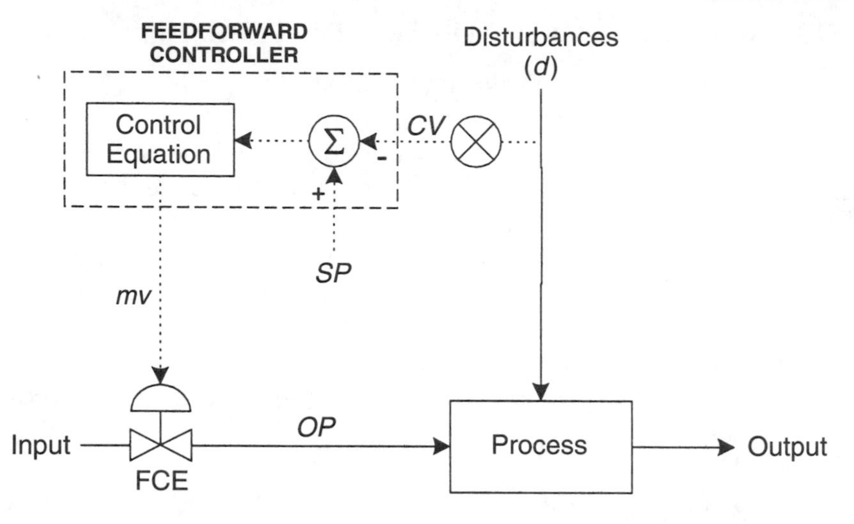

Figure 6.5 Feedforward control system.

the process disturbance measurement contains dead time, or lag, the feedforward dynamic compensation would have to be a predictor, which of course is impossible unless an exact and very fast dynamic model of the process is available. In practice, the feedforward dynamic elements are approximated by a lead-lag network [3, 4] that is adjusted to yield as much disturbance rejection as possible over as wide a range as possible.

When feedforward control is used equations are needed to calculate the amount of the manipulated variable needed in order to compensate for the disturbance. This sounds simple enough; however, the equations must incorporate an understanding of the exact effect of the disturbances on the process variable. Therefore, one disadvantage of feedforward control is that the controllers often require sophisticated calculations as even steady models can be nonlinear and thus need more technical and engineering expertise in their implementation.

Another disadvantage of feedforward control is that all of the possible disturbances and their effects on the process must be precisely known. If unexpected disturbances enter the process when only feedforward control is used, no corrective action will be taken and the errors will build up in the system. If all the disturbances were measurable and their effects on the process precisely known, a feedback control system for regulatory purposes would not be needed. However, such complete and error-free knowledge is never available, so feedforward is generally combined with feedback, as illustrated in Figure 6.6. The intent of this union is that the feedforward mitigates most of the effects of the principal disturbances and the feedback loops provide residual control and set point tracking.

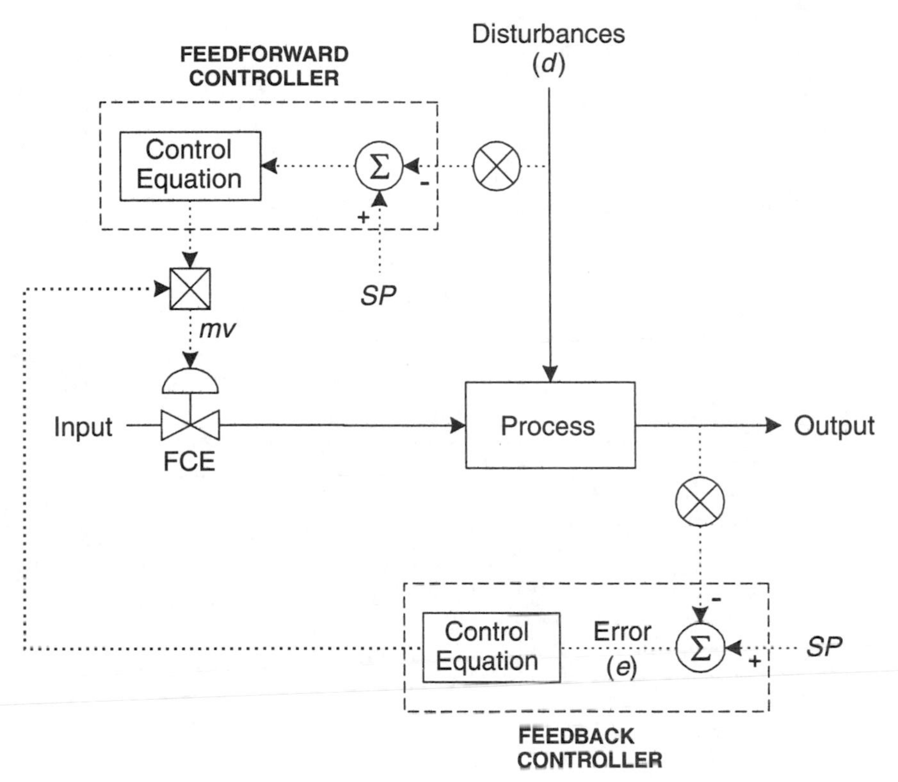

Figure 6.6 Feedforward/feedback control system.

Consider the following example of the feedforward control of a heat exchanger with cascaded feedback trim control (shown in Figure 6.7). The addition of feedback and cascade control serves to eliminate offset due to modelling inaccuracies and other non-measured disturbances.

At steady-state an overall heat balance can be written for the process as shown in Equations 6.1–6.3.

$$q_{in} - q_{out} = 0 \tag{6.1}$$

$$W_s\lambda - WC_p\,(T_2 - T_1) = 0 \tag{6.2}$$

Or,

$$W_s = \frac{C_P}{\lambda} W(T_2 - T_1) \tag{6.3}$$

where: λ = enthalpy transferred by the steam condensing to form condensate (kJ/kg)

C_p = heat capacity of the process fluid (kJ/kg.K)

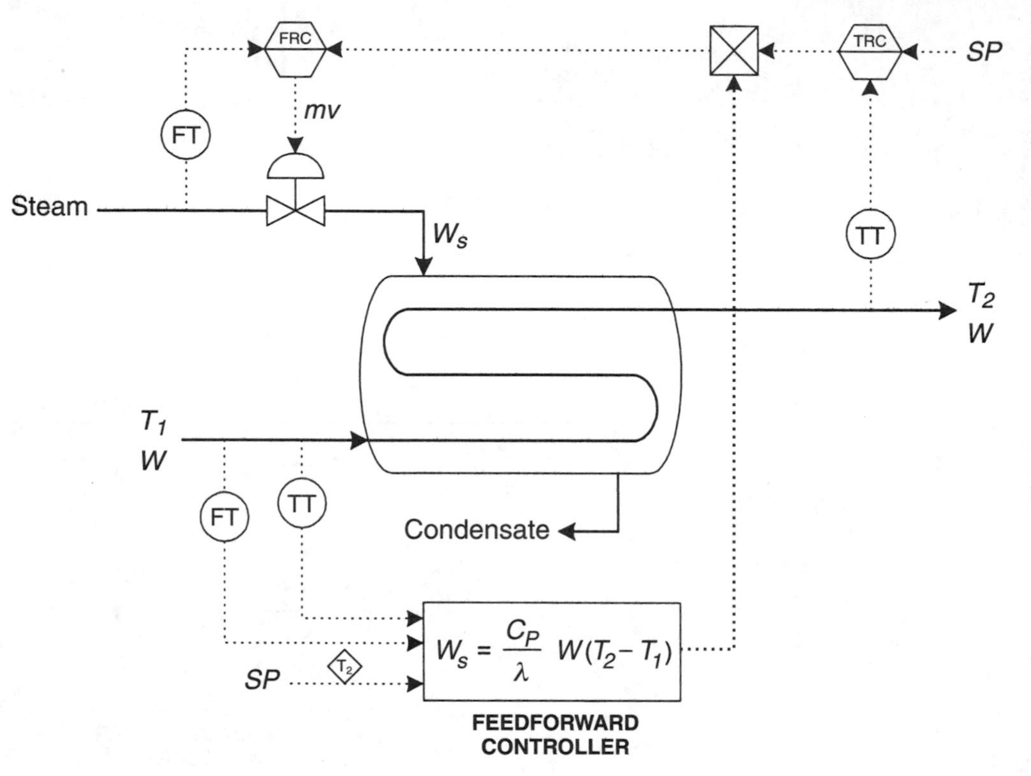

Figure 6.7 Combined feedforward and cascade control of a heat exchanger.

In this example, the inlet flow of liquid, W, and the temperature, T_1, are measured to determine the amount of steam required as per Equation 6.3. The desired outlet temperature, T_2, is the set point into the feedforward controller. The feedback temperature controller on the liquid stream measures T_2 to adjust for any disturbances that are not corrected by the feedforward controller.

Typical response curves for a load upset would appear as shown in Figure 6.8. Included in this figure for comparison is the response curve for feedback control only (with a PID controller) on the same process.

The response of the outlet temperature T_2 for the base case of the feedback control shows the type of improvement in control that can be achieved with even a simple steady-state feedforward controller. The lead/lag dynamic compensation [3, 4] shows further improvement over the steady-state feedforward control.

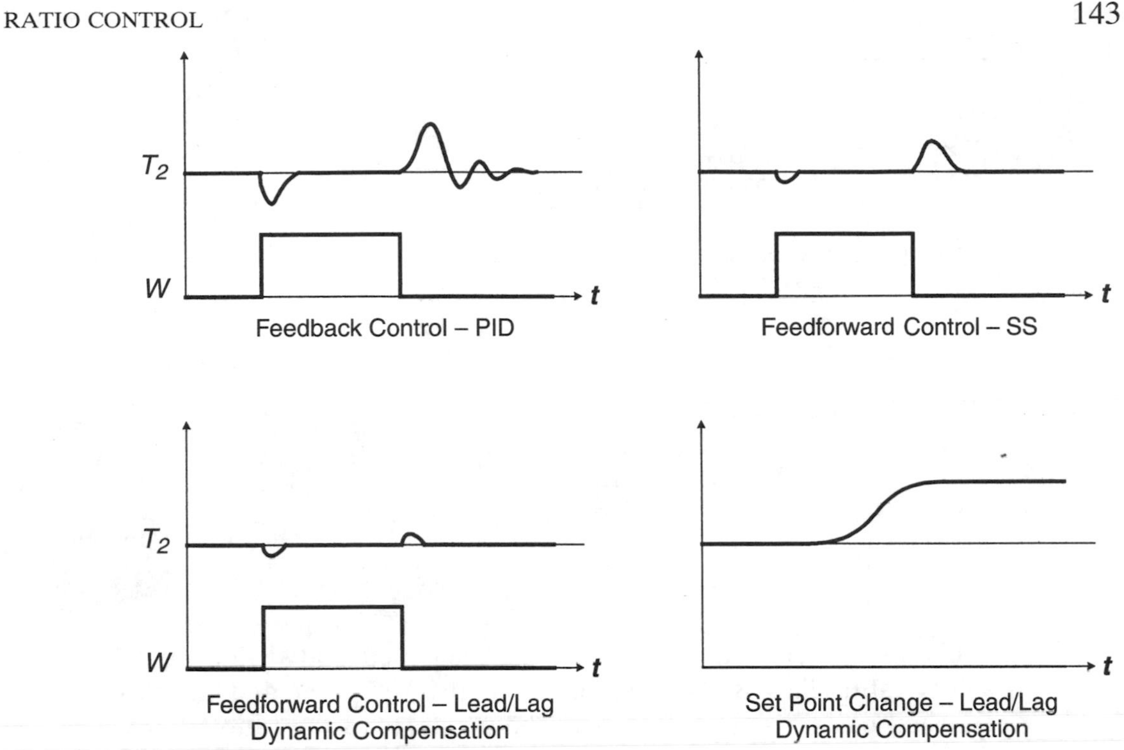

Figure 6.8 Typical responses of the heat exchanger.

6.3 Ratio control

Ratio control involves keeping the ratio between two variables fixed [2, 4], as illustrated in Figure 6.9. Typically these two variables, y and y_w, are flow rates, where y_w is the wild or uncontrolled flow rate and y is the manipulated or controlled flow rate. The wild flow rate is measured, and the controlled flow rate is then adjusted to maintain a fixed ratio between the two.

Ratio control can be considered to be a form of feedforward control. This is obviously true since in ratio control the process variable is measured upstream of the process, as is the case in feedforward control (Figure 6.4). Take for example a reactor with two liquid feed streams. Ratio control would ensure that these streams were being stoichiometrically fed to the reactor by measuring the flow of one stream and adjusting the other accordingly. The product stream would be of no real use in determining if the stoichiometric ratio was met.

There are two methods by which the ratio between the two variables can be fixed when only one stream is being manipulated. The first is shown in Figure 6.9.

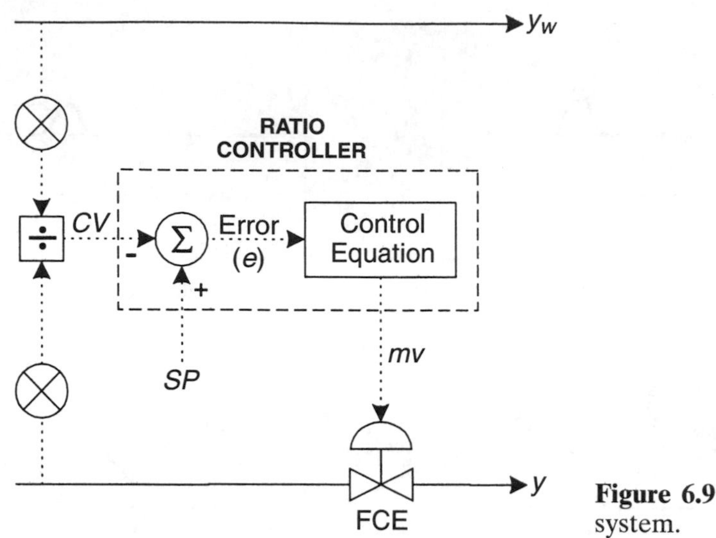

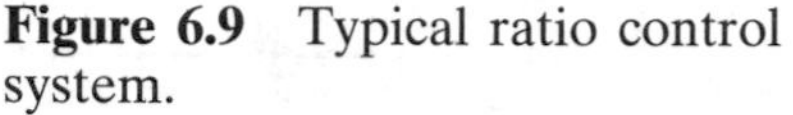
Figure 6.9 Typical ratio control system.

In this first scheme, both flows are measured and divided to obtain the actual ratio. This is then compared to the set point and the flow of y is adjusted based on the difference. The set point to the ratio controller is the desired ratio.

In the scheme shown in Figure 6.10, the flow of the wild stream, y_w, is measured and then multiplied by the desired ratio. The output from the multiplier is the set point for the controller, which compares it to the measured flow and adjusts the flow of y accordingly. In this scenario, the desired ratio is a constant variable in the multiplier and if a new value for the ratio is needed, it must be set in the multiplier.

A common example of ratio control is the case of an adsorption column where a fixed ratio of V/L is desired. The wild flow rate is the vapour feed to the column, V, while the controlled flow rate is the liquid flow rate, L. The ratio control seeks to maintain constant absorption factors in the column by keeping a constant V/L profile.

6.4 Override control (auto selectors)

Frequently a situation is encountered where two or more variables must not be allowed to exceed specified limits for reasons of economy, efficiency or safety. If the number of controlled variables is greater than the number of manipulated variables, a selection must be made for control purposes (Single Input / Single Output). A selector is used to accomplish this. Selectors are available in both electronic and pneumatic versions. The only difference between selectors is the number of inputs a particular hardware implemen-

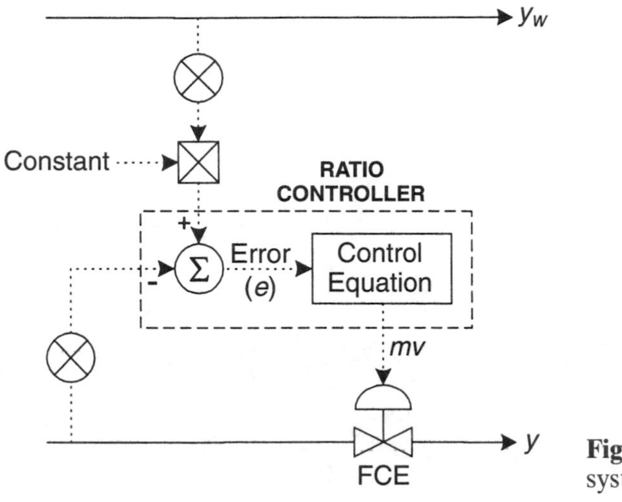

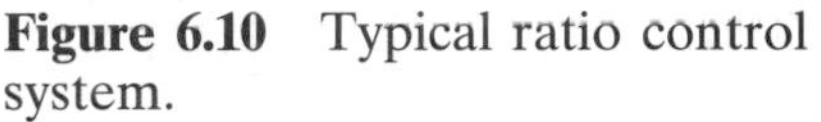

Figure 6.10 Typical ratio control system.

tation may be able to accommodate. In this section specific examples of such selectors will be discussed. It must be kept in mind that these are only a few examples of such auto selectors [4].

The two basic building blocks for selector systems are the high and the low selector. The high selector, shown in Figure 6.11, will pass the highest value of the multiple input to the output signal, ignoring all other inputs.

The low selector, shown in Figure 6.12, will choose the lowest of inputs to pass through as the output while ignoring all other inputs.

By using combinations of these basic building blocks it is possible to build other types of selectors, such as a median value selection, shown in Figure 6.13. The selector output for a median value selector is a signal that falls between the highest and lowest input.

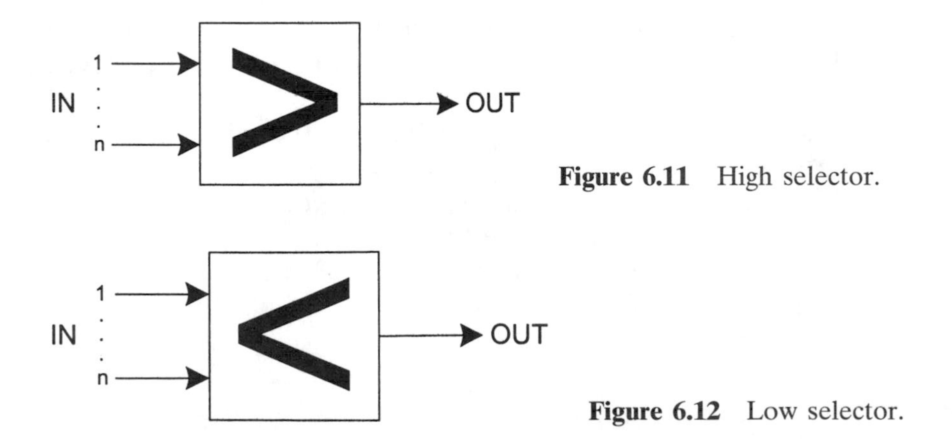

Figure 6.11 High selector.

Figure 6.12 Low selector.

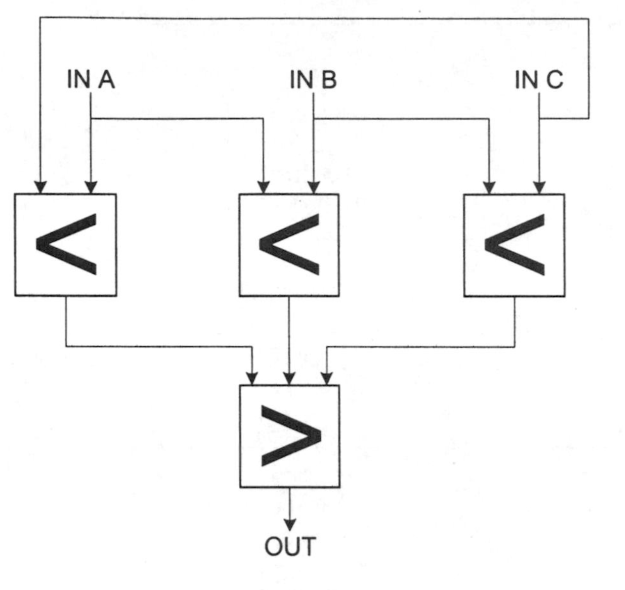

Figure 6.13 Median value selector.

Let us investigate some typical applications of these selectors in four areas:

1. Protection of equipment.
2. Auctioneering (choosing from several signals).
3. Redundant instrumentation (used commonly with process analytical equipment).
4. Artificial measurements (establishing artificial limits).

6.4.1 PROTECTION OF EQUIPMENT

To illustrate how selectors can be used to protect equipment, examine the pump system shown in Figure 6.14. The pump system demonstrates a situation where there are multiple measurements, multiple controllers and only one manipulated variable that can provide the following protection:

- Surge protection: when P_{in} drops below a certain minimum value, close the valve.
- High temperature: when the temperature of the pump exceeds a certain maximum temperature, close the valve.
- Excessive downstream pressure: when P_o exceeds a certain maximum pressure, close the valve. (Assume $P_o > P$ shut off.)

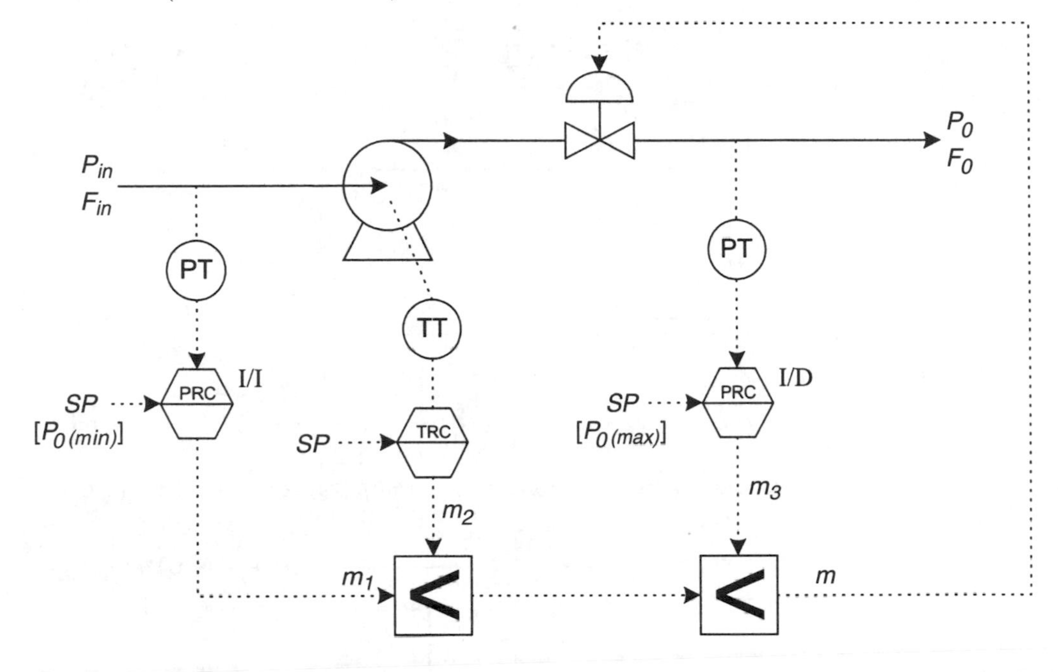

Figure 6.14 Protection of equipment – pump.

Surge protection
As P_{in} begins to drop, the output, m, will also decrease (note Increase/Increase action on pressure controller). The output, m, will be selected by the first and second low selector and will be passed through as the manipulated variable causing the valve to close.
High temperature and excessive downstream pressure
If either the pump temperature or outlet pressure begins to increase, both outputs m_2 and m_3 begin to decrease (note Increase/Decrease action on both of these controllers). The smallest value will be chosen and passed through to manipulate the valve. In general the smallest output from either of the controllers will always be operating the valve.

6.4.2 AUCTIONEERING

The objective of autioneering is to protect against the highest temperature sensed by one of many temperature transmitters. In the example shown in Figure 6.15 the control equipment consists of one controller, four transmitters and one FCE. The highest temperature will be selected by the high selectors and will be used as the measurement for controlling the fuel to the oven.

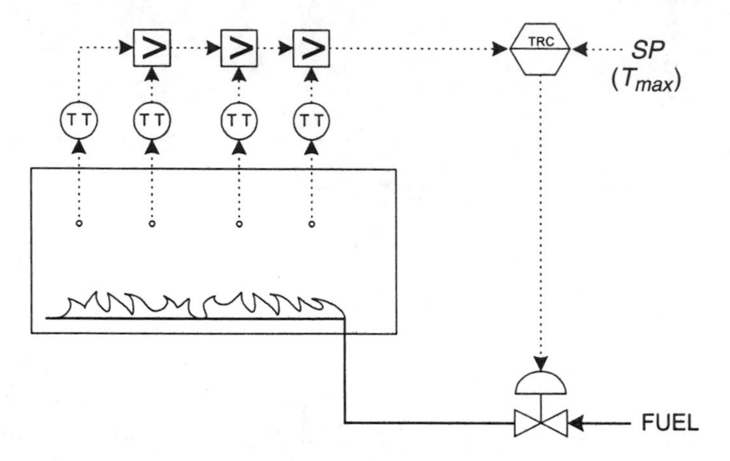

Figure 6.15 Auctioneering – temperature control in an oven.

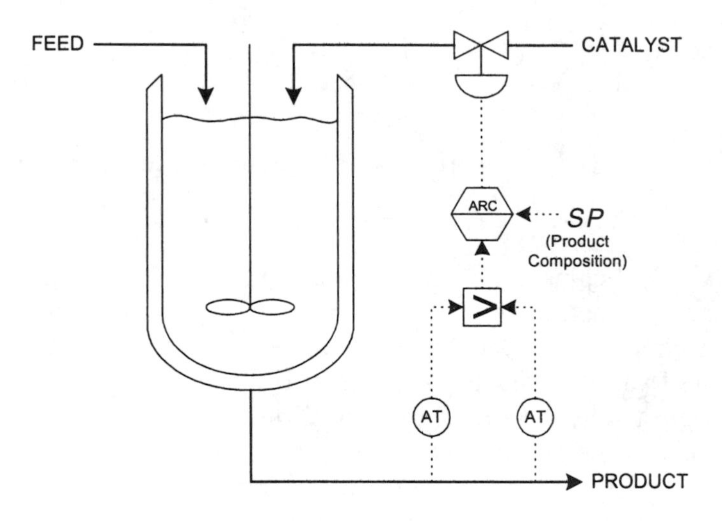

Figure 6.16 Redundant instrumentation – reactor.

6.4.3 REDUNDANT INSTRUMENTATION

For an exothermic reactor (shown in Figure 6.16) too much catalyst can prove disastrous. By implementing a fail-safe scheme that consists of two composition transmitters that are analysers and a high selector, the highest reading from the analysers will be utilised by the composition controller to control catalyst flow. The following actions will occur in the event of catastrophic failure of the analysers:

1. Down scale failure of analyser – If one analyser fails to zero, the other will be selected to control catalyst flow and production will not be interrupted.

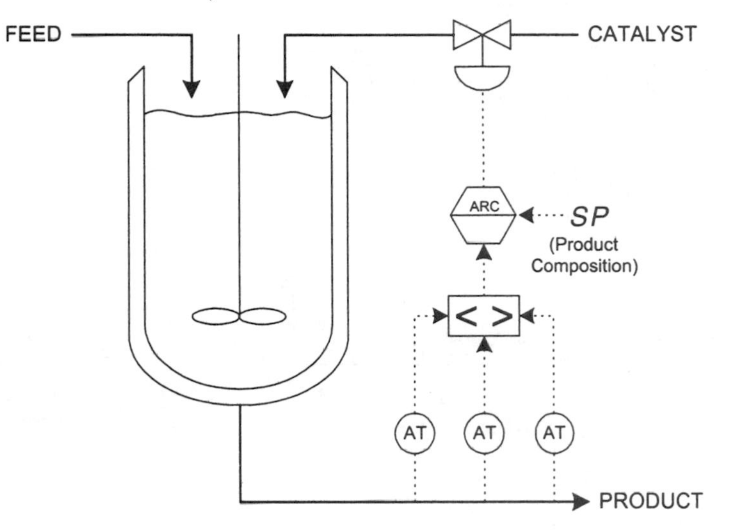

Figure 6.17 Redundant instrument – reactor/median selector.

2. Upscale failure of analyser – If one analyser fails to full scale it will be selected and the catalyst flow will be stopped. Production is stopped and a possible hazardous situation avoided.

An alternate scheme, shown in Figure 6.17, implements analysers with a medium selector that will keep the process operating regardless of the failure mode of one of the analysers. The measurement variable to the controller will always be the median transmitter output. If one of the analysers fails, either upscale or downscale, the selector will still choose the median value.

In summary, the amount or quantity of redundancy depends on the importance of the process unit (reactor, distillation column, etc.). This is because the higher the quantity of redundancy the higher the cost (capital/operating) becomes and therefore the cost must be justified.

6.4.4 ARTIFICIAL MEASUREMENTS

Some processes require certain operating constraints to be set. These are referred to as artificial measurements. These operating constraints can be set through the use of selectors. For example, consider a distillation column whose feed versus steam characteristic is shown in Figure 6.18.

Instead of operating the steam versus feed flow on a straight line, operating constraints are set. The operating constraints require a minimum stream rate of 10%, even if the feed rate drops to zero. This sets the low limit of the steam flow. Furthermore, at maximum feed rate the stream rate should

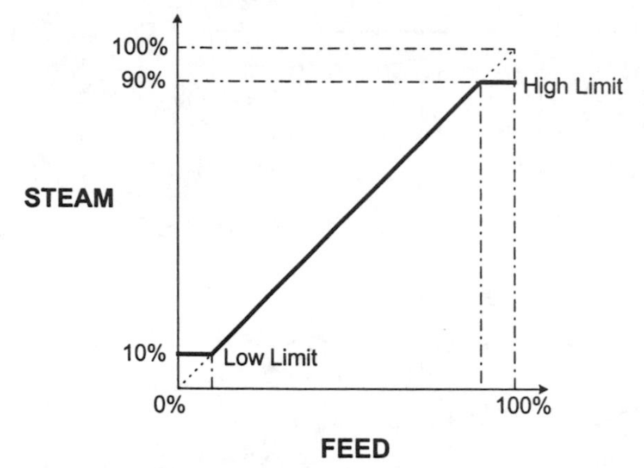

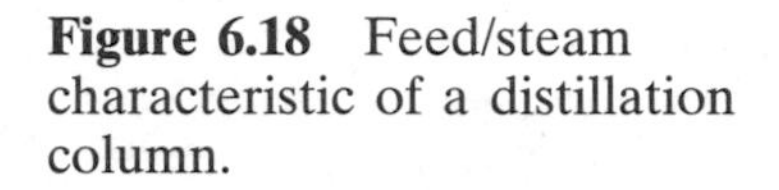
Figure 6.18 Feed/steam characteristic of a distillation column.

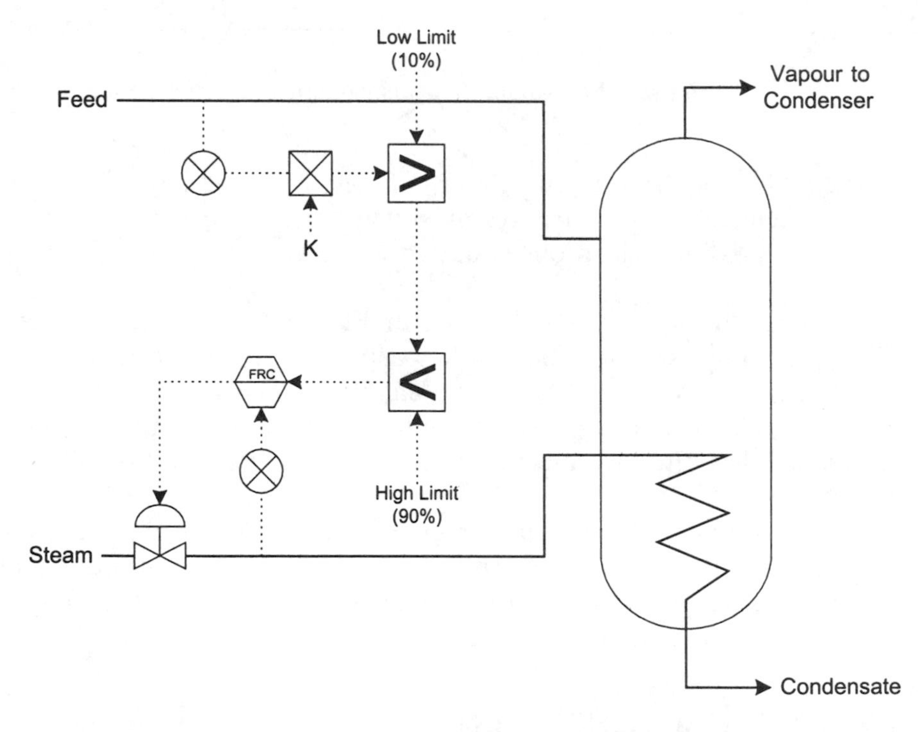

Figure 6.19 Artificial constraints.

not exceed a maximum flow of 90%, the high limit. These constraints can be implemented as shown in Figure 6.19.

If the feed flow is within the safe operating region, the signal from the multiplier will pass through the high selector since it is higher than the low limit. It will also pass through the low selector since it is lower than the high limit, and then act as the set point for the steam flow controller. If the feed signal falls below the low limit or above the high limit the proper limit will be selected and that limit will be a constant high or low signal to the steam flow controller. This prevents the high and low limits from being exceeded.

6.5 References

1. Seborg, D.E., A Perspective on Advanced Strategies for Process Control. *Modeling, Identification and Control*, 1984, **15(3)**: 179–89.
2. Eckman, D.P., *Principles of Industrial Process Control*. John Wiley & Sons, New York, 1945 (Metered and Ratio Control, pp. 194–9).
3. Murril, T.W., *Automatic Control of Processes*. International Textbook Company, Scranton, Penn., 1967 (Cascade/Ratio, pp. 431–44, Feedforward Control, pp. 405–25).
4. Shinskey, F.G., *Process Control Systems*. McGraw-Hill, New York, 1967 (Cascade/Ratio, pp. 154–60, Feedforward Control, pp. 204–29, Override Control, pp. 167–9).

7 COMMON CONTROL LOOPS

This chapter will describe some common loops encountered in process control. The loop characteristics, type of controller to use, response, tuning and limitations will all be examined.

7.1 Flow loops

A typical flow control loop is shown in Figure 7.1. This process is very fast responding and, even for long lengths of pipe, the dead time and the capacitance are very small. Typically the process response is limited by the valve response (time constant).

As shown in Figure 7.1, the flow sensor/transmitter is always placed upstream of the valve for several reasons. First, many flow measuring devices have upstream and downstream straight run pipe requirements. Usually the upstream straight run is longer than the downstream straight run. Therefore, the flow measuring device can be placed closer to the valve upstream than downstream, where there might be problems with additional pressure drop through piping if a head flow device is used. Some examples of head flow devices are orifice plates, venturi tubes and flow nozzles. Second, when the flow sensor is upstream from the valve there is a more constant inlet pressure since it is closer to the source. Finally, there might be pressure fluctuations introduced to elements installed downstream from the valve as a result of valve stroking. Valve stroking results when the valve moves up and down, causing pressure changes that can affect downstream units or elements.

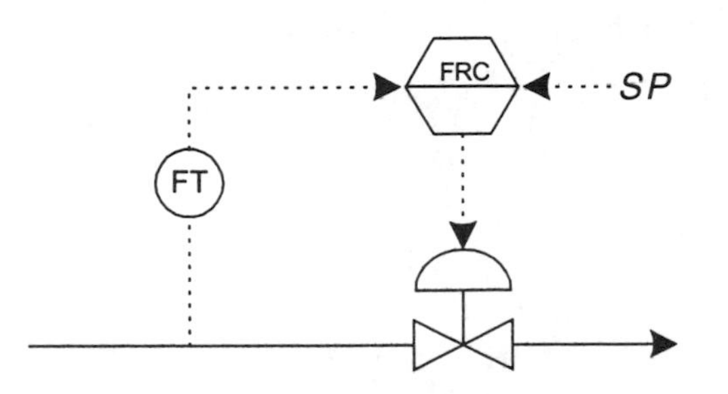

Figure 7.1 Flow control loop.

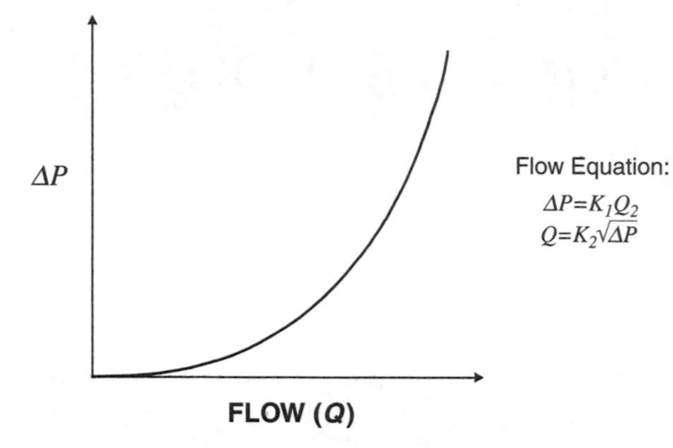

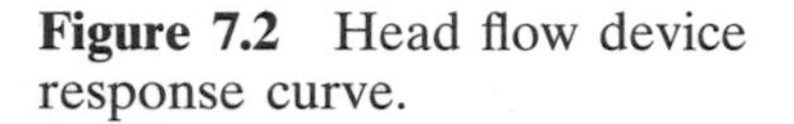
Figure 7.2 Head flow device response curve.

Another consideration when using head flow devices, besides additional pressure drop, is their nonlinear response, illustrated in Figure 7.2.

It is important to examine all the elements in the loop when determining what type of response is expected. For example, a differential pressure transmitter, also known as a d/p cell, has a linear response. However, when the head flow element, i.e. orifice, and the d/p cell are connected together the response is nonlinear, as shown in Figure 7.2.

A desirable attribute of a control loop is a response that is independent of the operating point, or a linear response. To this end, good practice requires offsetting nonlinearities in the loop to create an overall linear response. For example, when using a head flow device, a square root extractor is used to linearise the flow signal. The square root extractor is a device that simply takes the square root of the signal in order to linearise it.

Another factor that affects the linearity of the response is the characteristic of the valve selected. Three of the most common valve characteristics include quick opening, linear and equal percentage. If the majority of the system pressure drop is taken across the valve, a linear valve should be used since its installed characteristic will also be linear, giving the linear response desired. However, if the pressure drop across the valve is a small part of the total line drop and is not constant, an equal percentage valve can be used since its installed characteristic will be close to linear. Quick opening valves are most commonly used with on-off controllers where a large flow is needed as soon as the valve begins to open. More information on these valve flow characteristics can be found in Chapter 2.

Flow measurement by its very nature is noisy. Therefore, derivative action cannot be effectively used in the controller because the noisy signals cause the loop to become unstable. Flow control is one type of loop where an integral-only controller can be used. One drawback to I-only control is that it can greatly slow down the response of the loop, but the flow process is so fast that this slowing down may not be significant. To understand just how fast a flow loop is, consider again the heat exchanger cascade control scheme shown in Figure 6.4, where the primary loop may have a response period of several minutes. However, the secondary flow loop, even under I-only control, is fast enough for effective cascade control. Despite this, typically PI controllers are most often employed in flow loops because they are standard in the process control industry.

Tuning a flow loop for a PI controller is easy in comparison to other types of loops. The flow loop is so fast that quarter amplitude damping[1] cannot be observed. The objective is for the flow measurement to track the set point very closely. To achieve this, the gain should be set between 0.4 and 0.65 (PB ≈ 200%) and the integral time, T_i, between 0.05 and 0.25 minutes. If there is an instability limit over the operating range due to nonlinearities in the loop, the controller gain can be reduced, but the integral time should not. For a fast loop, such as flow control, an offset may persist because more gain is contributed from the proportional action than from the integral action.

7.2 Liquid pressure loops

A liquid pressure loop has the same characteristics as a flow loop. The objective of the loop is to control the pressure, P, at the desired set point by controlling the flow, F, as the needs of the process change (see Figure 7.3).

The pressure loop shown in Figure 7.3 is in fact a flow control loop, except that the controlled variable is pressure rather than flow. Since the flow is an incompressible fluid, the pressure, P, will change very quickly. The process behaves like a fixed restriction, i.e. an orifice plate, whose ΔP is a function of flow through the process. The process gain, K_p, can be determined from Equations 7.1–7.4.

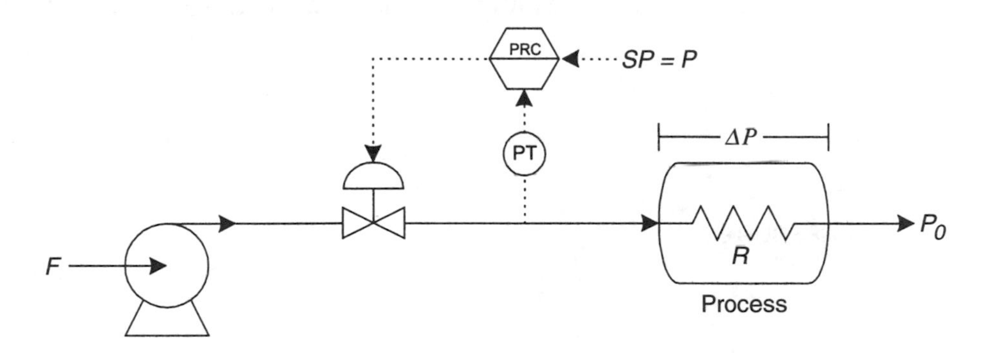

Figure 7.3 Liquid pressure control loop.

[1] Also called the quarter decay ratio (QDR); see Chapter 5 for more detail

$$P = \Delta P + P_o \tag{7.1}$$

P_o is the downstream pressure at zero flow.

Also,

$$\Delta P = \frac{F^2}{R^2} \tag{7.2}$$

So

$$P = P_o + \frac{F^2}{R^2} \tag{7.3}$$

As illustrated in Figure 7.3, R is the process flow resistance. The gain of the pressure loop is calculated as shown in Equation 7.4:

$$K_p = \frac{dP}{dF} \tag{7.4}$$

Substituting Equation 7.3 into Equation 7.4 results in Equation 7.5:

$$K_p = \frac{dP}{dF} = \frac{2F}{R^2} \tag{7.5}$$

Plotting $P = P_o + [F^2/R^2]$ results in Figure 7.4, where the slope of the curve at any point is the process gain, K_p, as calculated in Equation 7.4.

The response of pressure to flow is exactly the same shape as the head flow device response discussed previously and shown in Figure 7.2. Therefore, the same rules apply for a liquid pressure loop as for the flow loop. The only difference between the two is that the pressure varies from P_o to 100%, and not from 0% to 100% as for the head flow device. For this case, the process

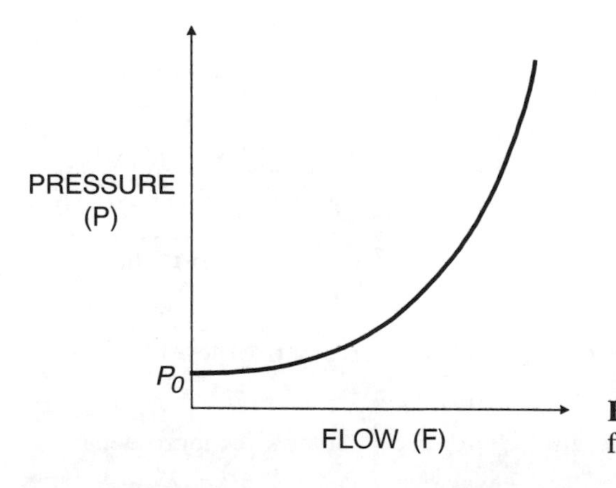

Figure 7.4 Process gain of pressure flow loop.

gain is somewhat smaller than that for the flow process and thus a higher controller gain can be used, i.e. between one and two.

Other considerations for the liquid pressure loop are as follows:

- The controller can be proportional plus integral (PI) or integral-only (I-only) and is tuned similarly to the flow controller.
- K_p is not constant, therefore a square root extractor or the highest loop gain should be used in tuning the controller. The reason behind using the highest loop gain is to prevent the loop from becoming unstable. This concept is explained in more detail in the liquid level control section in this chapter.
- Since the liquid pressure loop is similar to a flow loop, it is also noisy. Therefore, derivative action in the controller is not advisable.

7.3 Liquid level control

A liquid level control loop, shown in Figure 7.5, is essentially a single dominant capacitance without dead time. Typically, processes that are dominated by a single, large capacitance are the easiest to control. However, liquid level processes are not necessarily as simple as they first appear to be. In many liquid level control situations, considerable noise in the measurement is present as a result of surface turbulence, stirring, boiling liquids, etc. The fact that this noise exists often precludes the use of derivative action in the controller. Still, some applications use unique methods of level measurement to minimise the noise in the measurement in order to apply derivative action in the controller.

The first example of using a unique measurement method to minimise noise is using a displacer in a stilling well, shown in Figure 7.6. The intention of

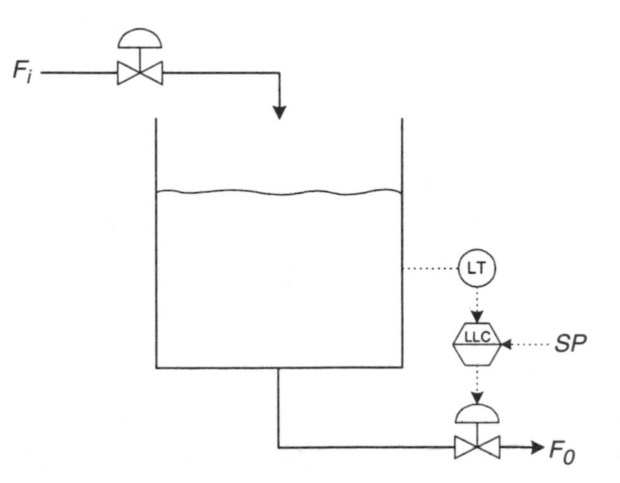

Figure 7.5 Liquid level control loop.

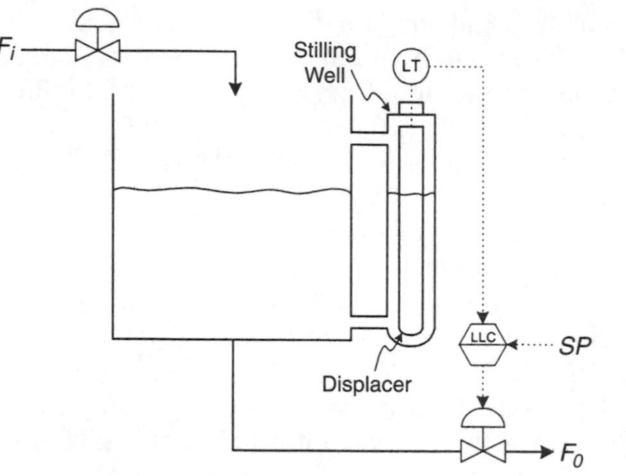

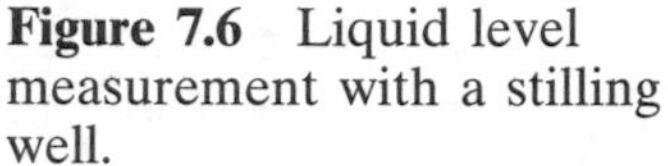
Figure 7.6 Liquid level measurement with a stilling well.

this arrangement is to effectively filter out high frequency noise due to turbulence in the measurement by using the stilling well. However, one caution is that the tank and stilling well form a U tube and the result could be low frequency movement of the liquid from the tank to the well and back. This will make the transmitter believe that the level is slowly moving up and down. If control action were taken, the controller would aggravate the situation.

Another example of noise filtering is to use ultrasonic level measurement with electronic filtering of the level signal, shown in Figure 7.7. This method works because the noise frequency is much higher than the period of response of the tank level. The electronic filtering in this case is a relatively simple matter.

Another technique that has been employed effectively to minimise noise is to use some sort of tank weighing method, illustrated in Figure 7.8. In this

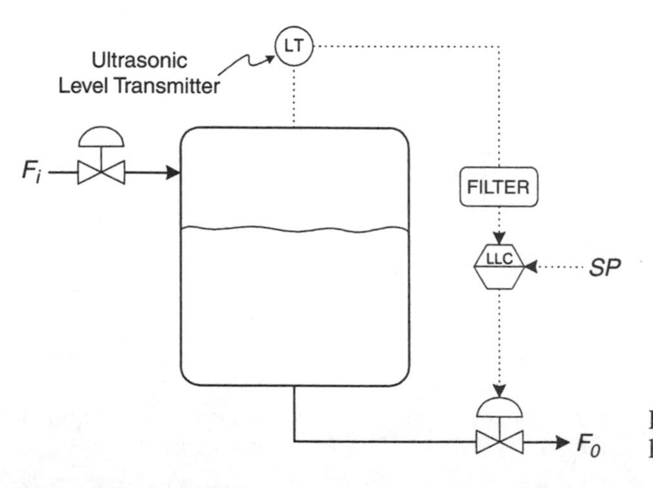

Figure 7.7 Ultrasonic liquid level measurement.

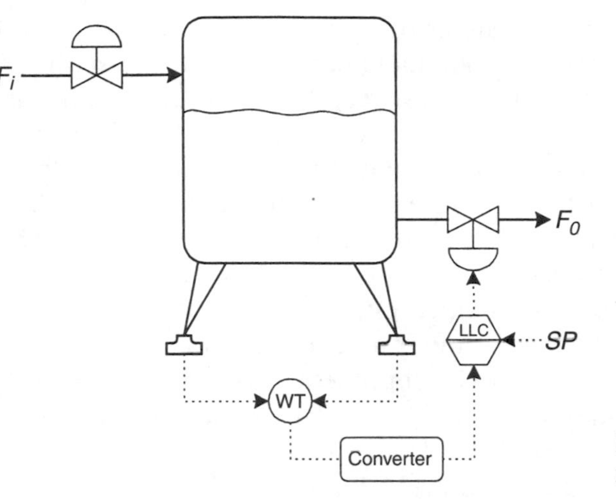

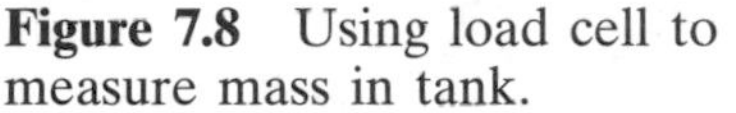

Figure 7.8 Using load cell to measure mass in tank.

application, a load cell is placed under each tank support in order to measure the mass of the tank. The outputs are sent to an averaging weight transmitter, and then to a weight/level converter before entering the controller. Obviously this method is effective in eliminating noise in the measurement because the turbulence in the tank does not affect the weight measurement.

The three methods suggested (Figures 7.6–7.8), are ways to minimise noise in the level signal so that derivative action can be used in the controller. Derivative action in the controller will overcome the sluggish response caused by the integral action. Integral control is required to maintain the level at the set point, which cannot be accomplished by a proportional only controller. Just how large an offset results from applying P-only action to a level process and is it small enough to justify use of a P-only controller for liquid level control? The equation for the error resulting from P-only control of a process is given in Equation 7.6:

$$e = \frac{mv - b}{K_c} \tag{7.6}$$

As seen in the above equation, increasing the controller gain, K_C, will minimise the error. If the process gain is low, the larger controller gain will result in a small error while still maintaining a stable loop.

Therefore, applying a P-only controller to control the liquid level in large tanks should definitely be considered. In many cases, an acceptably small error of only a few percent will result. The response will be just as fast as with a PID controller and noise in the measurement does not need to be a consideration. Effective application of P-only control is possible in this case because of the low process gain of the large capacity tank, which allows for a high controller gain and thus smaller error. P-only control should be considered whenever a single large dominant capacitance with very little or no dead time is present. Keep in mind, however, that tight level control is not always desirable. Deviations from the level set point can sometimes be tolerated in exchange for a smoother flow in the manipulated stream feeding other more sensitive equipment. This is termed averaging control [1, 2, 3].

Another interesting problem to be considered in a liquid level process is the dependence of the process gain on load, which is a problem that exists in any single dominant capacitance situation. Load is defined as anything that will affect the controlled variable under a condition of constant supply. Consider the open loop case of a liquid level process, shown in Figure 7.9. The process gain, K_P, for a constant outflow, F_o, is calculated using Equation 7.7.

$$K_p = \frac{\Delta \text{Out}}{\Delta \text{In}} = \frac{\Delta \text{Level}}{\Delta \text{Inflow}} = \frac{\Delta h}{\Delta F_i} \quad \text{for } F_o = \text{Constant} \tag{7.7}$$

Since the outflow is set at a constant value, the inflow is considered to be a load on the process. Figure 7.10 shows the effect that the inflow has on the process gain. In this figure, the process gain is the slope of the curve and F_i is the relative opening of the inlet valve.

In the first case, the outlet valve is closed, $F_{01} = 0$, and $F_i \neq 0$, causing the tank level to increase. The level will theoretically increase to infinity or $K_p = \infty$. However in reality, the tank will overflow and the level, h, will never saturate at the maximum tank capacity. Also, regardless of what the level is in the tank, if F_{01} is set to zero and then F_i is set to the original flow, the level will continue rising at the same rate. This is not the case if $F_0 \neq 0$. As F_0 gets larger, the steady-state level is a lower value. The reason that this occurs is because

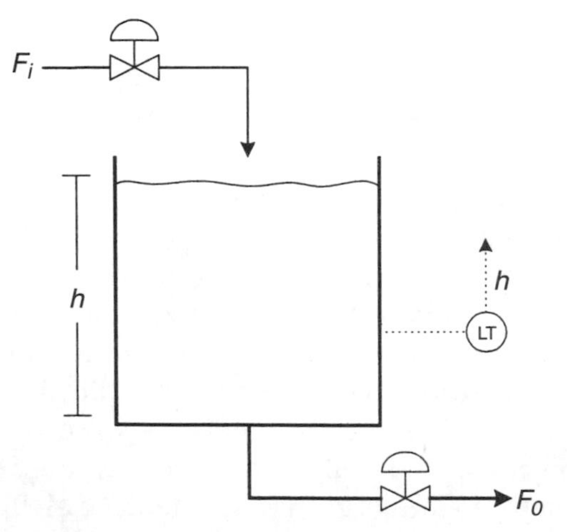

Figure 7.9 Liquid level process.

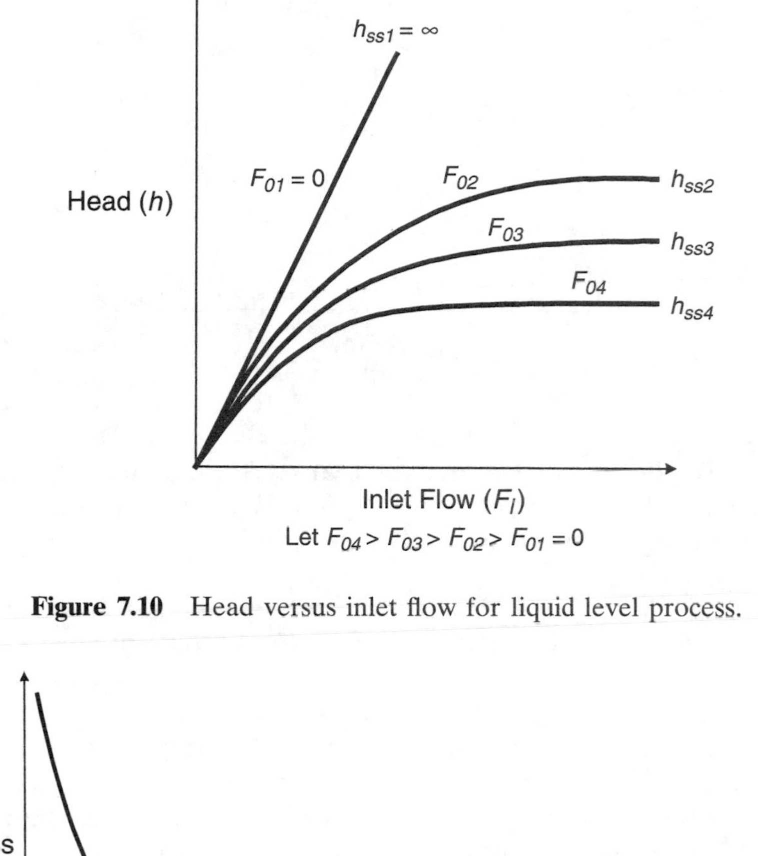

Figure 7.10 Head versus inlet flow for liquid level process.

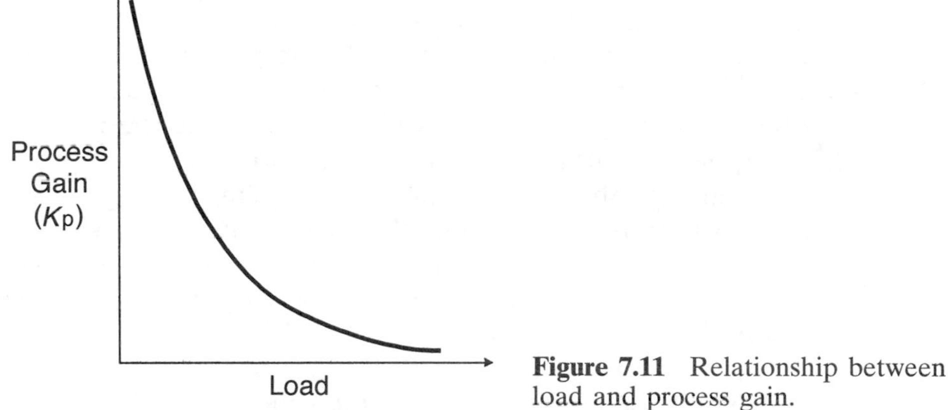

Figure 7.11 Relationship between load and process gain.

as F_0 is made larger, the head in the tank does not have to be as large to make the inflow equal to the outflow. Also, for any given outlet valve setting, i.e. F_{02}, F_{03}, etc, if $F_i = F_0$ and then F_i is moved to its original setting, the level in the tank will rise and Δh for a given ΔF_i will be less.

This behaviour is generally true of a class of processes dominated by a single capacitance. The process gain is a function of the load, with the gain decreasing as the load increases, as shown in Figure 7.11.

However, the inflow is not the only load on this process. The set point, i.e. fixed value of F_0, is also considered to be a load, as illustrated in Figure 7.12.

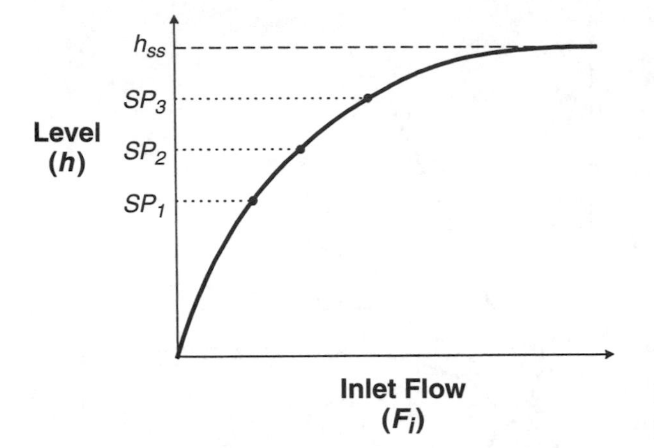

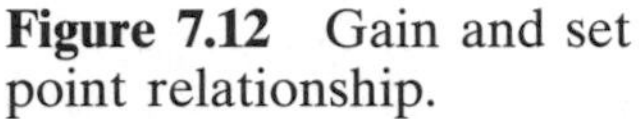
Figure 7.12 Gain and set point relationship.

As the set point is increased, $SP_3 > SP_1$, and the process gain, which is the slope of the curve at SP_1, SP_2, and SP_3, would decrease. Again, the process gain shows a reciprocal relationship to the load. In this case, the load is the set point.

Why is the dependence of process gain on load a consideration? The previous discussion shows that when a process contains a capacitance and the controller gain is adjusted to give a particular response at a given load, the response will change as the load changes. If the load on the process is reduced, the process gain rises and therefore the loop response tends to be more oscillatory. If the process gain increases enough, the loop could become unstable. On the other hand, if the load increases, the process gain decreases, and the loop responds sluggishly. The important fact to remember is that the loop must not become unstable, i.e. the loop gain, K_L, must be < 1. Therefore, for a situation where the process gain is a function of the load, the simplest thing to do is tune the loop at the highest process gain and live with a sluggish response for the situation of the process gain decreasing with increasing load.

Another approach to this situation would be to put a component in the loop that would have a complementary gain to the process gain. An example of this is using a square root extractor with a head flow meter in the flow control loop. If the pressure drop across the valve remained fairly constant, then the valve and installed characteristic would be almost the same. An equal percentage valve could be used to complement the process, and the product of the valve and process gain ($K_V \times K_P$) would almost be constant, as illustrated in Figure 7.13.

Yet another approach to this situation would be to adjust the controller parameters, the controller gain for example, with the variation in load of set point to compensate for the variation in process gain. This approach is termed gain scheduling [4, 5] or programmed adaptation [6, 7] and can be considered a form of adaptive control [4].

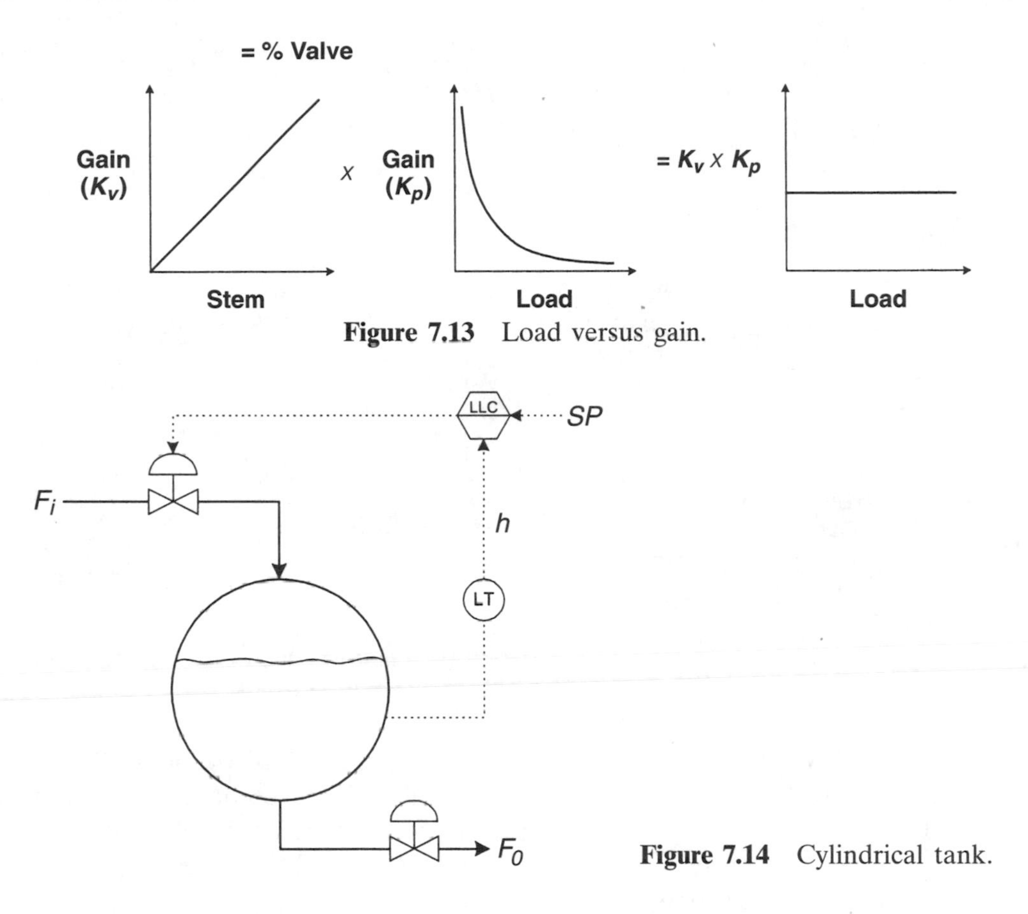

Figure 7.13 Load versus gain.

Figure 7.14 Cylindrical tank.

Another level control situation where a non-linear element might be introduced into the loop is the case of level control of a cylindrical tank lying on its side, shown in Figure 7.14. The previous liquid level processes were shown to have the outlet flow as the manipulated variable. Although this is the common method for liquid control, the inlet flow can also be used to control the level in the tank, as shown in this example.

The response of the level in the cylindrical tank shown above is given in Figure 7.15. Obviously the response for the horizontal, cylindrical tank differs from those previously discussed due to the differences in tank geometry.

Figure 7.16 represents the process gain qualitatively as it varies with load. In this case the load is the height in the tank, h. The gain of the tank is high at both low and high levels and is low at normal levels in the cylindrical tank.

To make the loop gain independent of the tank level, a signal compensator must be added to the liquid level loop. The signal compensator has a response which can be varied as shown in Figure 7.17, so that when the process gain is high, the compensator gain is low and vice versa, thus giving an overall linear response.

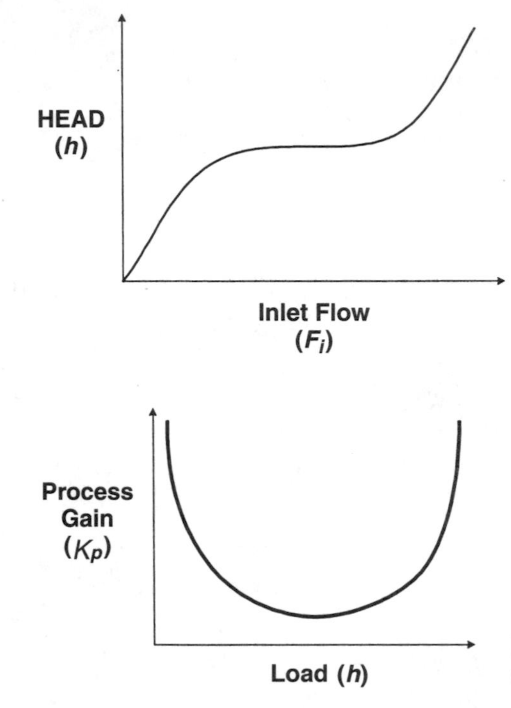

Figure 7.15 Cylindrical tank level response.

Figure 7.16 Process gain versus load for a cylindrical tank.

Now consider the case, shown in Figure 7.18, of a tank with a P-only controller and some valve hysteresis. All valves have some hysteresis, but excessive valve hysteresis typically occurs when the valve sticks as it tries to open and close. This can happen for a number of reasons, including over-tightened packing, etc.

The response for this process might not be as good as desired, since any misposition in the supply valve will show up as an incorrect level to the controller. If this were a large tank, the response might be very slow because the level would change very slowly due to the large surface area. By applying a valve positioner, V/P, a cascade system is set up that will effectively minimise the effects of valve hysteresis and improve loop response. In this cascade system the level controller acts as the primary loop and the valve positioner is the secondary loop. A valve positioner can be used in this case because the response of the level control loop is much slower than that of the valve positioner loop and therefore the rule $\tau_{01} > 4\tau_{02}$ is obeyed, giving an effective cascade system.

Another level control situation commonly encountered is that of using the capacity of a tank to prevent surging or pulsing of flow from upstream processes to downstream process units. This is the general case of an integrating process that does not require tight control. An example of an integrating process is a buffer tank with pumped outflow, as illustrated in Figure 7.19. The buffer tank provides exit flow smoothing in the face of

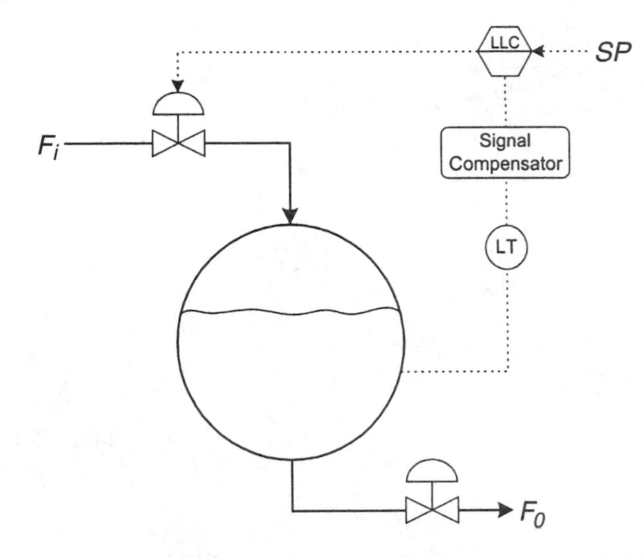

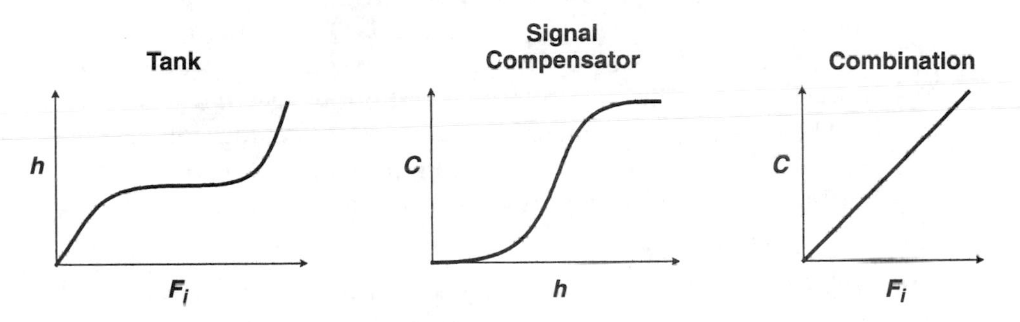

Figure 7.17 Cylindrical tank level compensator.

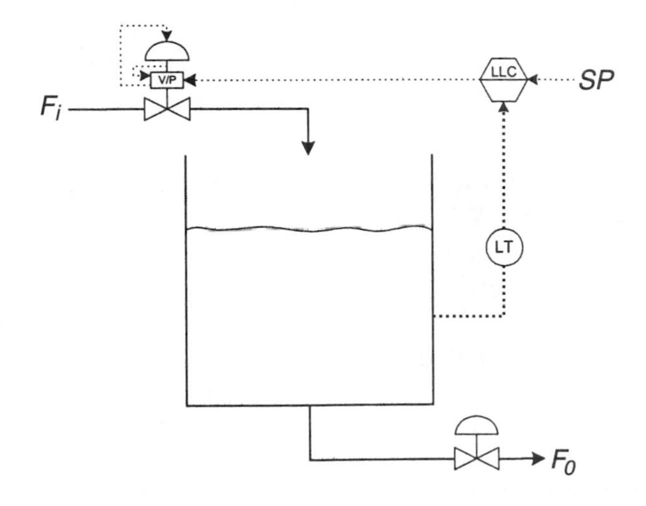

Figure 7.18 Liquid level control loop.

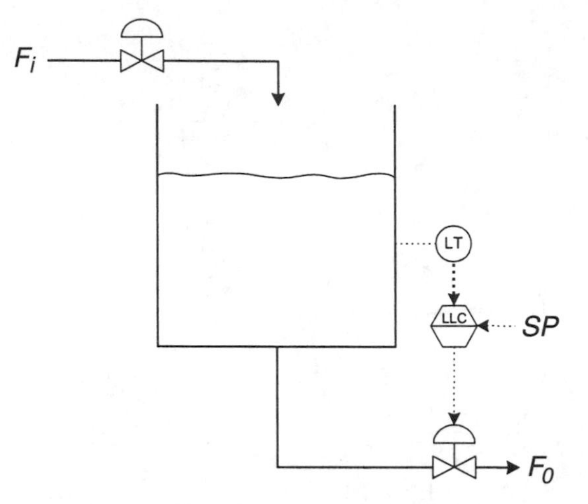

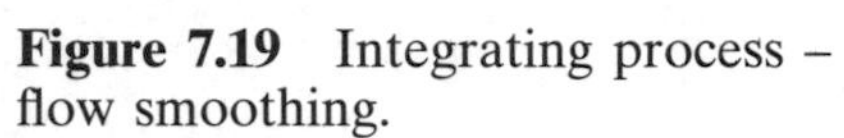

Figure 7.19 Integrating process – flow smoothing.

incoming flow disturbances. The objective is not to provide tight level control but to let the level swing. Since the tank and pump are in effect an integrator, if a PI level controller were used the result would be a double integrator with the potential for continuous cycling.

There are two popular approaches to tuning integrating processes to provide flow smoothing of the exit stream. One involves proportional only, and the other uses proportional plus integral action.

7.3.1 PROPORTIONAL ONLY CONTROL FOR INTEGRATING PROCESSES

The following defines the desired conditions for proportional only control of an integrating process with flow smoothing:

Proportional Controller Gain	= 2
Bias	= 50%
Set point	= 50%

To set the bias at 50%, put the controller into manual, set the output to 50%, and then switch to auto. With no integral action, the bias will remain at whatever the valve position was when it was last in manual mode.

These settings offer the following benefits:

1. Output flow will shut off before the level drops below 25%; *thus the tank cannot run dry.*
2. Output flow reaches the maximum before the level exceeds 75%; *thus the tank level will never exceed 75% for maximum throughput.*
3. Effective buffering is increased when compared to using a PI controller. In other words, at low flow the level is low and there is room for the

most likely change, a large dump of fluid into the tank. In fact, there is approximately 50% more room, i.e. 25/50 or 50%.

Using Equation 3.3 for a proportional only controller[2] and inserting the above values for the set point, *SP*, the bias, *b*, and the controller gain, K_C, and rearranging results you get:

$$mv = 2(PV) - 50 \tag{7.8}$$

Therefore, at a 25% level, the output is 0%, and for 75% level, the output is 100%. What may not yet be obvious is how the effective buffering is increased. Consider a lower-than-normal throughput, say $mv = 5\%$. From Equation 7.8, the level, *PV*, would line out at about 27.5%. For a higher-than-normal throughput, say $mv = 95\%$, the level would line out at about 72.5%. Thus at high throughput, the level runs high, and for low throughput, the level runs low, always providing maximum room for the most likely disturbance. The increased buffering comes from the fact that:

- at low throughputs there is more "headroom" to buffer sudden increases in throughput when the level is low; and
- at high throughputs, there is more inventory in the tank to buffer sudden decreases in throughput when the level is high.

PI controllers will all eventually drive the level to the set point value, and therefore lack this particular benefit. However, P-only level controllers have two important drawbacks:

1. The proportional gain of 2 required here is sometimes higher than desired and provides insufficient flow smoothing, i.e. small holdups. Lower gains offer improved flow smoothing; however, there is the risk of running the tank dry or overflowing the tank during maximum throughput.
2. Without integral action, P-only controllers typically never operate at the set point, so there is always an offset between the level and its set point. While this is actually what provides some of the benefits described above, many operators dislike seeing this sustained offset and resist its use.

7.3.2 PI CONTROLLER TUNING FOR INTEGRATING PROCESS

If the limitations of the P-only controller preclude its use, the following outlines a tuning procedure for a PI control scheme:

---

[2] For a direct acting controller, $e = CV - SP$

1. Select a value for controller gain that is less than 2. Try a gain between 0.5 and 1.0. Only if $K_C < 2$ does it make sense to use a PI controller at all, otherwise a P-only controller is used.
2. Determine the total holdup time, T_{HU}, of the tank by dividing the volume of the tank, as measured between the minimum and maximum level control points, the maximum flow through the control valve (Equation 7.9). It is important to note that the volume in Equation 7.9 is the volume of the tank between minimum and maximum controlled levels and not the total tank volume.

$$T_{HU} = \frac{\text{Volume [ft}^3]}{\text{Maxflow}\left[\dfrac{\text{ft}^3}{\text{min}}\right]} = \frac{V}{Q_{max}} \tag{7.9}$$

3. Calculate the integral time using Equation 7.10.

$$T_i = 4\left(\frac{T_{HU}}{K_C}\right) \tag{7.10}$$

7.4 Gas pressure loops

The characteristics of the gas pressure loop are almost the same as that of a liquid level control loop. A typical gas pressure loop is shown schematically in Figure 7.20.

Varying the flow of a compressible fluid controls the pressure in a large volume. This process is dominated by a single large capacitance with no dead time. The measurement is normally noise-free and, due to its capacitive

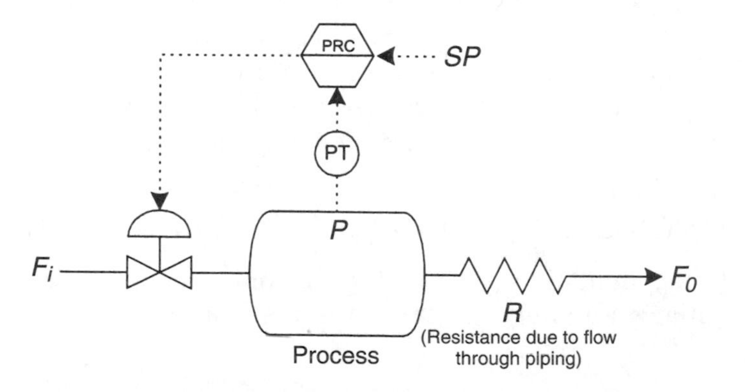

Figure 7.20 Gas pressure control loop.

nature, is characterised by a slow response and a small process gain. As shown for liquid level control, a proportional controller is more than adequate for gas pressure control.

The gas pressure loop is perhaps the easiest type of process loop to control. Due to the low gain in the process, a high controller gain will result in good control with very little offset and very little possibility of unstability. It is perhaps the only loop in the fluid processing industry that is very close to being unconditionally stable. As with the level control loop, a valve positioner can be used to improve loop response for a valve with hysteresis.

The gain of this process is a function of the load, F_0. However, since the loop is almost unconditionally stable it is not necessary to tune the controller at the highest process gain. The process gain changes but, even at the lowest load, it is stable. It is simply not possible to increase the controller gain to a high enough value to cause cycling.

7.5 Temperature control loops

Temperature loops may be divided into two main categories:

1. Endothermic – requiring heat energy.
2. Exothermic – generating heat energy.

Both of these processes have similar characteristics in that they are typically comprised of one large and many small capacities, i.e. valve actuator, transmitter, etc. The net result is a response indicative of a process with a dominant capacitance plus a dead time. Both of the above categories will be investigated and their specific differences and similarities will be identified.

For both of these processes, one of the following devices for measuring temperature is typically used.

- Thermocouple (TC);
- Filled thermal system (FTS); and
- Resistance temperature detector (RTD).

Although the overall loop response is characteristic of a large dominant capacitance plus dead time, care should be taken when installing temperature measuring elements. The temperature measuring devices should be selected so that the devices add a minimum lag to the process lag. It is common practice to insert the measuring element into a thermal well to protect it from the process fluid and to facilitate change out of the element if a problem should occur. Thermal wells are typically made of metal or ceramic depending on the environment. Figure 7.21 illustrates a thermal well.

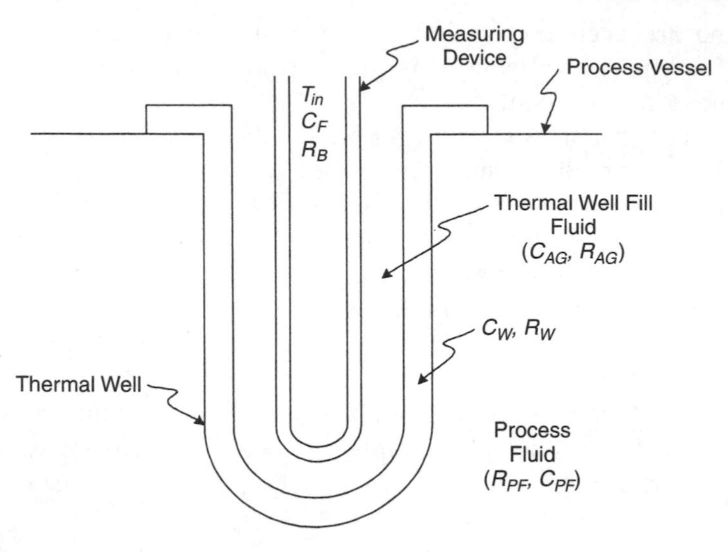

Figure 7.21 Typical temperature measurement device installation.

Every component in the measuring system, as shown in Figure 7.21, has an associated time constant, τ, where $\tau = RC$ (see Equation 3.28). So, each component in the measuring system will increase the measurement lag depending on the size of its time constant. Good practice dictates that the dominant time constant is the process fluid time constant, $\tau_{PF} = R_{PF}C_{PF}$. The other time constants: thermal well, $\tau_w = R_wC_w$, thermal well fluid, $\tau_{AG} = R_{AG}C_{AG}$, and measuring device, $\tau_B = R_BC_F$, should be as small as possible.

These time constants may be minimised in various ways. For instance, τ_w can be reduced by choosing a thermal well made from a material of low thermal resistance and also made as small as possible to reduce C_w. To minimise τ_{AG}, the air gap is filled with high thermally conductive fluid to decrease R_{AG} or the measuring device is attached via welding to the well. Using the smallest bulb and shortest capillary possible to reduce both R_B and C_F will minimise τ_B. With respect to short capillary runs in a filled thermal system (FTS) to increase response speed a transmitter is mounted close to the process and a gas filled bulb is used. This is done to minimise capillary length and gas is used because its thermal capacity, C_F, is low. This in turn results in a small measurement time constant since the pressure signal from the FTS is changed to an electronic signal of 4–20 mA sooner without a long capillary run.

When using a resistance temperature detector (RTD) or a thermocouple (TC) the time constant considerations are similar but the actual response times of the devices will vary. The FTS and RTD will have response times of nearly the same magnitude, while the TC is somewhat faster. For a thermocouple, τ_B varies with the device's construction and length of the extension wires. Hence, a TC made of small wire with short extension wires will give a fast response. A typical TC response is around 0.5 seconds.

Let us put some numbers on the response times of the other two temperature measurement devices. The filled thermal system (FTS) and resistance temperature detector (RTD) have similar response times except that the response of the RTD in water is generally longer than the FTS due to its greater internal resistance. But for the same size bulb, either a liquid-filled FTS or RTD has a time constant of about three seconds.

For a filled thermal system (FTS) the capillary has a time constant of 0.55 seconds per 10 feet, and thus it is obvious why a transmitter is often used since some capillaries can be up to 100 feet long. Using a short capillary, with a gas-filled system and a transmitter, results in a response that is twice as fast as the liquid-filled system with a long capillary.

The response time of RTD or FTS bulbs in a thermal well depends on the material of the well and the clearance between the bulb and the well. For a bulb in a dry well a typical time constant is one to two minutes, while for a bulb with thermal fluid in the well the following apply:

- *in a gas stream* – τ is the same as for the dry well due to the high thermal resistance of the gas (τ_{PF} is very large).
- *in a liquid stream* – τ should be about 2–3 times that of the bulb alone, since the thermal resistance of the liquid is very small (τ_{PF} is very small) and the response improves with the lowering of the thermal well fluid resistance, i.e. makes τ_{AG} as small as possible.

If it is necessary to use a thick metal or ceramic well due to corrosive process fluid, the response time can increase to 10 times that of the bulb alone. In addition, a large well creates a static error as a result of conduction along the wall of the thermal well. The addition of a large well can increase the measuring device plus well time constant by approximately 1.5 minutes. This increase can be detrimental in certain processes, i.e. exothermic reactor, but is not significant in others.

The following are general rules of thumb for reducing temperature measurement lag:

1. Use a small diameter bulb or thermal well.
2. Increase the velocity of flow past the measuring device by using a small pipe or a restriction orifice near the bulb. Be cautious though because there is the possibility of thermal well fracture as the velocity increases.
3. When measuring temperature in two phase flow situations, place the measuring element in the liquid phase, if possible, to gain the benefit of faster heat transfer from the process fluid to the measuring element.
4. Consider using a transmitter with derivative action. Some manufacturers make a gas-filled thermal system connected via a short capillary to a

temperature transmitter with derivative action in it. The derivative action acts to cancel out some of the lag in the measuring element. Reduction of the derivative gain in the controller is required to accommodate this added derivative gain in the transmitter.

The whole point of the preceding discussion on minimising temperature measurement lag in the temperature control loop is to make you aware that this is important in slow as well as fast loops.

7.5.1 THE ENDOTHERMIC REACTOR TEMPERATURE CONTROL LOOP

A good example of an endothermic process is a process heat exchanger being used to heat a fluid from the inlet temperature, T_1, to an outlet temperature, T_2, as shown previously in Figure 6.2. This heat exchanger's response will be that of a single large dominant capacitance with dead time. Typically either a PI or PID controller is used. Derivative action can be used since the temperature measurement is not noisy. The response of the loop under PID control will be equal to that of P-only control except the temperature will be maintained at the desired set point.

The steady-state gain of the heat exchanger is calculated from Equation 7.11:

$$K_p = \left(\frac{\Delta T_2}{\Delta F_s}\right)_{F_w = \text{Constant}} \tag{7.11}$$

The process gain, K_p, is a function of the load, F_w, as in the case of liquid level control previously discussed. However, F_w is not the only load; it is one of several. Other loads include the inlet and outlet temperatures of the cold fluid stream. The steady-state equation describing the behaviour of the heat exchanger is given by Equations 7.12 and 7.13.

$$F_S = KF_w(T_2 - T_1) \tag{7.12}$$

$$K = \frac{C_p}{\lambda} \tag{7.13}$$

where: C_p = specific heat of F_w
λ = heat of condensation
F_w = flow of cold fluid (load variable)
T_1 = inlet temperature of cold fluid (load variable)
T_2 = outlet temperature of cold fluid (load variable)

The heat exchanger energy balance equation can be solved for the heat exchanger gain, K_p, as shown in Equation 7.14.

$$K_p = \frac{dT_2}{dF_s} = \frac{1}{KF_w} = \frac{K^t}{F_w} \tag{7.14}$$

As expected the gain is inversely proportional to the load, F_w. This result is identical to that for the liquid level process and may be minimised with similar approaches.

A valve positioner, V/P, can also be added to the flow valve as described in the case of liquid level control. However, if there is a chance of a supply upset to the heat exchanger, a temperature on flow cascade is used instead, as shown in Figure 6.4. Another approach to minimising supply upsets is to use a pressure regulator ahead of the steam. The pressure regulator is used to make the steam supply pressure constant. This scheme negates the need for another flow loop while still providing protection against supply upsets.

There are many methods used in controlling heat exchangers. Figures 7.22–7.25 show several methods in addition to the basic feedback loop, in which the flow of steam was directly throttled by the temperature controller.

Figure 7.22 shows a situation in which F_s is a wild flow and T_2 is controlled by controlling the condensate level in the heat exchanger, i.e. overhead condenser. When the temperature is too high, the valve closes, which causes condensate to cover more tubes and reduce heat transfer to the cold fluid. Because the condensation time is large, the response is slower than for other systems. Also, due to condensate splash, T_2 can show significant fluctuations.

Figure 7.23 shows a scheme employed when temperature control is critical and the response time, τ_1, of the heat exchanger is very long.

In this approach a sidestream of the input F_w, is bypassed and mixed with the outlet F_w. This gives a fast response, with an energy penalty of first heating up and then cooling down F_w. It is also necessary to ensure good mixing at the output and to ensure a fast response in the temperature measurement,

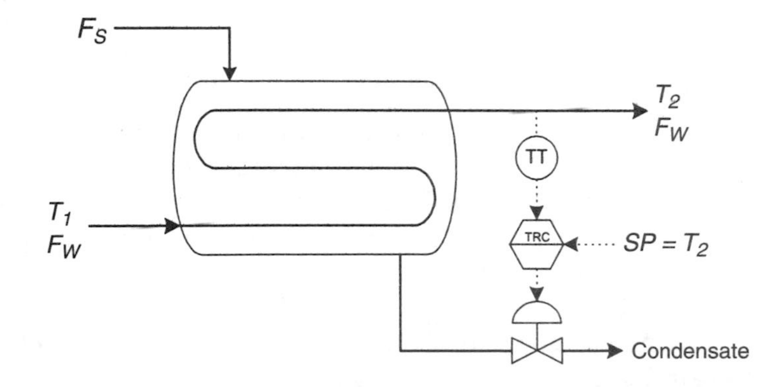

Figure 7.22 Temperature control via level control.

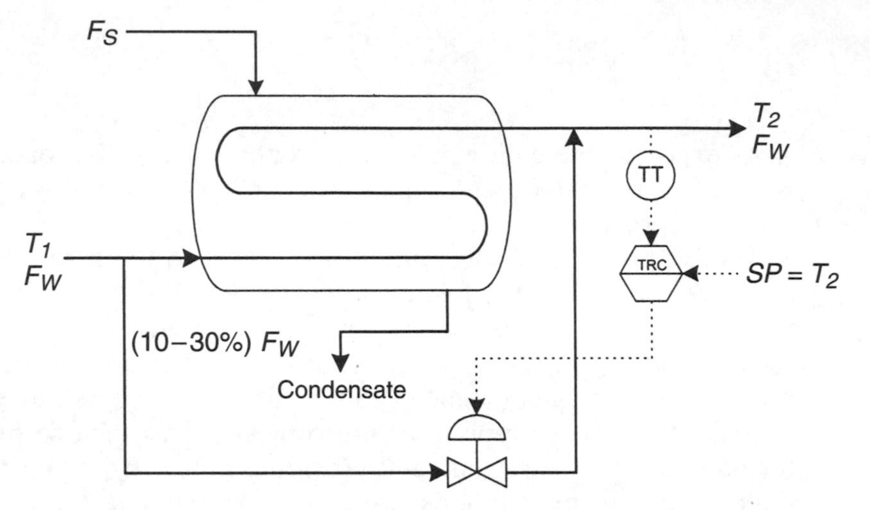

Figure 7.23 Control scheme for critical temperature control.

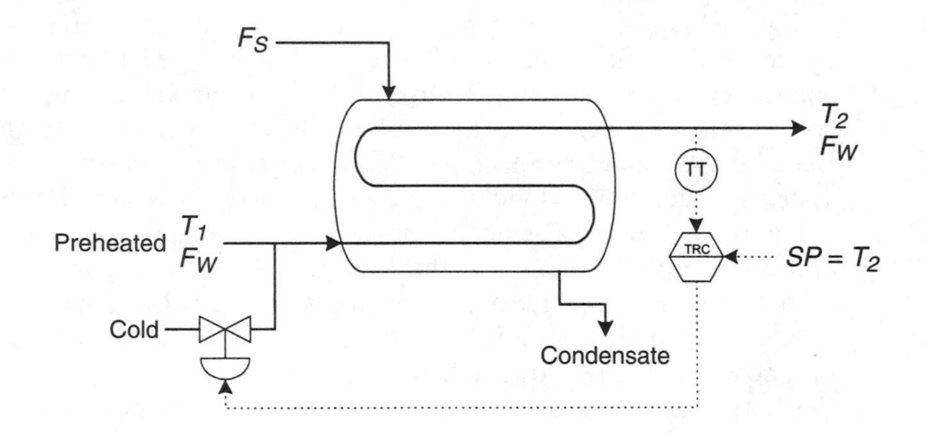

Figure 7.24 Variation of critical temperature control scheme.

since a flow loop is being used. Another variation on the control scheme shown in Figure 7.23 is shown in Figure 7.24.

The scheme shown in Figure 7.24 provides control over a wider range of F_w and gives a nearly constant process gain. Hot and cold F_w are blended together to maintain a uniform T_1. The energy penalty for this scheme is the cooling of the stream which energy has been expended in heating up.

The scheme shown in Figure 7.25 throttles F_w, to maintain T_2 and is usually used on heat exchangers that are capacity-limited. It is more important to maintain T_2 at the set point than to maintain F_w at a given demand.

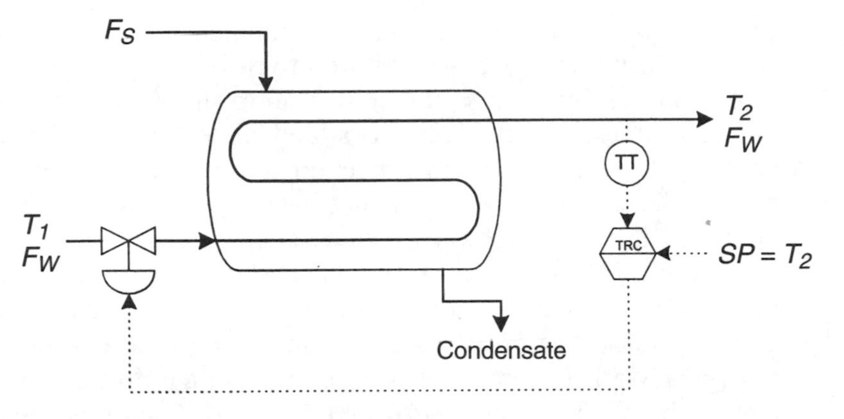

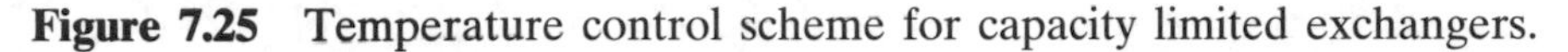

Figure 7.25 Temperature control scheme for capacity limited exchangers.

A detailed discussion of tube and shell, aerial coolers and fired heaters is found in the hydrocarbon processing articles by W.C. Driedger [8,9].

7.5.2 THE EXOTHERMIC REACTOR TEMPERATURE CONTROL LOOP

The exothermic reactor is perhaps the most difficult process to control due to its instability and extreme nonlinear response [7]. A chemical reactor is quite often an exothermic process where some feed stock and catalyst are mixed together, and the temperature must be controlled at a specific set point. A typical temperature control scheme is illustrated in Figure 7.26.

The degree of stability that can be achieved in this temperature control loop depends on the rate at which the heat can be removed from the reactor.

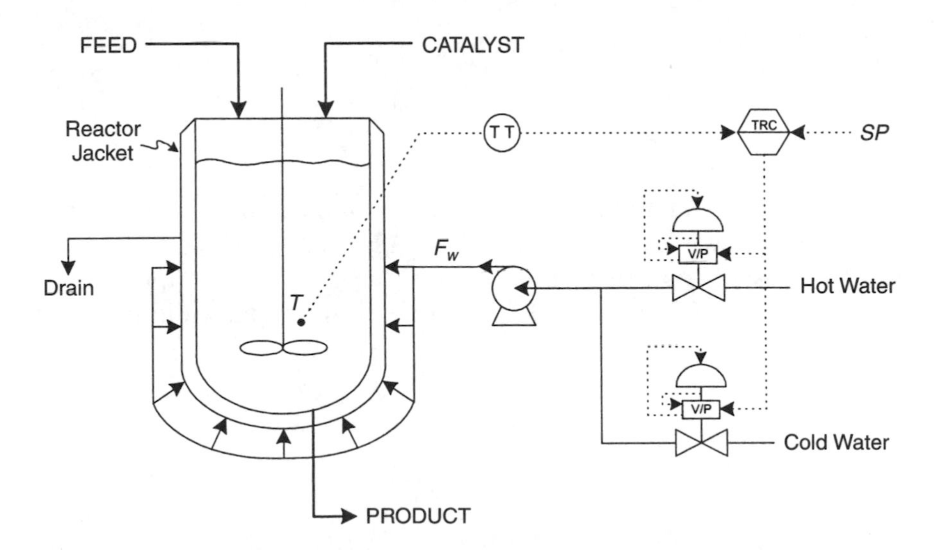

Figure 7.26 Control scheme for an exothermic reactor.

In other words, the reactor can be stabilised if the reaction temperature changes fairly slowly when compared to the rate at which the jacket temperature changes. The idea of the control loop shown in Figure 7.26 is that once the feed stock and catalyst are added, hot water in the jacket is used to initiate the reaction. As the reaction temperature increases, the controller output decreases, closing the hot water valve, which is air-to-open and opening the cold water valve, which is air-to-close. A valve positioner, V/P, is used to minimise valve hysteresis. The pump and multiple water inputs to the jacket are used to minimise dead time and to change the jacket temperature as quickly as possible, i.e. minimise the time constant of the jacket.

Typically a PID controller is used, but using a proportional only controller may stabilise the reactor provided that the reactor is the dominant single capacitance in the loop and there is no appreciable dead time[3].

Extremely fast reactions are sometimes carried out in semi-batch fashion to prevent a runaway temperature. In this case, one reagent is continuously added to a cooled reactor containing the other reagent or catalyst, and the addition rate is controlled to maintain a given batch temperature or a given heat removal rate. For safe operation, the temperature is kept high enough to ensure that a low concentration of the added reactant is required.

Often an emergency control system[4] is implemented to stop the reaction by dumping the charge or stopping the catalyst flow in case the main control fails to halt a runaway reaction temperature. Other typical reactor control schemes can be found in [3,7].

In addition to these problems there is also the problem of the extreme nonlinear process response. As before, the gain is a function of the temperature operating level. For the control system shown, the gain is defined by Equation 7.15.

$$K_p = \frac{\Delta T}{\Delta F_W} \tag{7.15}$$

In this equation, T is the reactor temperature and F_w is the cooling/heating water flow to the reactor jacket. Assuming perfect mixing in the reactor and a constant rate of heat evolution, the gain can be approximated as shown in Equation 7.16 where Q is the rate of heat generation:

$$K_p \approx \frac{Q}{F_W^2} \tag{7.16}$$

[3] Similar to a gas pressure loop
[4] Override controls

Due to this nonlinearity as well as to the problems mentioned earlier, some exothermic reactors are controlled with advanced control techniques such as feedforward, model reference or adaptive control [17].

7.6 Compressor control

Simply stated, compressors are employed whenever a gas at a certain pressure in one location is required to be at a higher pressure at another location. However, this belies the fact that compressors are major ticket items in the capital cost of a chemical or petroleum plant. For example, a large centrifugal compressor with a gas turbine is an investment of many million dollars.

Two major types of compressors are commonly used in chemical and petroleum plants; reciprocating and centrifugal compressors.

7.6.1 RECIPROCATING COMPRESSOR CONTROL

Control of reciprocating compressors [10] involves the control of compressor capacity, engine load and speed; the control of auxiliary items on the compressor package; and the control of compressor safety.

Control of compressor capacity, engine load and speed

Compressor capacity is controlled by varying driver speed, opening or closing fixed or variable volume clearance pockets, activating pneumatic suction valve unloaders, bypassing gas back to suction, or varying suction pressure. Driver speed control is not always possible with synchronous AC motors although solid state devices are now available for varying input frequency and speed. It is outside the scope of this book to go into the details of these electromechanical control mechanisms. The interested reader is directed to the Gas Processors Suppliers Association, *Engineering Data Book* [11].

Control of auxiliary items on the compressor package

Oil, water and gas temperatures, oil, water and scrubber liquid levels and fuel and starting gas pressures need to be controlled.

Control of compressor safety

Safety shutdown controls must also be provided in case of harmful temperatures, pressures, speed, vibration, engine load and liquid levels.

7.6.2 CENTRIFUGAL COMPRESSOR CONTROL

As mentioned previously, a large centrifugal compressor with a gas turbine as a driver is typically a multi-million dollar investment. A dedicated computer control system is usually employed to monitor multiple operating parameters and this is specifically designed for the purpose. Such a control system is shown in Figure 7.27.

The control of a centrifugal compressor involves the control of capacity, the prevention of surge and the protection of equipment.

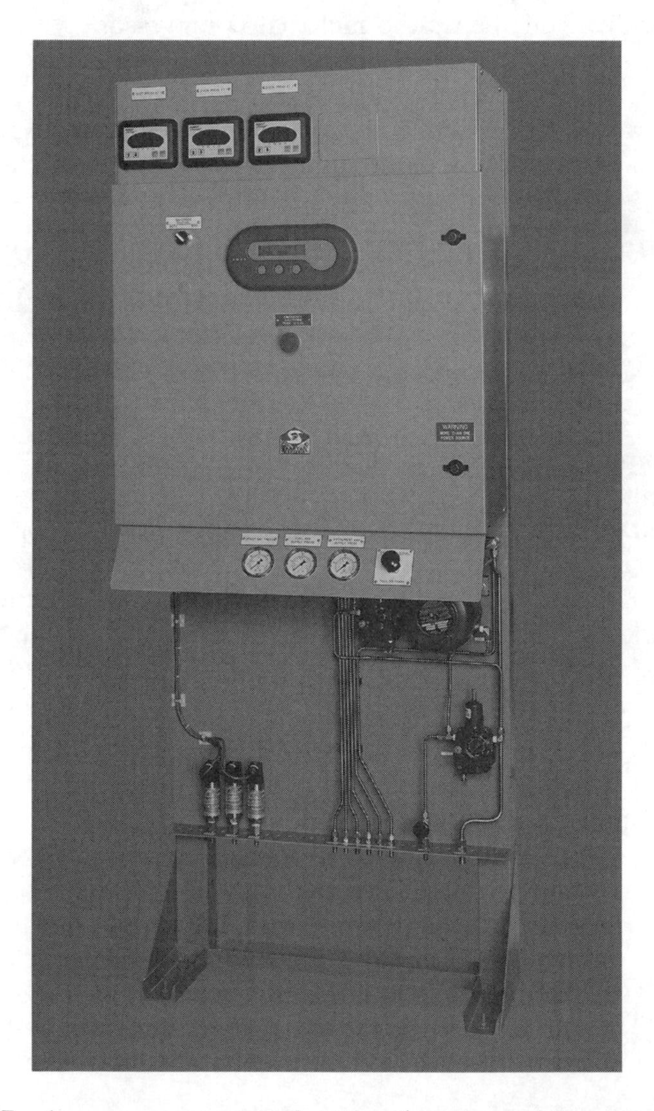

Figure 7.27 A compressor control station (courtesy of Spartan Controls).

Capacity control

The means available for controlling compressor capacity are suction throttling, discharge throttling, recirculation, variable guide vanes and motor speed control. Most of these controls are used in practice and which is best depends on the application. Some of the pros and cons of these alternative control methods are as follows:

1. *Suction throttling*: The spinning vanes of a centrifugal compressor sling the gas outward. The centrifugal force develops a pressure proportional to the density and to the speed squared. Suction throttling reduces the density and hence the ΔP; thus the machine operates as a constant pressure ratio machine.
2. *Discharge throttling*: This is much like suction throttling but it is less efficient because the increased temperature means that the volume is more than P_1/P_2, therefore $V\Delta P$ loss is greater at discharge. Therefore discharge throttling is never done.
3. *Recirculation*: Recirculation has a much lower efficiency for similar reasons, but is essential for low turndown.
4. *Variable guide vanes*: These work by directing the gas flow with respect to blade rotation. Theoretically there is no efficiency loss but they tend also to act as inlet throttlers. Although excellent, this approach involves extra expense.
5. *Motor speed control*: This is cheap on turbine and engine drives but expensive for electric motors.

Each of the methods of control affects the compressor curve to produce a set of curves called the compressor map, shown in Figure 7.28.

The various curves show the compressor characteristic at different values of the parameter being varied, such as inlet valve setting or speed. At each crossing of a compressor curve and a system curve, specific operating points occur which collectively establish the controllable range.

If the load on the compressor changes – that is the system curve changes – as shown in Figure 7.29 the operating point moves along the compressor curve. The range within the compressor curve is called the operating range (shown as points 1 to 2 on Figure 7.29).

Returning to Figure 7.28, it can be seen that each compressor curve shows two end points. The lower right end point is relevant when discussing capacity control. Beyond the lower right end point the volume is so great that the internal flow velocity approaches sonic. A further drop in discharge pressure cannot affect the inlet flow therefore the flow rate no longer increases – this phenomenon is known as "stonewalling".

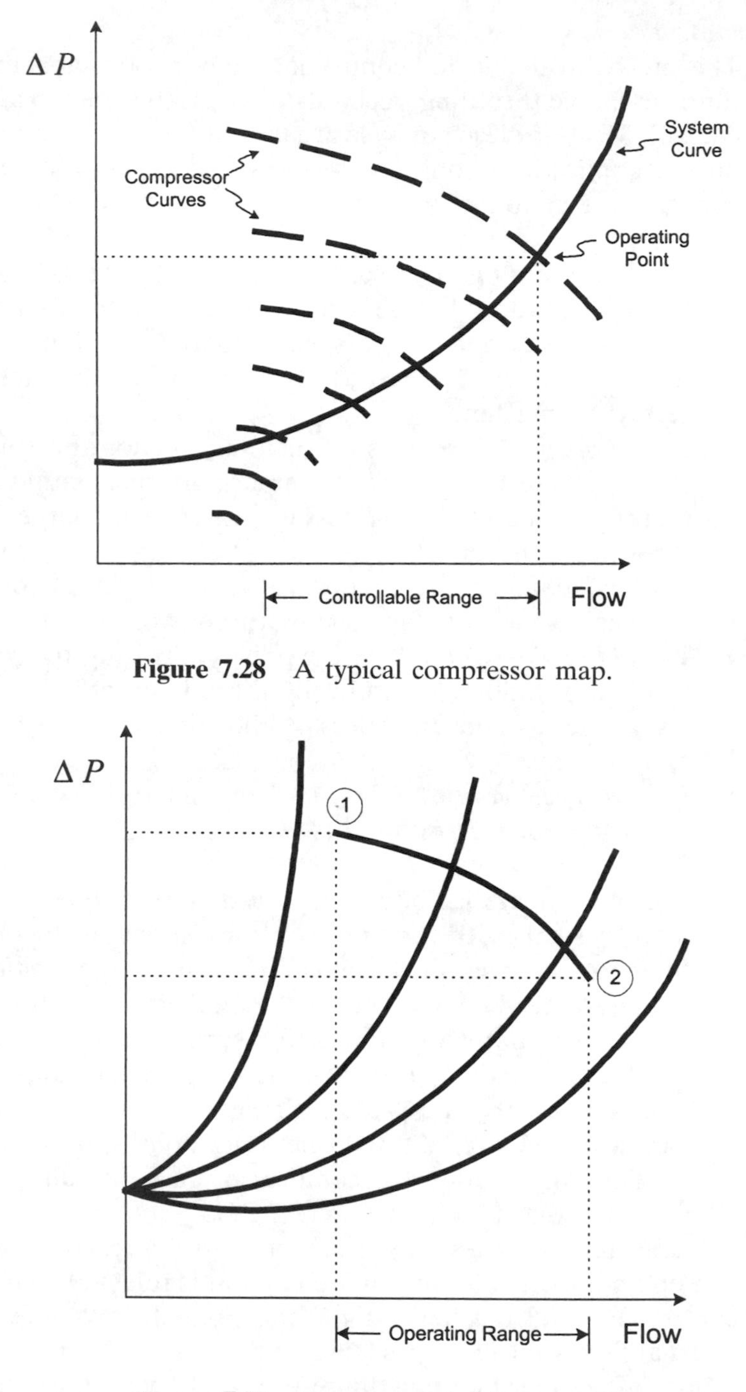

Figure 7.28 A typical compressor map.

Figure 7.29 Compressor load changes.

Surge prevention

Beyond the upper left end point of Figure 7.29 the ΔP drops to a minimum and then rises again. This causes severe oscillations known as "surge". As the pressure rises up the curve it eventually reaches a maximum. The pressure cannot fall unless some of the gas flows out of the discharge volume or into the inlet volume.

The symptoms of surge are pulsating pressure, rapid flow reversals, a drop in motor current and a jump in turbine speed. Continuous, rapid flow reversals can cause severe damage to the compressor. In axial compressors the blades may touch, resulting in instant destruction. However, centrifugal compressors are more rugged and only seal damage results [12].

The frequency of surge varies from 5–50 Hz. Suction and discharge volumes also influence surge. Minimising the volume that has to be depressured can mitigate surge. Preventing ΔP from getting too high also prevents surge. Surge protection involves the determination of the surge limit line, i.e. the limiting values of ΔP versus throughput that can initiate surge. Surge control keeps the compressor from crossing a surge control line that is arbitrarily set at a safe distance from the surge limit line [13].

The above compressor control theory is applied in the following example [12]. For more details on centrifugal compressors and their control, the interested reader is again directed to the Gas Processors Suppliers Association, *Engineering Data Book* [14] or the ISA Instructional Resource Package on Centrifugal and Axial Compressor Control [15].

Application example

The application to be considered is a plant with a compressor drawing vapour from the top of a distillation column and moving this vapour to downstream processing units. The plant also has a considerable amount of waste heat in the form of steam, therefore it is economically worthwhile to use steam turbines as drivers with superheated steam as the motive force. The schematic of the example plant is shown in Figure 7.30.

In order to control the compressor, its purpose in terms of a process variable needs to be known. The purpose of the compressor in this application example is to control the pressure at the top of the column. A suitable measuring instrument would be a pressure transmitter located at the knock out (KO) drum. The compressor throughput is controlled by speed control on the steam turbine. Steam turbines generally have special control valves that are an integral part of the machine and usually have their own governors. The pressure controller provides a set point for the governor.

Excess flow control (stonewall) protection is not needed as long as the compressor is not grossly oversized and the downstream process will provide sufficient backpressure to prevent excess flow. The process fluid is a light hydrocarbon and is never vented directly to atmosphere.

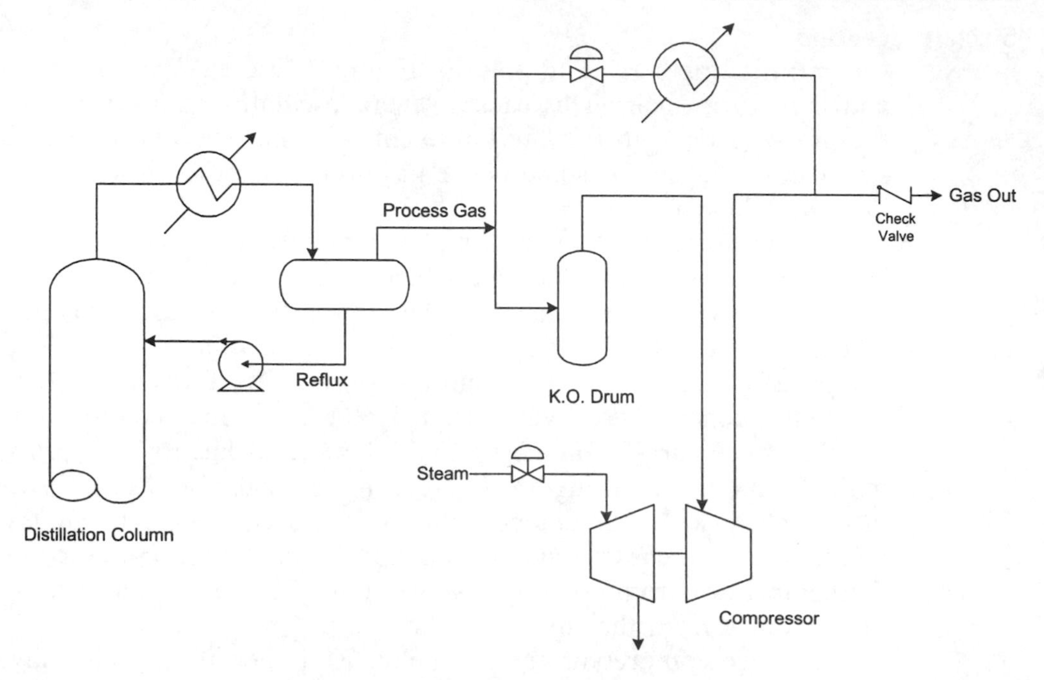

Figure 7.30 Compressor control example process schematic.

However, minimum flow (surge) protection is needed as every compressor needs surge protection. The surge loop is placed as close as possible to the discharge. A check valve is placed downstream of the recycle "tee" to prevent recycling the entire downstream process flow.

The recycle line returns to the suction KO drum. A cooler must also form part of the recycle loop as there is no other way of removing the energy that accumulates as heat of compression.

In order to control surge the compressor map must be known. From the Fan Laws we know that flow varies proportionally to speed and ΔP varies with speed squared.

$$Q \propto n \tag{7.17}$$

$$\Delta P \propto n^2 \tag{7.18}$$

From this we can calculate a family of curves based on the original compressor curve. These curves can be well fitted by a cubic equation. Surge occurs at the maximum or flat part of the curve. Applying the Fan Laws and solving for the maxima results in a quadratic equation called the "surge line". To avoid surge, the compressor never operates to the left of the surge line, so the square of the flow must be greater than the proportionality constant times the ΔP.

$$Q^2 > k\,\Delta P \tag{7.19}$$

In order to provide surge control, suction flow and ΔP must be measured. Suction flow must be in terms of actual, not standard, volume units at the inlet. The effects causing surge are based on gas velocity, not mass flow. These measurements are made as follows. ΔP is measured across the compressor. A venturi, which has by definition output proportional to the square of the flow, is placed in the compressor suction. It is important that the flow transmitter does not apply a square root to provide a linear signal so that it may be used directly in the surge controller without further squaring.

In order to apply these process measurements, the compressor map is cast into a new form, ΔP versus the square of the flow, which results in a straight line for the surge line. However, it is not a good idea to use the surge line as the set point to the surge controller because of instrument error, transmitter, controller and valve delays, compressor variations with time and molecular weight variations. Instead a surge control line is established perhaps 5% to the right of the actual surge line, as a safety factor.

The resulting complete compressor control system with pressure/speed and surge loops is shown in Figure 7.31.

As always, it is important to verify the control scheme dynamically with the use of a suitable dynamic simulator. Other application examples that are

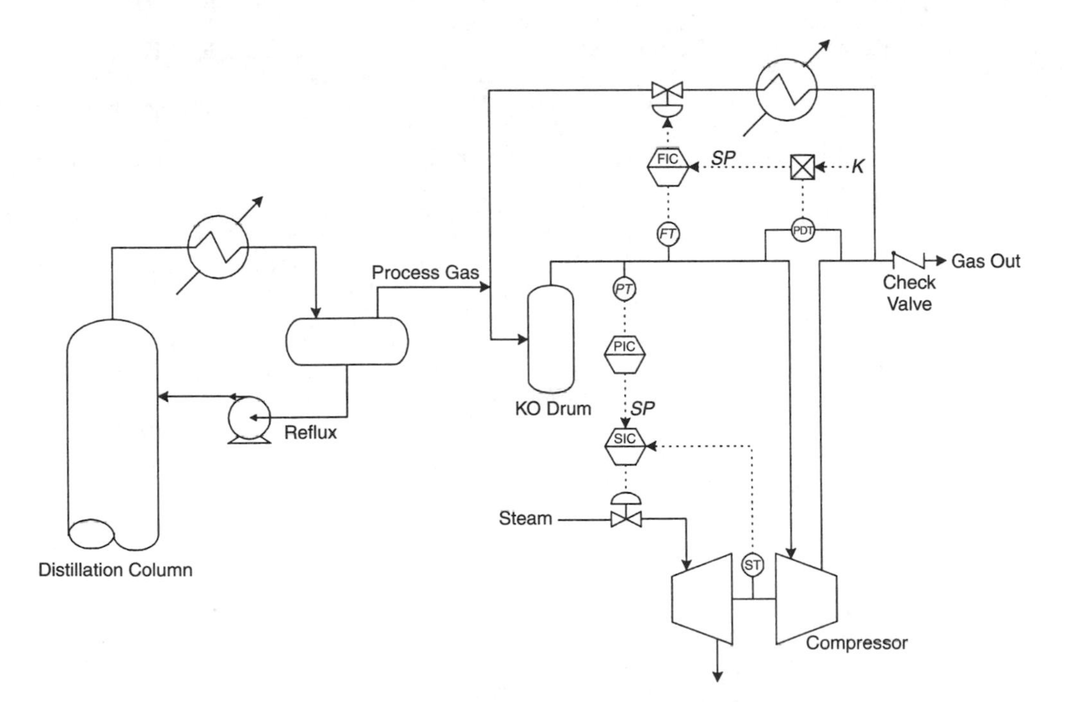

Figure 7.31 Complete compressor control system schematic.

documented in the literature [16,17,18] include a substantial emphasis on the importance of dynamic simulation for control scheme design validation and performance evaluation.

7.7 References

1. Buckley, P.S., *Techniques of Process Control*. John Wiley & Sons, New York, 1964, pp. 167.
2. Shinskey, F.G., "Averaging Level Control". *Chemical Processing*, 1997, **60** (9) Sept.: 58.
3. Shinskey, F.G., *Process Control Systems*. McGraw-Hill, New York, 1967, p. 147.
4. Åström, K.J., Wittenmark, B., *Computer Controlled Systems: Theory and Design*. Prentice Hall, New Jersey, 1984, pp. 351–2.
5. Marlin, T.E., *Process Control: Designing Processes and Control Systems for Dynamic Performance*. McGraw-Hill, New York, 1995, pp. 538–40.
6. Ogunnaîke, B.A. and Ray, W.H., *Process Dynamics, Modeling, and Control*. Oxford University Press Inc., New York, 1994, pp. 628–9.
7. Seborg, D.E., Edgar, T.F. and Mellichamp, D.A., *Process Dynamics and Control*. John Wiley & Sons, New York, 1989.
8. Driedger, W.C., "Controlling Fired Heaters". *Hydrocarbon Processing*, 1997, **April**: 103–16.
9. Driedger, W.C., "Controlling Shell and Tube Exchangers". *Hydrocarbon Processing*, 1996, **March**: 111–32.
10. Manning, F.S. and Thompson, R.E., *Oil Field Processing*, Volume 2. PennWell Books, Tulsa, OK, USA, 1995, Chapter 15, pp. 340.
11. Gas Processors Suppliers Association, *Engineering Data Book*. GPSA, FPS Version, 11th Edn, Volume I, Section 13, 1998, pp. 14–21.
12. Driedger, W.C., "Controlling Compressors". Lecture notes, ENCH 529 Process Dynamics and Control, University of Calgary, March 16, 1998.
13. Magliozzi, T.L., "Control System Prevents Surging in Centrifugal-Flow Compressors". *Chemical Engineering*, May 8, 1967, pp. 139–42.
14. Gas Processors Suppliers Association, *Engineering Data Book*. GPSA, FPS Version, 11th Edn, Volume I, Section 13, 1998, pp. 33–5.
15. McMillan, G.K., *Centrifugal and Axial Compressor Control*. Instrument Society of America, Research Triangle Park, NC, USA, 1983, pp. 57–74.
16. Campos, M.C.M.M. and Rodrigues, P.S.B., "Practical Control Strategy Eliminates FCCU Compressor Surge Problems". *Oil and Gas Journal*, 1993, **91** (2): 29–32.
17. Muhrer, C.A., Collura, M.A. and Luyben, W.L., "Control of Vapour Recompression Distillation Columns". *Ind. Eng. Chem. Res.*, 1990, **29** (1): 59–71.
18. Van der Wal, G., Haelsig, C.P. and Schulte, D., "Minimising Investment with Dynamic Simulation". *Petroleum Technology Quarterly*, 1996/97, **Winter:** 69–75.

8 DISTILLATION COLUMN CONTROL

Steady-state simulation and design methods for separation processes, with emphasis on distillation, have been presented in detail in many references, a few of which are listed in the references for this chapter. This chapter will present a discussion of the basic control schemes for distillation columns. Let us start by stating the obvious: the amount of literature on separation processes, particularly distillation is colossal. Particularly readable books and references are those by Buckley [1, 2], King [3], Tyreus [4], Seborg *et al.* [5], Shinskey [6], Smith and Corripio [7], Svrcek and Morris [8], and Wilson and Svrcek[9].

8.1 Basic terms

When determining the control system design for a multivariable process, the terms control strategy, control structure and controller structure are used interchangeably. In this context, the meaning is the selection and pairing of manipulated and controlled variables to form a complete, functional control system. However, the three terms can also have individual meanings. Control strategy can describe how the control loops in a process are configured to meet a given overall objective such as the purity of a given stream. Control structure, on the other hand, is the selection of controlled and manipulated variables from a set of many choices. Finally, controller structure means the specific pairing of controlled and manipulated variables by way of feedback controllers.

This chapter will describe a methodology for designing a multivariable control system that includes elements of control strategy considerations, control structure selection and variable pairing. The methodology is largely empirical and based on general principles for distillation control. The methodology for control system design assumes that the process configuration is fixed and that changes are not possible. This is the case in many instances where control engineers are asked to design the control system for process configurations in an existing plant or a plant well into the design phase. The task for the control engineer is to select appropriate variables to be controlled and design controllers that will tie these variables to the control valves (manipulated variables) in such a way that the resulting controller structure meets the desired objectives. The final assumption is that the controller structure will be built up around conventional PID controllers, ratio, feedforward and override control blocks found in all commercial distributed control systems (DCS).

8.2 Steady state and dynamic degrees of freedom

When a process engineer works with a detailed steady-state simulation of a distillation column, a certain number of variables have to be specified in order to converge to a solution. The number of variables that need to be specified, or degrees of freedom, can be determined through the concept of the description rule as stated by King [3]:

"In order to describe a separation process uniquely, the number of independent variables which must be specified is equal to the number which can be set by construction or controlled during operation by independent, external means."

Applying the description rule to a distillation column with a total condenser and two product streams gives two steady-state degrees of freedom. In this case the column would require two specifications, i.e. a composition and a component recovery. The steady-state simulator will then manipulate two variables, such as reboiler and condenser duties, in order to satisfy the specifications and close the steady-state material and energy balances. If a partial condenser is added to the column, another degree of freedom is added to the steady-state column. Likewise, for each additional side draw added to the column, a new degree of freedom is added, requiring another specification.

When the same two-product distillation column is viewed in dynamics, the number of degrees of freedom increases from two to five. These three new dynamic degrees of freedom correspond to three new manipulated variables needed to control the integrating, inventory variables within the column that are not fixed by the steady-state material and energy balances alone. The inventory variables for this column are condenser level, reboiler level and the column pressure.

One control valve (or degree of freedom) must be used for each controlled variable. This relationship between controlled variables and degrees of freedom (or control valves or manipulated variables) is known as variable pairing and is an important concept in control system design.

When the five manipulated variables, which correspond to five valve positions as shown in Figure 8.1, are viewed, it can be seen that the two steady-state manipulated variables are a subset of the overall five. However, there is nothing about the heat duties that make them exclusive steady-state manipulators and prevent them from being used for inventory control. In many control schemes, the condenser duty is used for pressure control rather than composition control. For the same reason, any three of the five manipulated variables can be used to control the column inventories.

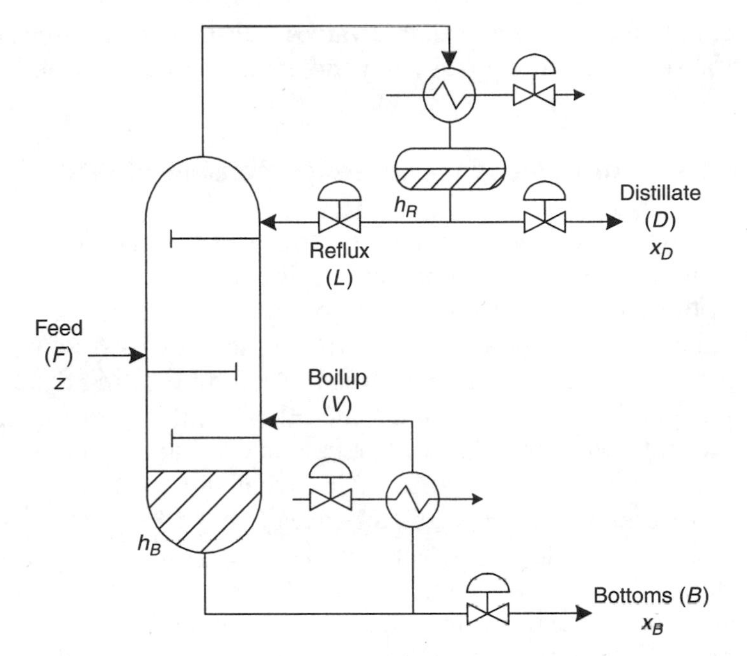

Figure 8.1 Basic distillation column schematic.

Although the above paragraph describes the manipulated variables as control valves, there are many choices available other than just the individual valves. For example, many columns have reflux ratio as a manipulated variable for either inventory or composition control. When ratios and linear combinations of variables are included, the choice of a manipulator for a given loop broadens considerably for a simple two-product column. However, the steady-state and dynamic degrees of freedom remain unchanged as two and three, respectively, totalling five.

One must take care in determining the number of steady-state and dynamic degrees of freedom for more complex columns. Tyreus [4] describes the determination of the degrees of freedom for an extractive distillation system and for an azeotropic column with an entrainer. In the case of an extractive distillation system, recycle streams reduce the dynamic degrees of freedom through an increase in the steady-state degrees of freedom if the recycle contains a component that neither enters nor leaves the process. As well, if it is important to control the inventory of a "trapped" component, such as an entrainer for azeotropic distillation, it is necessary to provide extra control valves to account for the loss of degrees of freedom. This loss comes from the addition of a side stream.

In summary, the total degrees of freedom for actual plant operation equal the number of valves available for control in that section of the plant. To find out how many integrating variables, i.e. pressures and levels, are to be controlled with the available valves, subtract the degrees of freedom required for steady-state control from the total degrees of freedom.

8.3 Control system objectives and design considerations

Defining and understanding the control system objectives should be a collaborative effort between process engineers and control engineers. Left to either of these contributors alone, the objectives can be severely biased. The control engineer might be tempted to make the control system too complex in order for it to do more than is justified based on existing disturbances and possible yield and energy savings. On the other hand, a process engineer might underestimate what process control can achieve and thus make the objectives less demanding. It is crucial to define what the control system should do as well as to understand what disturbances it has to contend with.

Process understanding is another key, but often overlooked, activity for successful control system design. In practice, more time is spent designing and implementing algorithms and complex controllers than on analysing process data and understanding how a process really works. Modelling and simulation are integral parts in the process understanding step.

Rigorous dynamic simulation is the third important activity in control system design. A flexible dynamic simulator allows for rapid evaluation of different control structures and their response to various disturbances. In choosing a control scheme there are several design considerations to take into account. First it is important to remember that a distillation column performs two basic functions:

1. Feed split.
2. Fractionation.

The feed split is the primary point of separation between the overhead and bottoms product. Fractionation is determined by the number of separation stages in the column and the energy input. Figure 8.2 illustrates these concepts with a mixture of a low boiling point component (light shading) and a high boiling point component (dark shading). The boiling point of the distillation products is determined by how much of each component is present in each product. As the distillation feed split changes, the line will shift left or right. As the fractionation changes, the slope of the line will change with a steeper slope representing better separation. It is important to realise that fractionation increases the purity of both products simultaneously while changing the feed split will make one product more pure and the other less pure.

Once the inventory variables are controlled there are two degrees of freedom left in the case of the column shown in Figure 8.1. One degree of freedom should be used to control the feed split while the last available

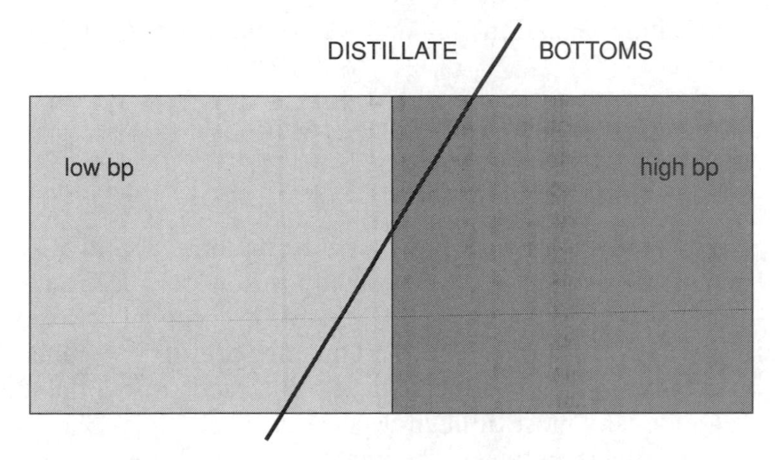

Figure 8.2 Feed split and fractionation.

degree of freedom controls fractionation. Feed split has a much more significant effect on the product compositions than fractionation. Therefore after the inventory and capacity variables have been paired, the primary controlled variable is normally used to set the feed split while the secondary controlled variable is used to set fractionation.

The following equations describe how the various manipulated variables are related and show that virtually any variable pairing can be used to achieve the desired control objectives. However, some pairings will provide significantly better sensitivity and responsiveness. Better sensitivity means that the control scheme will react with smaller changes, whereas a more responsive control scheme reacts more quickly.

$$\text{Overall material balance: } F = D + B \tag{8.1}$$

$$\text{Component balance: } Fx_f = Dx_d + Bx_b \tag{8.2}$$

where: F = feed
B = bottoms
D = distillate
x_I = concentration of a particular component in the feed, distillate or bottoms
Q_{reb} = reboiler duty
Q_{cond} = condenser duty

$$\text{Energy balance: } FH_f + Q_{reb} = DH_d + BH_b + Q_{cond} \tag{8.3}$$

Where: H_F = enthalpy of the feed
H_B = enthalpy of the bottoms
H_D = enthalpy of the distillate

Combining Equations 8.1 and 8.2 to eliminate B or D gives:

$$D = F\,(x_f - x_b)/(x_d - x_b) \tag{8.4}$$

$$B = F\,(x_f - x_d)/(x_b - x_d) \tag{8.5}$$

The control system must satisfy Equations 8.1 and 8.3 at all times. For particular values of x_d and x_b (i.e. composition specifications), Equation 8.4 or 8.5 also has to be satisfied. D, B, Q_{reb} and Q_{cond} can all be fixed or adjusted dynamically by control valves on the flow rate or utility streams. The reflux flow can also be adjusted dynamically and will directly affect the energy balance.

One of the most difficult aspects of distillation column control is the interaction effects between the material and energy balance and composition controls. Depending on the inventory controls, heat input or removal can alter both the material draws and the compositions. This interaction can work for us or against us depending on the control strategy.

Another point to consider when choosing a column control scheme is that typically the process gains from a high purity separation are very nonlinear. This can be verified by simply using the component balance equations. For example, Equation 8.4 can be rearranged and differentiated at constant x_b to give:

$$\left(\frac{\partial x_d}{\partial D}\right)_{xb} = -\frac{F(x_f - x_b)}{D^2} \tag{8.6}$$

Equation 8.6 shows that changes in the distillate rate, D, will have a much larger effect on the distillate composition, x_d, when the distillate rate is relatively low compared with cases when the distillate rate is relatively high.

A final and important consideration to keep in mind is the dead time that may be present in the column. In Chapter 3 dead time was described as being generated by a series of lags (material or energy capacitances). It is easy to see how a distillation column with its multiple stages can generate dead times. The control scheme on a distillation column should be set up to minimise dead times with respect to the process lags and disturbances.

The steps for determining a suitable controller structure are as follows:

1. Define the objectives of the control system and the nature of the disturbances.
2. Understand the principles of the process in terms of its dynamic behaviour.
3. Propose a control structure consistent with the objectives and process characteristics.

4. Assign controllers and evaluate the proposed control structure with anticipated disturbances through the use of dynamic simulation.

Ultimately the importance of process control is seen through increased overall process efficiency allowing the plant engineer to get the most from the process design. This is especially true of distillation control. Most distillation columns are inherently flexible and a wide range of product yields and compositions can be obtained at varying levels of energy input. A key requirement of any control system is that it relates directly to the process objectives. A control system that does not meet the process objectives or produces results that conflict with the process objectives does not add value to the process.

8.4 Methodology for the selection of a controller structure

The economic performance of a distillation system is linked to its steady-state degrees of freedom. In other words, the economic benefits of a column control scheme depend on how well it controls composition, recovery or yield and not on how well it holds integrating variables such as levels and pressures. The integrating variables must obviously be controlled, but their control performances do not directly translate into profits. However, inventory controls can be the most troublesome of all loops and can preoccupy the operators to the point where the economically important composition and recovery are neglected. This problem has been resolved by designing the level and pressure controls before dealing with the composition controls [2]. However, one must be careful in the selection of the manipulated variables for inventory control as they can significantly impact the control performance of the composition loops.

The following methodology [4] can be employed to define a control structure for a simple distillation column shown in Figure 8.1.

1. Count the control valves in the process to determine the overall degrees of freedom for control.
2. Determine from a steady-state analysis the steady-state degrees of freedom.
3. Subtract the steady-state degrees of freedom from the overall degrees of freedom to determine how many inventory loops can be closed with available control valves.
4. Design pressure and level controls and then test for disturbance rejection.
5. Design composition controls based on the product stream requirements.
6. Design optimising controls with the remaining manipulative variables.

For the simple distillation column in Figure 8.1, there are five degrees of freedom, which translates into five independent valves from a control point of view. In this 5×5 system, there are 120 possible single input/single output (SISO) control combinations of controlled and manipulated variables. Fortunately, most of these combinations are not useable due to various constraints, such as economics. From a steady-state degree of freedom analysis there are only two degrees of freedom, since a total condenser is assumed. If the column had a partial condenser there would of course be three degrees of freedom instead of two. Inventories that must be controlled are the reflux drum level (h_R), level in column base or reboiler (h_B), and the column pressure (vapour hold-up). The remaining two variables are used to control the feed split and the fractionation.

The feed split is simply the amount of feed that leaves as distillate versus the amount that leaves as bottoms. The other variable, fractionation, is the amount of separation that occurs per stage. The overall column fractionation depends on the number of stages, the energy input and the difficulty of separation. A typical control scheme for this column is shown in Figure 8.3.

The most convenient method of verifying the operability of a purposed multivariable control scheme is through dynamic simulation. However, to effectively use dynamic simulation it is first necessary to define the objectives of the control system, define the nature of the expected disturbances and develop a basic understanding of the process both in terms of its steady-state and dynamic behaviour.

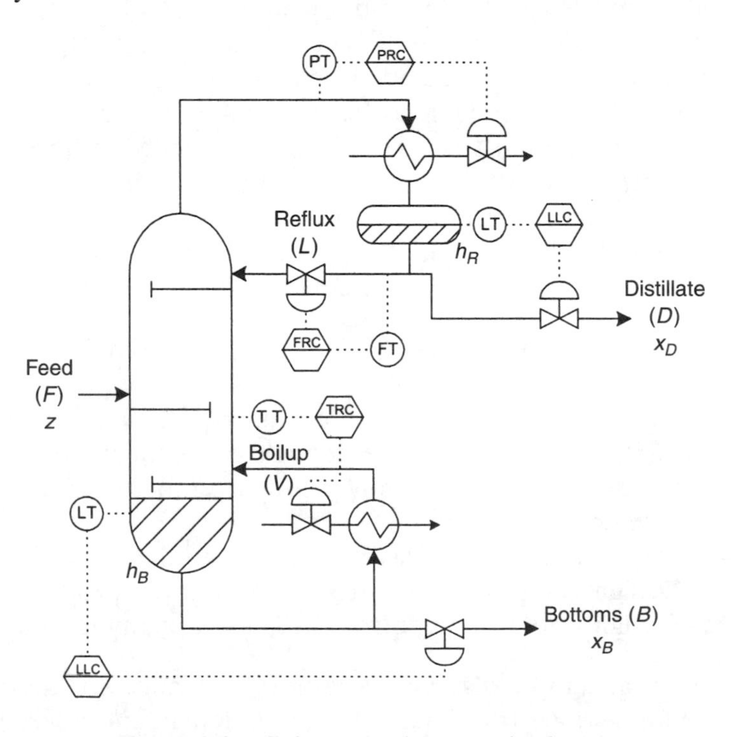

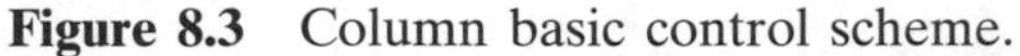
Figure 8.3 Column basic control scheme.

8.5 Level, pressure, temperature and composition control

Measurement of fractionating column variables must be within certain tolerances of accuracy, speed of response, sensitivity and dependability; and they also must be representative of the true operating conditions before successful automatic control can be realised. The instrument equipment selected, the installation design and the location of the measuring points determine these requirements.

This section is concerned with selecting the specific location in a fractionating column of the measuring points that will provide the best automatic control under variable process operating conditions. Specifically, level, temperature, pressure and composition measuring points in conventional fractionating columns are discussed. It should be clearly understood that this discussion, which is general in nature, is intended only to serve as a guide from which detailed recommendations may be formulated and tested through dynamic simulation.

Locating temperature, pressure, flow and composition measuring points for automatic control systems depends on the control scheme used and the static and dynamic interdependence of these variables. The control scheme utilised is usually determined by the source of energy or process stream to be manipulated to control a particular variable. It is therefore important to consider the static measuring sensitivity of the instrument selected to measure the controlled variable. Measuring sensitivity should generally increase with requirements of control precision by the use of narrow span suppressed range instruments. In addition, the location of the measuring element with respect to the energy source and the time lag involved for it to sense effects of changes in manipulated variables will determine dynamic measuring lags introduced by changing process conditions. The dynamic measuring lags will determine the quality and stability of the control scheme.

The interaction of temperature, pressure and composition will differ with location in the column. The selection of a temperature control point in a fractionating column, which is determined assuming that the pressure and composition are constant, may be unsatisfactory when these variables are permitted to vary with changing process conditions, i.e. feed composition changes. The complex effect of all of the sources of disturbances in the form of changing process conditions on the measuring point must be considered for dynamic stability via dynamic simulation.

8.5.1 LEVEL CONTROL

Level control was discussed in detail in Chapter 7 in the liquid level control section. From a common sense point of view, to assign manipulative variables for level control, simply choose the stream with the most direct impact. For example, in a column with a reflux ratio of 100, there are 101 units of vapour entering the condenser and 100 units of reflux leaving the reflux drum for every unit of distillate leaving. Therefore, the reflux flow or vapour boilup should be used to control the drum level. If this assignment principle is not followed and distillate flow is selected for level control, it would only take a change of slightly more than 1% in either vapour boilup or reflux flow to saturate the drum level controller and saturate the distillate valve. The Rule of 10 can be applied. This rule states that if there is a 10 to 1 or greater difference, say reflux versus product, then the larger stream must be used to control the level.

8.5.2 PRESSURE CONTROL

Pressure control is a primary requirement for all towers because of its direct influence on the separation process. Columns are typically designed to operate at sub-atmospheric, atmospheric or above atmospheric pressure. Tower pressure control configurations can also be required to vent varying amounts of inerts from the overhead accumulator. The venting of inerts or maintaining the desired operating pressure is often the crux of the control problem.

The same general principle is followed when finding manipulated variables for pressure control as for level control. Column pressure is generated by boilup and is relieved by condensation and venting. To find an effective variable for pressure control, it is necessary to determine what affects pressure the most. For example, in a column with a total condenser either the reboiler heat or the condenser cooling are good candidates for pressure control. On a column with a partial condenser, it is necessary to determine whether removing the vapour stream affects pressure more than condensing the reflux. Sometimes the dominating effect is not obvious. If the vent stream is small, it might be assumed that the condenser cooling should be manipulated for pressure control. However, if the vent stream contains noncondensables, these will blanket the condenser and affect the condensation significantly. In this situation the vent flow, although small, is the best choice for pressure control.

Figure 8.4 shows a typical pressure control scheme for sub-atmospheric column operation used for total condensing service. The eductor is not controlled by regulating the motivating steam because the turndown on the jets is very limited. Rather, the capacity is controlled by regulating the addition of non-condensable gas. This method provides a smooth and rapidly responding control system.

Figure 8.5 shows a typical control scheme used for an atmospheric or above atmospheric tower in a total condensing service with little or no inerts. In this situation the pressure is controlled by regulating the flow of the coolant which in turn changes the condensing surface temperature and the vapour

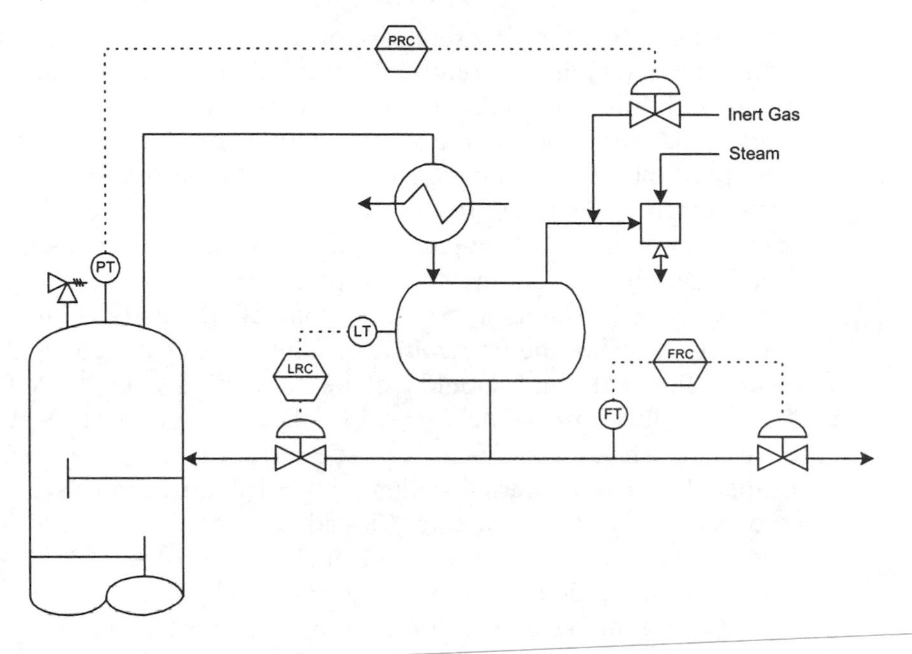

Figure 8.4 Pressure control for sub-atmospheric operations.

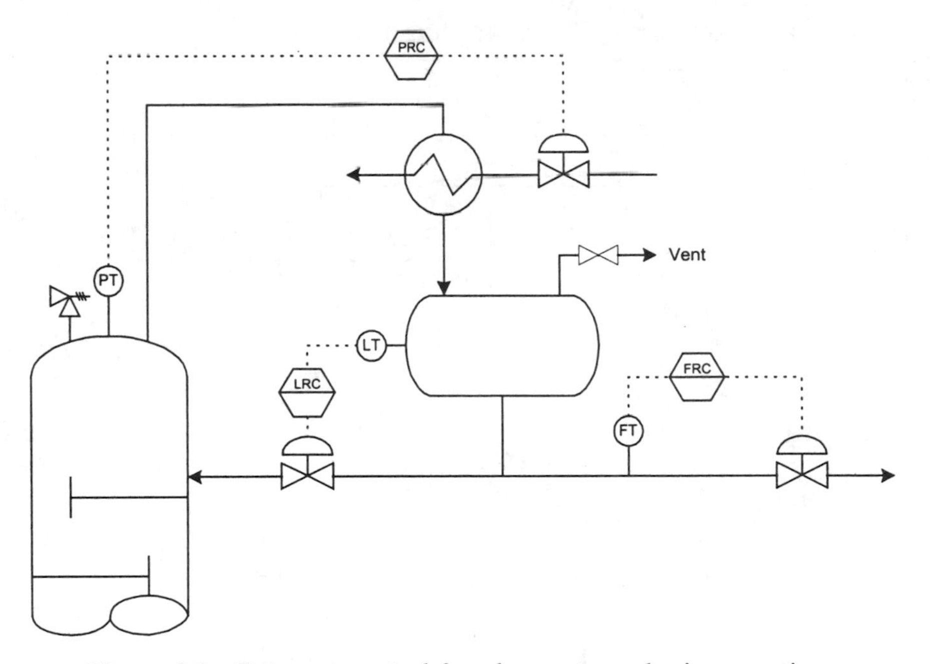

Figure 8.5 Pressure control for above atmospheric operation.

condensing rate. The pressure response of this scheme to changes in the coolant flow rate is inherently slow in comparison to methods regulating vapour withdrawal directly and/or condenser surface area control.

Figure 8.6 shows a scheme where column pressure is controlled by regulating the flow of the vapour product from the accumulator. The reflux is on flow control. A level controller is required to control the coolant flow in order to maintain accumulator liquid inventory. This method provides a smooth, rapidly responding column pressure control.

In Figure 8.7 column pressure is controlled by regulating the inert and vapour flow from the accumulator. The condenser coolant is fixed at a constant flow rate and should not be subject to change. A flow controller fixes the reflux rate while a cascade (level to flow) is used to adjust the product overhead rate. This cascade arrangement isolates the overhead product flow from internal column pressure disturbances that could affect the overhead product flow rate. Cascade control is only used if minimisation of overhead product flow rate is critical to downstream unit operations.

For a total condensing service the column pressure can be controlled by varying the condenser level or the condenser surface area exposed to the column overhead vapours, as shown in Figure 8.8. The accumulator pressure and reflux temperature can also be controlled by providing a condenser vapour by-pass or by controlling the coolant flow rate.

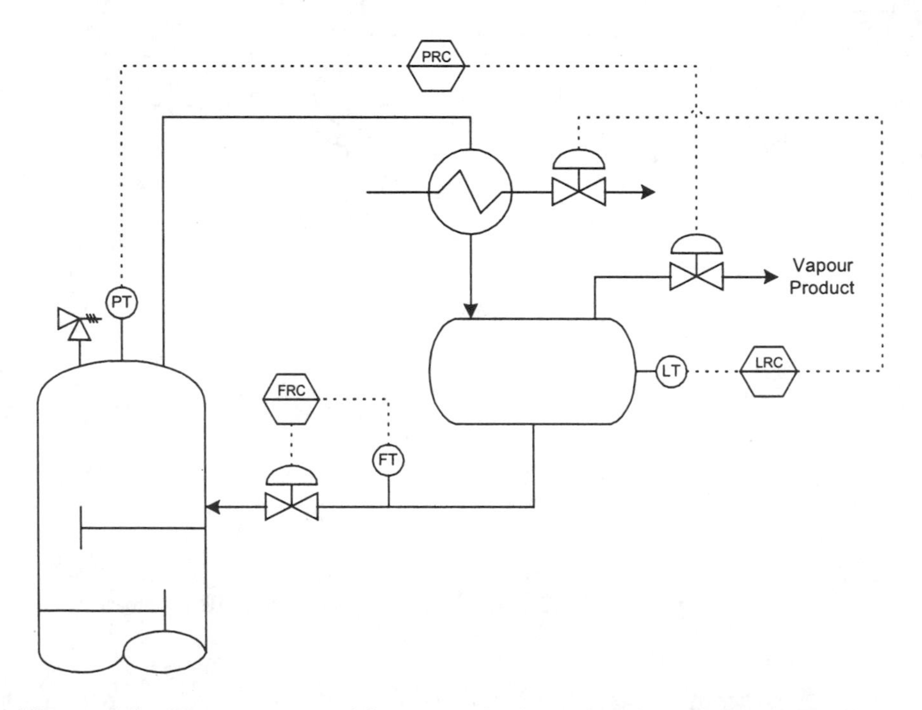

Figure 8.6 Pressure control by control of overhead product vapour flow.

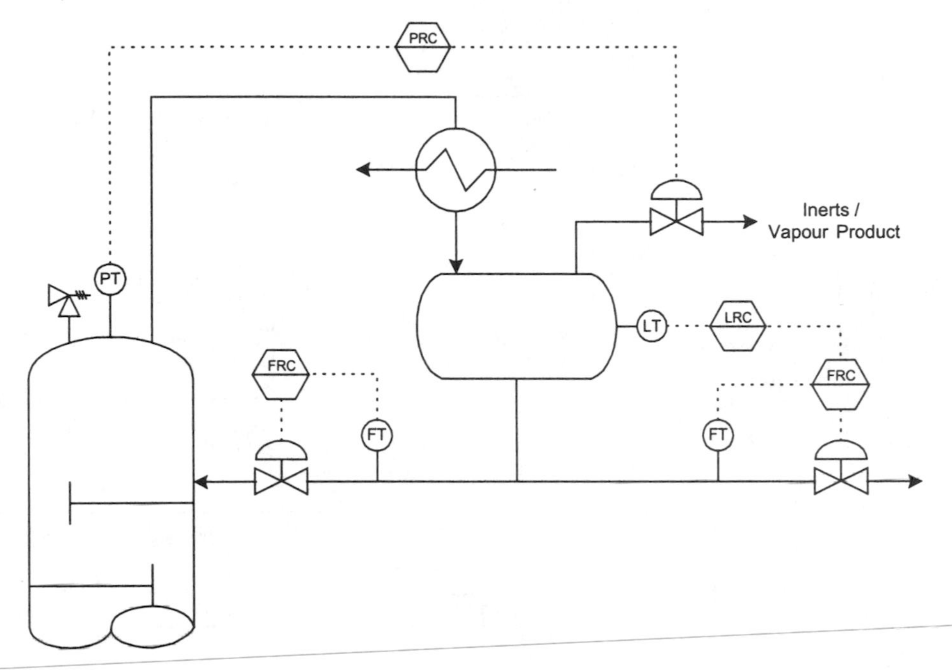

Figure 8.7 Pressure control by venting of inerts.

In a total condensing service, when varying quantities of inerts are present in a pressurised tower, it is often necessary to vent or alternatively inject a blanketing inert gas. This is normally accomplished using a split range control scheme, as shown in Figure 8.9. The column pressure is controlled by either injecting or venting blanketing gas from the accumulator/condenser.

The location of a column pressure control point is not restricted by dynamic considerations. The response time of pressure changes in a column and the dynamic measuring lag has been found to be equally fast for any location in the column when the manipulated energy source used to control pressure is either condenser cooling or reboil heat. Pressure is regulated at a constant value and is rarely used as a variable to control a product specification.

Some of the factors that should be considered in locating the pressure measuring point are:

1. When columns are operating near the relief valve pressure setting, the pressure measurement should be located near the relief valve.
2. Bottom pressure control of an atmospheric column via reboiler heat input will in effect control the column differential pressure and thereby the column vapour flow and tray vapour loading.
3. Providing temperature on a tray is a good indication of composition; pressure at this tray should be measured and controlled. This concept is explained in greater detail in the next section.

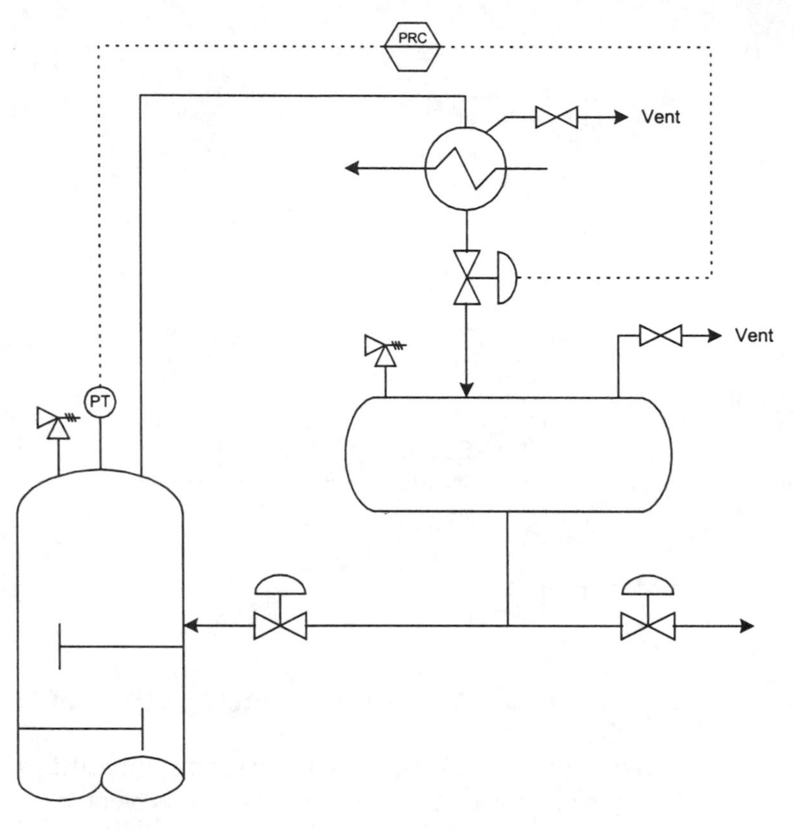

Figure 8.8 Pressure control by condenser level control.

8.5.3 TEMPERATURE CONTROL

Composition control of products from a column is usually realised via temperature control. Temperature sensors are inexpensive, continuous, and fast compared to composition sensors. The measurement lag is particularly important in dynamic considerations. For temperature it is a fraction of a minute whereas composition measurement by gas chromatography is of the order of 15–30 minutes. Infrared analysers are available that produce continuous composition estimates but they are not yet cost-effective. Periodic checks of product composition by analytical means provide information which is used in setting the temperature control point. The accuracy of the correlation of column temperature to product composition depends on the sensitivity of the controlled temperature to composition changes and pressure variations at the temperature measuring point.

The sensitivity of the temperature measurement to key or major component composition changes for each tray can be determined if tray-by-tray composition changes are large and the other component changes are small.

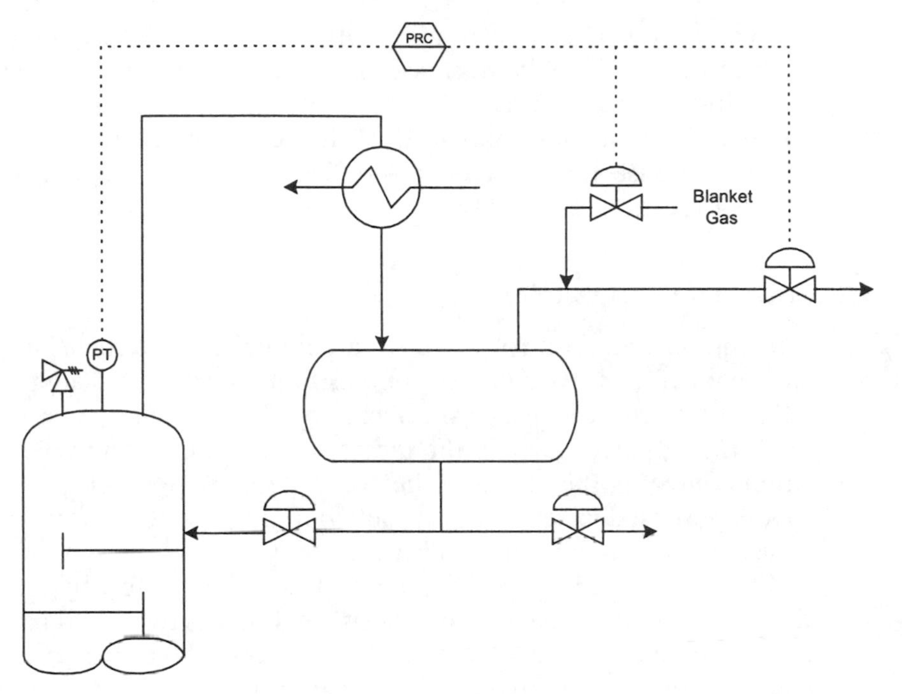

Figure 8.9 Pressure control by regulating condenser surface area with blanket gas.

A temperature measurement in this area will give a good indication of composition provided that the effects of the pressure variations are small. Controlling pressure at the point or tray where temperature is controlled can eliminate pressure variations that have large effects on composition.

The temperature-composition correlations of key components are often affected by changes in the concentration of other components, i.e. column feed composition changes. If the magnitude of these changes can be estimated, a calculation using equilibrium constants can be made to determine the effect on the temperature-composition correlation. Then a control tray can be selected where the effect of non-key component variations is small.

Stable column temperature control, from the tray selected by the foregoing static considerations, depends on the dynamic measuring lag or response of the tray temperature with respect to the manipulated energy source used to control the temperature. Based on experimental tests, the following observations are cited for use as guides:

1. Temperature control is made less stable by thermowell and measuring instrument lag or response times.
2. The speed of response and control stability of tray temperature, when controlled by reboil heat, is the same for all tray locations.

3. The speed of response and control stability of tray temperature, when controlled by reflux, decreases in direct relation with the number of trays below the reflux tray.
4. When pressure is controlled at the temperature control tray, the speed of response of the temperature instrument can vary considerably with tray location, and is normally slower.

8.5.4 COMPOSITION CONTROL

The composition control loops on a column are the most important steady-state controls. The purpose of composition control is to satisfy the constraints defined by product quality specifications. These constraints must be satisfied at all times, particularly in the face of disturbances. The objective of composition control is then to hold the controlled composition as close as possible to the imposed constraint without violating the constraint. This objective translates to on-aim, minimum variance control.

To achieve good composition control, two things must be examined: process dynamics and disturbance characteristics. Process dynamics includes measurement dynamics, process dynamics and control valve dynamics. Tight process control is possible if the equivalent dead time in the loop is small compared to the shortest time constant of a disturbance with significant amplitude. To ensure small overall dead time in the loop, it is necessary to find a rapid measurement along with a manipulated variable that gives an immediate and appreciable response. In distillation, a rapid measurement for composition control often translates into a tray temperature. A good manipulated variable is vapour flow, which travels quickly up through the column and usually has a significant gain on tray temperatures and indirectly on composition.

In situations where the apparent dead time in the composition loop cannot be kept small compared to significant disturbances, the disturbances themselves must receive the attention. Sometimes important disturbances can be measured or anticipated in which case feedforward control is a candidate. In other situations, the control loop structure can be rearranged to influence the way the disturbance affects the composition variable. Several researchers have proposed numerous algorithms for determining the disturbance sensitivity for different control structures. Tyreus [4] states that, in his opinion, direct dynamic simulation of the strategies resulting from assignment of the manipulated variables for pressure and level control gives the best insight into the viability of a proposed composition control scheme.

8.6 Optimising control

After the inventory and composition controls have been assigned, there are typically a few manipulated variables remaining. These variables can be used for process optimisation. Because process optimisation should be performed on a plant wide scale, in-depth discussion of this topic will be delayed until Chapter 10.

8.6.1 EXAMPLE: BENZENE COLUMN WITH A RECTIFYING SECTION SIDESTREAM

To better illustrate how the described control strategy design method is put into practice, consider the case of a liquid side draw benzene column.

Figure 8.10 shows the flowsheet configuration of a column with a rectifying section liquid side draw. The multi-component feed comes from an upstream unit in the process. The benzene liquid side draw is the product stream, and has a purity specification in terms of benzene. The distillation removes *n*-pentane from the feed mixture and the heavies (toluene, naphthalene and biphenyl) are purged from the reboiler. A small overhead purge stream is connected to the condenser for pressure relief.

The control scheme objective of this column is to operate close to the quality constraint of the liquid side draw product. The major disturbances are changes to the overall feed flow rate, as well as individual component feed flow rates.

The column has seven control valves and requires four degrees of freedom for steady-state control. The remaining three dynamic degrees of freedom are used to control the column inventories. Column pressure is controlled by

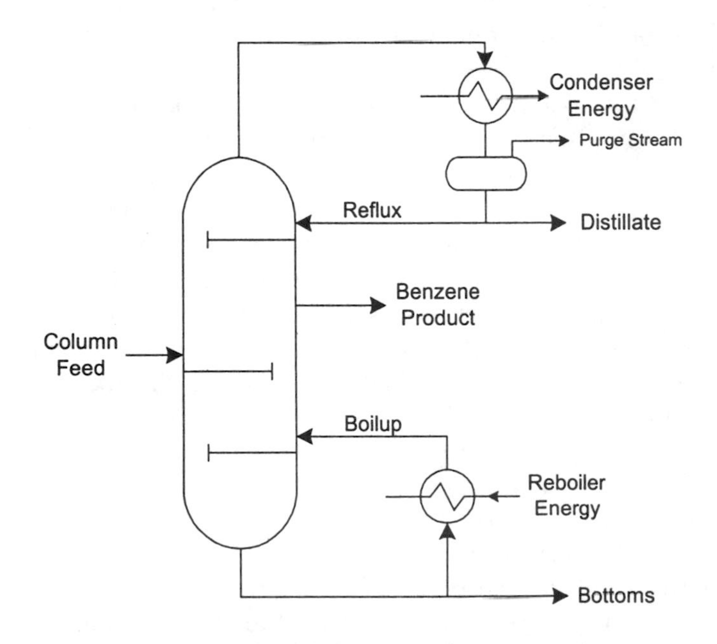

Figure 8.10 Liquid side draw benzene column.

manipulating the condenser duty. However, if there were non-condensables in the column, the overhead vapour stream would have been a more suitable choice as a manipulated variable. Non-condensables in a column tend to accumulate in the condenser and significantly reduce the dew point of incoming vapours. The low dew point reduces heat transfer because of small temperature driving force. Because the vent stream is rich in non-condensables, vent flow rate is an effective manipulator for removing the non-condensables and thereby quickly increasing heat transfer whenever needed.

Control of the reflux drum is fairly straightforward. Because the reflux ratio is very high, with a steady-state value of 145, reflux flow is the only reasonable manipulator for drum level. However, there is a potential loss of one dynamic degree of freedom unless it is ensured that the material balance for the distillate product is satisfied. This can be achieved by ratioing the distillate flow to the reflux flow. The effective manipulator is now the distillate flow and the reflux flow combined instead of just reflux flow. Control of the base level in the column is restricted to the use of reboiler steam due to the large vapour boilup to bottoms ratio.

At this point, the inventories in the system have been placed under control and composition control can be considered. However, first the side stream material balance must be considered. The condenser level control refluxes any disturbances in vapour flow rate back down into the column as liquid. On the other hand, the base level controller sends any disturbances in liquid flow back up the column as vapour. To prevent a buildup of side stream material in the column, a route must be provided for it to escape. This can be accomplished by ratioing the liquid side draw flow to the reflux flow.

Finally, a temperature controller can be added to provide a method of controlling the composition of the liquid side draw. This controller can have its temperature sensor on the bottom tray of the main tray section and use the bottoms flow rate as a manipulated variable. Temperature sensitivity analysis can be performed using the steady-state model to ascertain the proper location for the temperature sensor. Using the bottoms flow rate allows a method for excess heavies to be removed from the system in the event of a disturbance while retaining the target composition of the liquid side draw. The resulting control scheme for the liquid side draw benzene column is shown in Figure 8.11.

How does this control scheme respond to disturbances? The major control objective is to produce a side stream of essentially pure benzene, approximately 99%. To test this control scheme, the column was subjected to two different types of disturbances. The first disturbance was that of an increase in the total volumetric flow rate of the feed introduced to the column. A strip chart of the feed flow rate, the three product flow rates and the side stream benzene purity is shown in Figure 8.12. A step change in the feed

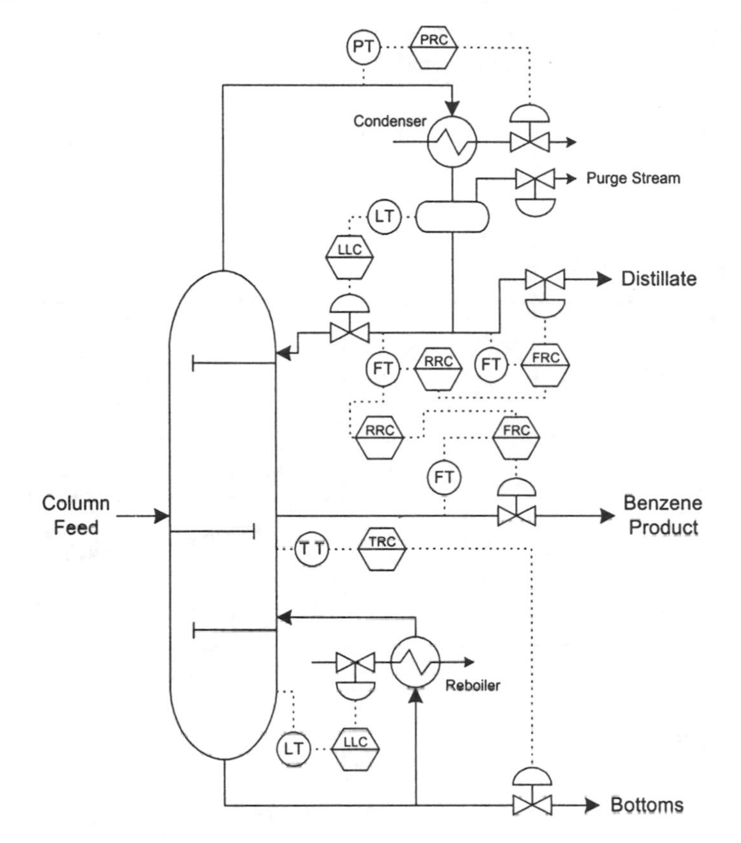

Figure 8.11 Liquid side draw benzene column control scheme.

flow rate is introduced, increasing the flow rate from 11.5 m^3/h to 12.075 m^3/h. This corresponds to an increase of 5%.

An increase in the overall volumetric flow rate of the liquid feed adds considerable liquid to the column system. Because the feed is primarily benzene, it is expected that the benzene side draw flow rate will increase. The flow rate overshoots and then assumes its new steady-state value. A similar shaped curve exists for the distillate. This, too, is expected due to the ratio control between the distillate stream and the reflux stream and the ratio control between the benzene side draw and the reflux stream. Throughout the overshoot in flow rates, the benzene purity in the side draw remains relatively constant. What is interesting to note is the response of the bottoms flow rate. Here, an inverse response is exhibited. The flow rate first decreases, then increases, overshoots, and finally assumes its new steady-state value. Why does the bottoms flow rate behave in such a manner? The introduction of more liquid feed means that more liquid benzene is cascading down the trays. As liquid reaches the bottom tray, the bubble point of the liquid on that tray decreases. The temperature controller reduces the valve opening

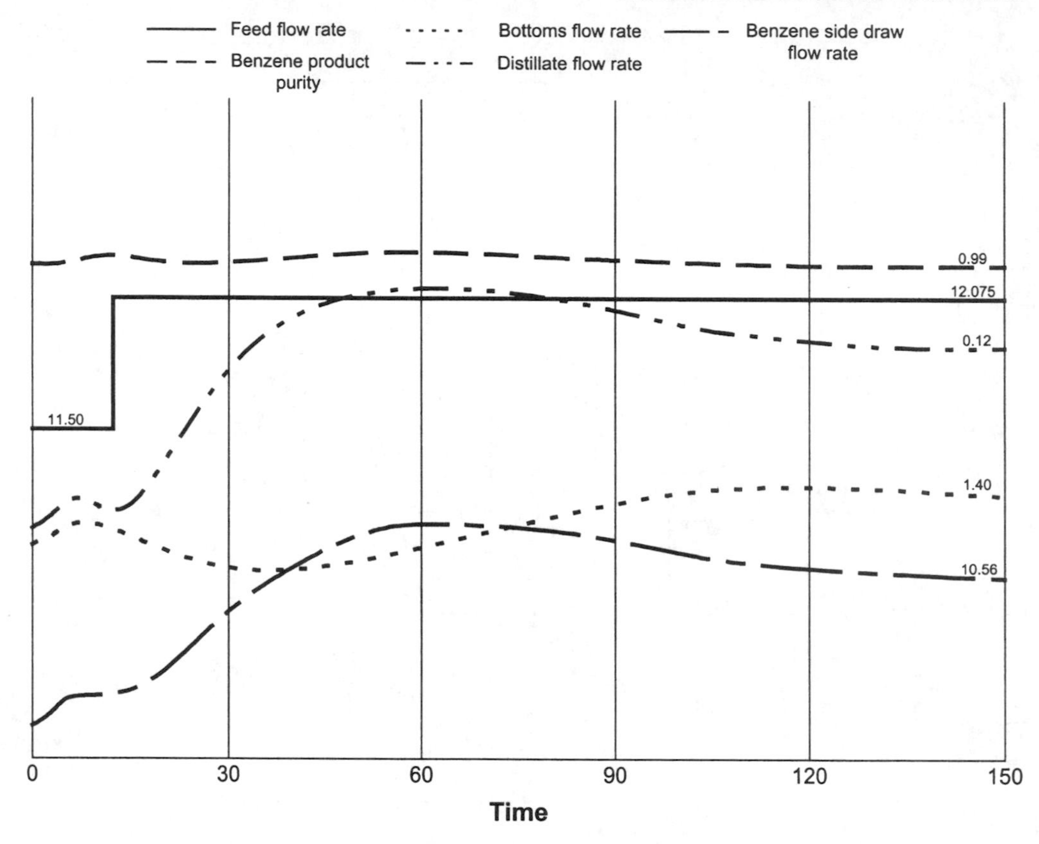

Figure 8.12 System response to a step change in feed flow rate.

on the bottoms stream to compensate. The reboiler level rises resulting in more steam being introduced by the action of the level controller. The benzene and pentane are vaporised and move back up the column. As the column adjusts to the increased feed flow rate, the temperature profile in the column rises. The bottoms flow rate is then increased and settles back down to its new steady-state value.

To test the control structure against changes in the composition of the feed stream, sinusoidal disturbances were introduced to the feed compositional flow rates. Each compositional flow rate was varied ±10% over periods ranging from 20 to 30 minutes. This example of a disturbance is a little unrealistic, but it demonstrates how the control structure would respond to compositional upsets. The strip chart of the same four flow rates and the benzene side draw purity is shown in Figure 8.13.

Each of the product flow rates responded in a similar fashion to that for a step change in the feed volumetric flow rate. As the feed flow rate increased, the products increased as well. The inverse response in the bottoms flow rate

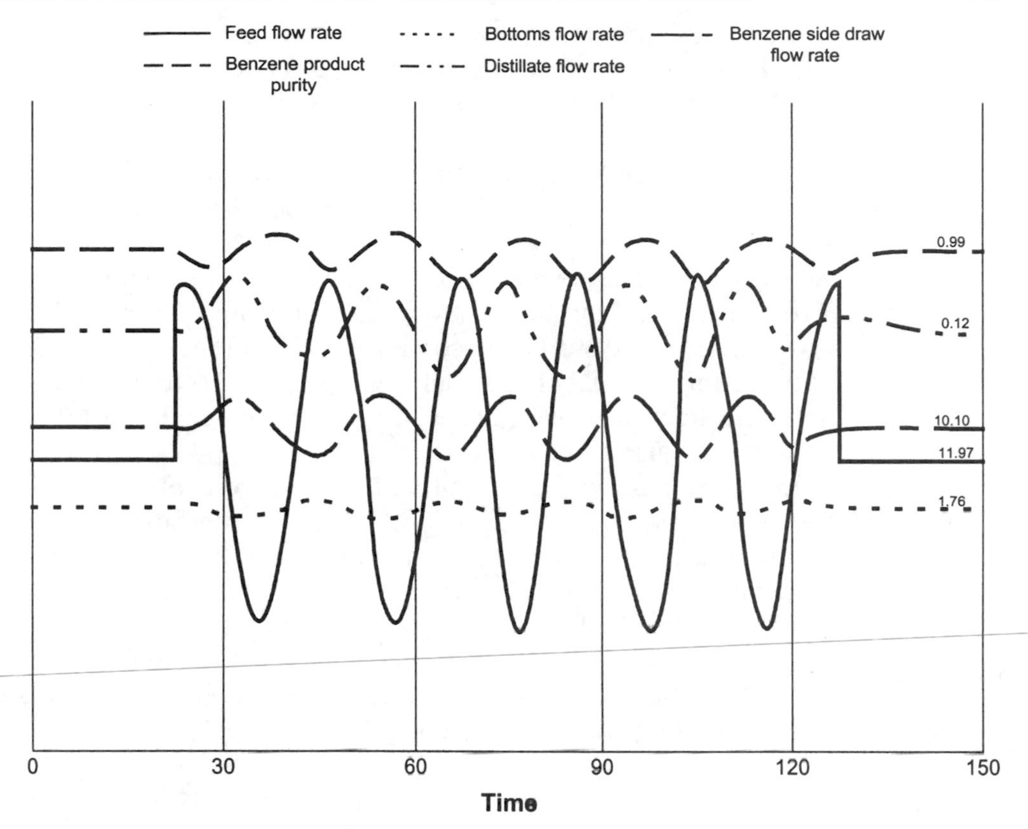

Figure 8.13 System response to sinusoidal disturbances in the component flow rates.

is not as observable now. The variable of interest is the benzene purity on the side draw. Although the feed composition is varying continuously, the variance in the benzene purity is much less. The effect on the benzene purity for drifting compositions is considerably damped.

8.7 Distillation control scheme design using steady-state models

Steady-state simulation of distillation columns has become routine. The use of these simulations has been restricted to use for heat and material balance and sizing purposes. Freuhauf and Mahoney [10] have shown that steady-state calculations [11] can be used to screen candidate control schemes, to provide a means for tray temperature location and to calculate static gains.

Steady-state models are easily manipulated and are robust. This allows for the efficient generation of a large number of case studies necessary for steady-state design procedures. The obvious disadvantage of this procedure is that nothing is known about the dynamic response, and hence the dynamic disturbance rejection capability of alternative control schemes is also not known. These need to be evaluated using a dynamic simulator.

The design procedure consists of the following five steps:

1. *Develop a design basis*: Here there is a need to define product composition specifications, disturbance type and size, constraints and original column design basis.
2. *Select a candidate control scheme*: The literature abounds with alternative control configuration. Consider as an example a typical column that has feed as the disturbance stream. As was pointed out in the previous section, for such a column, only two degrees of freedom remain, feed-split and fractionation. The resulting best feed-split control schemes are shown in Figures 8.14 and 8.15.

   In Figure 8.14, we have a direct feed-split control scheme, because the distillate is manipulated directly to control composition. Product compositions are controlled by fixing a column temperature. The temperature controller manipulates the distillate flow. This control scheme is often selected when the heat input is limited or must be fixed.

   Figure 8.15 illustrates the second common choice, indirect feed-split control, where the distillate flow is increased indirectly by increasing the steam flow. The compositions are controlled by a temperature controller that

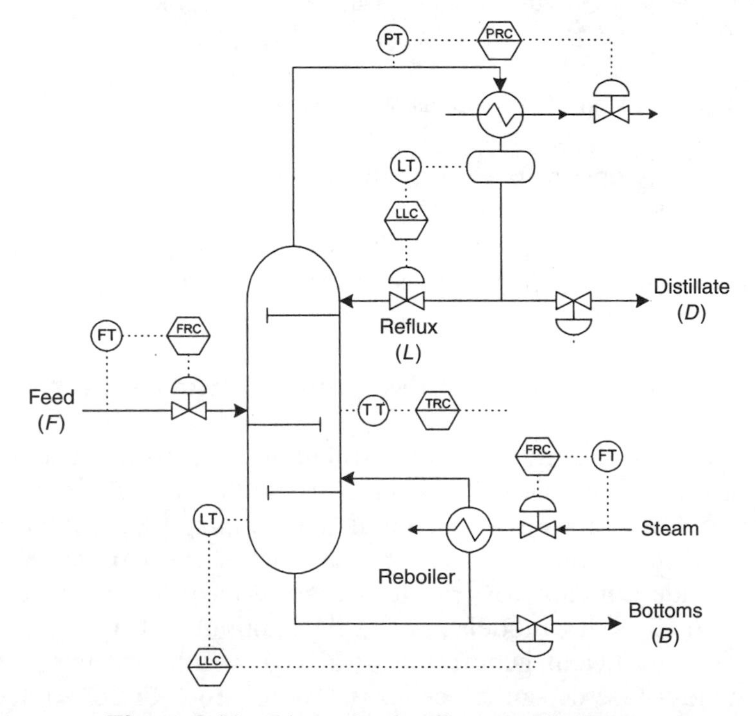

Figure 8.14 Direct feed-split control scheme.

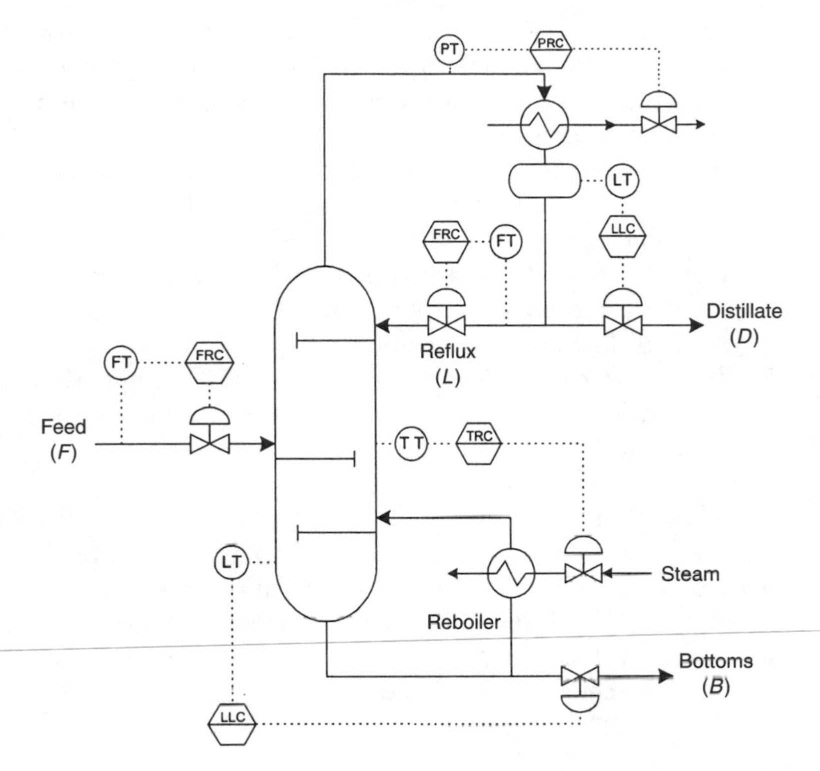

Figure 8.15 Indirect feed-split control scheme.

manipulates the steam flow. This alternative has two advantages. One is that the temperature loop has faster closed-loop response[1] and therefore provides better disturbance rejection. The second is that, because the reflux drum level sets the distillate flow, the reflux drum can be used to smooth flow disturbances to other downstream unit operations. To achieve flow smoothing, the level controller must have averaging level controller tuning.

The last part of the control strategy selection process is to select a ratio-control alternative that might use less energy than the primary alternative. One example of a ratio-control alternative for the scheme illustrated in Figure 8.15 would be a controller that keeps a constant reflux to feed flow ratio. This scheme likely will consume less energy than the non-ratio alternative because, as the feed flow to the column decreases, the amount of reflux will decrease. Less reflux will require less heat input.

3. *Conduct "open loop" testing*: The purpose of this step is to use the steady-state model to identify a suitable tray for the temperature sensor for composition control. The procedure consists of using a candidate control

[1] Shorter natural period

scheme such as in Figure 8.14 varying, in this case, the distillate flow and observing the change in column temperature profile. A good tray is one on which the temperature change was significant and nearly equal when the flow is increased or decreased.

4. *Run "closed loop" testing*: In this step the steady-state model is used to simulate the candidate control scheme and to test its robustness to feed flow and feed composition changes. This step consists of a series of runs (sensitivity studies) aimed at locating a set of operating conditions that meet or exceed the product specifications for all expected disturbances in feed flow and composition.
5. *Confirm the objectives have been met*: If the objectives have been met the procedure is complete. If not, the procedure is repeated with another candidate control scheme.

As mentioned previously, this methodology can be used to screen a large number of candidate control schemes quickly and efficiently. However, it is necessary to then evaluate the candidate control schemes using a dynamic simulator to check the dynamic disturbance rejection capabilities of the alternative control schemes. A case study that demonstrates the application of this design scheme is available in the literature [12].

8.8 References

1. Buckley, P.S., Luyben, W.L. and Shunta, J.P., "Design of Distillation Control Systems". Instrument Society of America, Research Triangle Park, NC, 1985.
2. Buckley, P.S., *Techniques of Process Control*. John Wiley & Sons, New York, 1964.
3. King, C.J., *Separation Processes*. McGraw-Hill, New York, 1971.
4. Tyreus, B.D., "Chapter 9: Selection of Controller Structure" In Luyben, W.L. (ed.), *Practical Distillation Control*, Van Nostrand Reinhold, 1992, pp. 178–91.
5. Seborg, D.E., Edgar, T.F. and Mellichamp, D.A., *Process Dynamics and Control*. John Wiley & Sons, New York, 1989.
6. Shinskey, F.G., *Distillation Control For Productivity and Energy Conservation*. McGraw-Hill, New York, 1977.
7. Smith, C.A. and Corripio, A.B., *Principles and Practice of Automatic Process Control*, 2nd edn. John Wiley & Sons, New York, 1997.
8. Svrcek, W.Y. and Morris, C.G., "Dynamic Simulation of Multi-component Distillation". *Canadian Journal of Chemical Engineering*, June 1981, pp. 382–7.
9. Wilson, H.W. and Svrcek, W.Y., "Development of a Column Control Scheme: Case History", *Chem. Eng. Prog.*, 1971, **67**(2): 45.
10. Freuhauf, P.S. and Mahoney, D.P., "Distillation Column Control and Design Using Steady-State Models: Usefulness and Limitations". *ISA*, Paper # 92–0279, 1992, p. 93–120.
11. Tolliver, T.L. and McCune, L.C., "Distillation Column Control Design Based on Steady-State Simulation". *ISA Transactions*, 1978, **17**(3): 3–10.
12. Young, B.R. and Svrcek, W.Y., "The Application of Steady-State and Dynamic Simulation for Process Control Design of a Distillation Column with a Side Stripper". Proceedings of Chemeca '96, 24th Australian and New Zealand Chemical Engineering Conference, Sydney. Published by the Institution of Engineers, Australia, Sept. 1996, Vol. 2, pp. 145–50.

9 SELECTION OF MULTIPLE SINGLE-LOOP CONTROL SCHEMES USING STEADY STATE METHODS

When a process has only one input variable to be used in controlling one output variable, the controller design problem can be handled fairly easily. However, in several important situations the system under consideration has multiple inputs and multiple outputs, making it multivariable. In fact, the most important chemical processes are often multivariable. This chapter outlines a procedure for designing control schemes for multi-input/multi-output processes using steady state methods.

9.1 Variable pairing

When designing controllers for multi-input/multi-output processes, it seems natural to start out by considering the possibility of pairing the input and output variables and assigning a feedback controller to each input/output pair, resulting in several single-loop controllers. With this approach, the issues to be resolved are twofold:

1. How to pair the input and output variables.
2. How to design the individual single-loop controllers.

Consider, as an example, the 2×2 system shown in block diagram form in Figure 9.1. This figure shows how each input variable is capable of influencing both output variables. The effect of the inputs on the outputs may also be represented by an input-output function or transfer function as shown by Equations 9.1 and 9.2.

$$y_1 = g_{11}m_1 + g_{12}m_2 \tag{9.1}$$

$$y_2 = g_{21}m_1 + g_{22}m_2 \tag{9.2}$$

where: y_i = the controlled or output variable "i"
m_j = the manipulated or input variable "j"
g_{ij} = the input-output relationship or transfer function between y_i and m_j

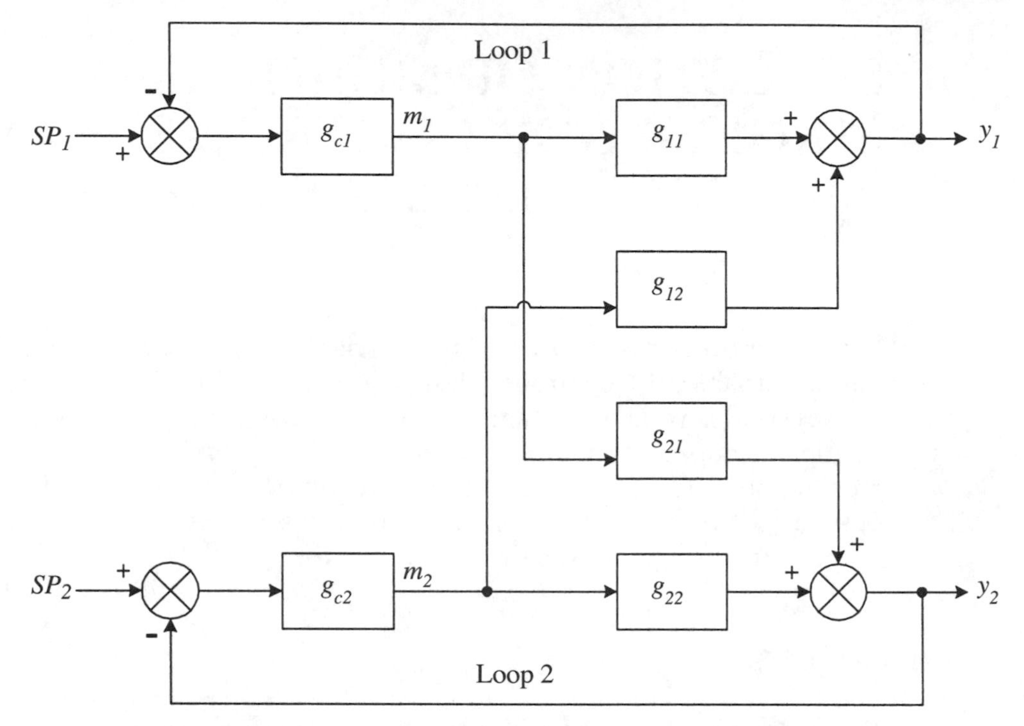

Figure 9.1 Loop interactions for a 2×2 system.

If m_1 has been assigned to y_1 (and, by default, m_2 to y_2), each of these control loops will experience interactions from the other loop. Loop 1 (the $m_1 - y_1$ loop) will experience interactions coming from Loop 2 (the $m_2 - y_2$ loop) via the g_{12} element. In the same manner, Loop 2 experiences interactions coming from Loop 1 through the g_{21} element.

There is a second configuration to be considered. If the input/output pairing were switched, each loop would still be subject to interactions from the other loop but, this time, the interaction would come about via the g_{11} and g_{12} elements in the equations shown above.

It seems reasonable that, as long as the intent is to use two single-loop controllers for this 2×2 system, pairing of the input and output variables needs to be such that the resulting interaction is at a minimum. In other words, since there are two possible input/output pairing configurations for this particular system, it would be advantageous to choose the configuration that results in the smaller net interaction among the variables. This strategy works well for the majority of processes; however, one should be aware that there are more complex control system designs that try to use the interactions rather than eliminate them.

In order to logically determine which input/output pairing configuration to use for the multiple, single-loop control strategy, a means of discriminating between the $n!$ different configurations possible with a $n \times n$ system is required. To accomplish this, there must be a method available to quantify the interactions experienced by the process under each configuration.

9.1.1 QUANTIFYING CONTROL LOOP INTERACTIONS

Consider the 2×2 system shown in Figure 9.1 with two outputs, y_1 and y_2, and two inputs, m_1 and m_2. When both loops are open, m_1 and m_2 can be manipulated independently and the effect of the inputs on the outputs is represented by the input-output function or transfer function shown in Equations 9.1 and 9.2.

Let us now consider m_1 as a candidate input to pair with y_1. To evaluate this choice against the alternative of using m_2, it is necessary to perform two experiments on the system.

Experiment 1: Unit step change in m_1 with both loops open

First, we will test the direct influence of m_1 on y_1. When a unit step change is made in m_1 with all the loops open, the output y_1 will change. y_2 will also change but for now the primary interest of this experiment is in the response of y_1. The change in the steady-state value of y_1 is equivalent to the steady-state gain between y_1 and m_1, as shown in Equation 9.3.

$$\Delta y_{1m} = k_1 \tag{9.3}$$

Experiment 2: Unit step change in m_1 with Loop 2 closed

In this next experiment the same step change is made in m_1. However, this time, Loop 2 is closed and the controller g_{c2} manipulates m_2 to reject upsets occurring in y_2 due to the change in m_1. Figure 9.1 shows that m_1 affects y_1 both directly and indirectly. In particular, the following happens to the process due to the change in m_1:

1. Obviously y_1 changes because of g_{11}, but so does y_2 and that is because of interactions via the g_{12} element.
2. Under feedback control, Loop 2 wards off this interaction effect on y_2 by manipulating m_2 until y_2 is restored to its initial state.
3. The changes in m_2 now affect y_1.

Thus, the changes observed in y_1 are from two different sources:

1. The direct influence of m_1 on y_1, which is already known from Experiment 1 to be Δy_{1m}.
2. The indirect influence precipitated by the retaliatory action of controller in Loop 2 while it is rejecting the effect of m_1 on y_2, which is Δy_{1r}. The

term Δy_{1r} indicates the amount of interaction m_1 is able to provoke from the other control loop in its attempt to control y_2.

After steady-state has been achieved, there will be a net change observed in y_1, $\Delta y_1{}^*$, given by Equation 9.4.

$$\Delta y_1{}^* = \Delta y_{1m} + \Delta y_{1r} \tag{9.4}$$

It can be shown [1] that:

$$\Delta y_1{}^* = k_{11}\left(\frac{k_{12}k_{21}}{k_{11}k_{22}}\right) = k_{11}{}^* \tag{9.5}$$

where $k_{11}{}^*$ is the steady-state gain between m_1 and y_1 obtained in the second experiment, and k_{ij} is the open-loop steady-state gain between input i and output j. The ratio of these two gains can be used as a good measure of how well the process can be controlled if m_1 is used to control y_1:

$$\lambda_{11} = \frac{\Delta y_{1m}}{\Delta y_1} \tag{9.6}$$

or

$$\lambda_{11} = \frac{\Delta y_{1m}}{\Delta y_{1m} + \Delta y_{1r}} \tag{9.7}$$

λ_{11} provides a measure of the extent of interaction in using m_1 to control y_1 and using m_2 to control y_2. Similar experiments can be performed to investigate the candidacy of m_2 as the input to control y_1. Since λ_{11} quantifies the degree of steady-state interaction, it can be used to select a control configuration that minimises the steady-state interaction.

9.2 Relative gain array (RGA)

The quantity λ_{11} introduced in the last section is known as the *relative gain* between the output y_1 and the input m_1. The relative gain provides a measure of the extent of the influence of process interaction when m_1 is used to control y_1. Although the example introduced this quantity, λ_{ij}, is for a 2×2 system, it can be generalised to any other multivariable system of arbitrary dimension. The relative gain, λ_{ij}, between output y_i and input m_j is defined as the ratio of two steady-state gains, shown in Equation 9.8.

$$\lambda_{ij} = \frac{\left(\dfrac{\partial y_i}{\partial m_j}\right)_{\text{all loops open}}}{\left(\dfrac{\partial y_i}{\partial m_j}\right)_{\text{all loops closed except the } m_j \text{ loop}}} \tag{9.8}$$

$$= \left(\frac{\text{open loop gain}}{\text{closed loop gain}}\right)_{\text{for loop } i \text{ under control of } m_j}$$

Calculation of the relative gains for all combinations of input/output pairs in a multivariable system leads to the following matrix known as the relative gain array (RGA) or the Bristol Array:

$$\Lambda = \begin{bmatrix} \lambda_{11} & \lambda_{12} & \cdots & \lambda_{1n} \\ \lambda_{21} & \lambda_{22} & \cdots & \lambda_{2n} \\ \vdots & \vdots & \vdots & \vdots \\ \lambda_{n1} & \lambda_{n2} & \cdots & \lambda_{nn} \end{bmatrix} \tag{9.9}$$

The RGA was first introduced by Bristol [1] in 1966 and has become the most widely used measure of interaction.

9.2.1 PROPERTIES OF THE RGA

The most important properties of the RGA are as follows:

1. The elements of the RGA across any row, or down any column, sum up to 1:

$$\sum_{i=1}^{n} \lambda_{ij} = \sum_{j=1}^{n} \lambda_{ij} = 1 \tag{9.10}$$

2. λ_{ij} is dimensionless; therefore, neither the units nor the absolute values taken by the variables m_j and y_i affect it.
3. The value of λ_{ij} is a measure of the steady-state interaction expected in the ith loop of the multivariable system if its output y_i is paired with m_j. In particular, $\lambda_{ij} = 1$, implies that m_j affects y_i without interacting with, and/or eliciting interactions from, the other control loops. By the same token, $\lambda_{ij} = 0$ implies that m_j has absolutely no effect on y_i.
4. Let k_{ij}^* represent the ith loop steady-state gain when all the other loops except this one are closed, whereas k_{ij} represents normal, open-loop gain. By the definition of λ_{ij}:

$$k_{ij}^* = \frac{k_{ij}}{\lambda_{ij}} \qquad (9.11)$$

In fact, $1/\lambda_{ij}$ tells us by what factor the open-loop gain between output y_i and input m_j will be altered when the other loops are closed.

5. When λ_{ij} is negative, it indicates that changing m_j in the closed-loop situation has the opposite effect on y_i compared to its effect on y_i when other loops are open. Such input/output pairings are potentially unstable and should be avoided.

9.2.2 CALCULATING THE RGA

Depending on the availability of the process model, RGA can be calculated using the following methods:

- *Process model is available*: The transfer function matrix, $\hat{G}$, of the process can be derived based on the process model. The transfer function matrix is defined as follows:

$$\hat{y} = \hat{G}\hat{m} \qquad (9.12)$$

where: $\hat{y}$ = Vector of output variables.
$\hat{m}$ = Vector of the input variables.
$\hat{G}$ = Transfer function matrix of the process.

Let the steady-state gain matrix of the process be defined as follows:

$$\hat{G}_{ss} = \hat{K} \qquad (9.13)$$

where "ss" indicates steady-state. Then let $\hat{R}$ be the transpose of the inverse of the steady state gain matrix, $\hat{K}$, i.e.,

$$\hat{R} = (\hat{K}^{-1})^T \qquad (9.14)$$

The elements of the RGA can be obtained as follows:

$$\lambda_{ij} = k_{ij} r_{ij} \qquad (9.15)$$

It is important to note that the above equation indicates an element by element multiplication of the corresponding elements of the two matrices $\hat{K}$ and $\hat{R}$. It has nothing to do with the standard matrix product.

- *Process model is not available*: When a process model is not available, it is still possible to obtain the RGA from experimental process data. One may adopt either of the following two approaches in such situations:

  1. Determine the steady state gain matrix, $\hat{K}$, by implementing step changes in the process inputs one at a time and observing the ultimate change in each of the process outputs. The steady-state gain between the ith output and the jth input can be found using Equation 9.16:

  $$k_{ij} = \frac{\Delta y_i}{\Delta m_i} \tag{9.16}$$

     The elements of the $\hat{K}$ matrix are thus obtained. Once the gain matrix is known we can easily generate the RGA.
  2. It is also possible to determine each element of the RGA directly from experiment. Each RGA element λ_{ij} can be determined by performing two experiments. The first experiment determines the open-loop steady-state gain by measuring the response of y_i to input m_j when all the other loops are opened. In the second experiment, all the other loops are closed (using PI controllers to ensure that there will be no steady-state offset) and the response of y_i to a change in input m_j is predetermined. By definition the ratio of these gains gives us the desired relative gain element. This was shown in the previous sections of this chapter.

The second method is more time consuming, involves too many upsets to the process, and may include experimental error: hence, it is rarely used.

9.3 Loop pairing using the RGA

Having established the RGA as a reasonable means of quantifying steady-state interactions, we will now consider how it may be used as a guide for selection of input/output pairs that lead to minimum interactions among control loops.

Using RGA in loop pairing needs a proper interpretation of the RGA elements. This interpretation can be obtained by classifying all the possible values that each element of RGA can take into the following five categories:

1. $\lambda_{ij} = 1$, indicating that open-loop gain between y_i and m_j is identical to the closed-loop gain:
   Implication for loop interactions: Loop i will not be subject to retaliatory actions from other control loops when they are closed. Thus, m_j can control y_i without interference from other control loops.
   Pairing recommendation: Pairing y_i and m_j will therefore be ideal.

2. $\lambda_{ij} = 0$, indicating that the open-loop gain between y_i and m_j is zero:
   Implication for loop interactions: m_j has no direct influence on y_i.
   Pairing recommendation: Do not pair y_i with m_j. Pairing m_j with some other output will however be advantageous, since at least y_i will be immune to interaction from this loop.
3. $0 < \lambda_{ij} < 1$, indicating that the open-loop gain between y_i and m_j is smaller than the closed-loop gain:
   Implication for loop interactions: Since the closed-loop gain is the sum of the open-loop gain and the retaliatory effect from the other loops, the loops are definitely interacting; and they do so in such a way that the retaliatory effect from the other loops is in the same direction as the main effect of m_j on y_i. Thus, the loop interactions "assist" m_j in controlling y_i. The extent of the "assistance" from the other loops is indicated by how close λ_{ij} is to 0.5.
   When $\lambda_{ij} = 0.5$, the main effect of m_j on y_i is identical to the retaliatory effect it provokes from other loops but complementary. When $\lambda_{ij} > 0.5$ but < 1, this retaliatory effect from the other interacting loops is lower than the main effect of m_j on y_i. If however $0 < \lambda_{ij} < 0.5$, then the retaliatory effect is more substantial than the main effect.
   Pairing recommendation: If possible, avoid pairing y_i with m_j whenever $\lambda_{ij} < 0.5$.
4. $\lambda_{ij} > 1$, indicating that the open-loop gain between y_i and m_j is larger than the closed-loop gain:
   Implication for loop interactions: The loops interact, and the retaliatory effect from the other loops acts in opposition to the main effect of m_j on y_i, thus reducing the loop gain when the other loops are closed. However, the main effect is still dominant, otherwise λ_{ij} will be negative. For large values of λ_{ij}, the controller gain for loop i will have to be chosen much larger than when all the other loops are open. This could cause loop i to become unstable when the other loops are open.
   Pairing recommendation: The higher the value of λ_{ij}, the greater the opposition m_j experiences from the other control loops in trying to control y_i. Thus, where possible, do not pair m_j with y_i if λ_{ij} takes a very high value.
5. $\lambda_{ij} < 0$, indicating that the open-loop and closed-loop gains between y_i and m_j have opposite signs:
   Implication for loop interactions: The loops interact, and the retaliatory effect from the other loops is not only in opposition to the main effect of m_j on y_i, it is also the more dominant of the two effects. This is a potentially dangerous situation because opening the other loops may cause loop i to become unstable.
   Pairing recommendation: Because the retaliatory effect that m_j provokes from the other loops opposes to its main effect on y_i, and it is dominant, avoid pairing m_j with y_i.

The foregoing discussion leads to the following rule for pairing of the input/output variables:

RGA Rule 1: Pair input and output variables that have positive RGA elements that are closest to 1.0.

9.3.1 NIEDERLINSKI INDEX

Even though pairing Rule 1 is usually sufficient in most cases; it does not consider the stability of the resulting control structure. Therefore, it is necessary to check the stability of control structure obtained through Rule 1. This can be done according to the following theorem originally developed by Niederlinski [2], and later modified by Grosdidier *et al.* [3].

Consider the $n \times n$ multivariable system whose input and output variables have been paired as follows: $y_1-m_1, y_2-m_2, \ldots, y_n-m_n$, resulting in a transfer function model of the form:

$$\hat{y} = \hat{G}\hat{m} \tag{9.17}$$

In this transfer function model, each element of $\hat{G}$, g_{ij}, is rational and open loop stable. Furthermore, assume there are no individual feedback controllers with integral actions, and each controller is stable when all the other $n - 1$ loops are open. When all loops are closed, the system will be unstable for all possible values of controller parameters, i.e. it will be "structurally monatomic unstable", if the Niederlinski index, N, [4] defined in Equation 9.18 is negative.

$$N = \left. \frac{|\hat{G}|}{\prod_{i=1}^{n} g_{ii}} \right|_{SS} = \frac{|\hat{K}|}{\prod_{i=1}^{n} k_{ii}} \tag{9.18}$$

The following important points help us to use the theorem properly:

1. The theorem is both necessary and sufficient only for 2×2 systems. For higher dimensional systems, it provides only sufficient conditions. In other words, if Equation 9.18 holds, the system is definitely unstable, but if it does not hold, the system may or may not be unstable. The stability in this case will depend on the values taken by the controller parameters.
2. The theorem is for systems with rational transfer function elements. This assumption technically excludes time delay systems. Since Equation 9.18 depends only on steady-state gains, which are independent of time delay, the results of this theorem provide useful information about time delay

systems as well. However, the analysis is no longer rigorous. Thus, Equation 9.18 should be applied with caution when time delays are involved.

This leads to the second rule for input/output pairing:

RGA Rule 2: Any loop pairing is unacceptable if it leads to a control system configuration for which the Niederlinski Index is negative.

The strategy for using the RGA in loop pairing may be summarised as follows:

1. Obtain the steady-state gain matrix, $\hat{K}$, and from this obtain the RGA. Find the determinant of $\hat{K}$ and the product of the elements on its main diagonal.
2. Use Rule 1 to obtain tentative loop pairings.
3. Use Niederlinski's Index to verify the stability of the control structure obtained in previous step. If the pairing is unacceptable select another. After selecting a potential candidate control scheme, dynamic simulation should then be used to verify its suitability before adoption.

9.4 Controller tuning for multiloop systems

The main obstacle to proper controller tuning is presented by the interactions that exist between the control loops of a multi-loop system, which change the effective gain of each loop. Thus, the controllers in a multi-loop system cannot be tuned individually, i.e. when other loops are open. In practice, the controllers of a multi-loop system are tuned through the following procedure:

1. With the other loops on manual control, each control loop is tuned independently until satisfactory closed-loop performance is obtained.
2. All the controllers are restored to automatic mode and the tuning parameters are readjusted until the overall closed-loop performance is satisfactory for all loops.

When the interactions between the control loops are not too significant, this procedure can be quite useful. However, for systems with significant interactions, the readjustment of the tuning parameters in Step 2 can be difficult and time-consuming. It is possible to reduce the amount of trial and error that goes into such a procedure by noting that, in almost all cases, the controllers will need to be made more conservative when all the loops are closed, i.e. the controller gains reduced, and the integral times increased. Dynamic simulation can drastically reduce and simplify the above controller tuning procedure.

9.5 RGA example: conventional distillation column

In this section, RGA analysis is used to find the appropriate pairing for a conventional distillation column. There are typically two control schemes for distillation columns: single and dual composition control. The single composition control scheme maintains the composition of one of the products at a desired value, whereas in dual control the compositions of both products are regulated.

If column pressure is being controlled (using coolant flow rate in the condenser), then the following variables can be used as manipulating variables:

1. Reboiler duty (Q_R)
2. Reflux flow (L)
3. Distillate flow (D)
4. Bottom product flow (B)

The reasons why feed flow rate and reflux ratio are not considered as manipulating variables are as follows:

- The feed stream of the column is usually downstream of other units. Thus, its characteristics are usually set based on the operating condition of upstream units.
- Using reflux ratio as one of the manipulating variables results in a shock to the column whenever the distillate flow rate changes.

The variables usually considered as the process outputs for a distillation column are liquid levels at the base of the column and reflux drum and product compositions in dual composition control. Since there are four inputs that can be used to control four outputs, there are four different combinations. These combinations are shown in Table 9.1.

A preliminary screening of these 24 alternatives based on the dynamic response of the manipulated variable to the measured variable results in the three alternatives; 4, 10 and 18, which are shown in bold face in Table 9.1. The reasons for discarding other pairings are as follows:

- Schemes 1, 3, 5, 7, 9, 11, 13, 15, 19, 20, 23 and 24 are discarded since they involve control of base level by reflux flow or distillate flow.
- Schemes 6, 8, 14 and 19 are discarded since they involve manipulating flow rate of bottom product or reboiler heat to control the liquid level in the reflux drum.
- Schemes 21 and 22 are discarded since they do not regulate the material balance.

Table 9.1 Pairings in dual composition control

Case	Reflux drum	Column base	Top comp.	Bottom comp.
1	D	L	B	Q_R
2	D	Q_R	B	L
3	L	D	B	Q_R
4	***L***	***B***	***D***	***Q_R***
5	B	L	D	Q_R
6	B	Q_R	D	L
7	Q_R	D	B	L
8	Q_R	B	D	L
9	D	L	Q_R	B
10	***D***	***Q_R***	***L***	***B***
11	L	D	Q_R	B
12	L	B	Q_R	D
13	B	L	Q_R	D
14	B	Q_R	L	D
15	Q_R	D	L	B
16	Q_R	B	L	D
17	D	B	Q_R	L
18	***D***	***B***	***L***	***Q_R***
19	B	D	L	Q_R
20	B	D	Q_R	L
21	L	Q_R	D	B
22	L	Q_R	B	D
23	Q_R	L	B	D
24	Q_R	L	D	B

- Schemes 2, 12 and 17 are discarded since each involves the control of one (or both) product composition(s) at the end of the column using a manipulated variable at the other end of the column.

9.5.1 BASE CASE STEADY-STATE SOLUTION

The best pairing among these three alternatives, 4, 10, and 18, can be found through RGA analysis for a water-ethanol distillation column. Consider a conventional distillation column used in the separation of a water-ethanol mixture. A process simulation package can be used to determine the necessary gains from a steady-state solution for RGA analysis and the Niederlinski Index to determine the appropriate pairings for dual composition control. For this example we have used HYSYS. Process™ [4]. The condenser and reboiler levels will be assumed to be under perfect control. For this type of system we will use the NRTL activity model with the ideal gas vapour model. The column is fed by the stream shown in Table 9.2.

Table 9.2 Characteristics of the column feed

Conditions and composition	
Temperature, °C	20.0
Pressure, kPa	101.3
Comp Molar Flow of Water, kmol/h	60.00
Comp Molar Flow of Ethanol, kmol/h	40.00

The distillation column has 20 stages and a total condenser. A steady-state solution for the distillation column can be found using the information in Table 9.3.

Table 9.3 Distillation column data

Column characteristics		Column pressure	
No. of stages	20	Condenser pressure, kPa	95
Feed stage	10	Condenser ΔP	0
Condenser type	Total	Reboiler pressure, kPa	105

Table 9.4 Distillation column specifications for base case steady-state

Specification	Value
Reflux ratio	2.0
Distillate rate, kmol/h	30

The steady-state solutions for the column yield the following results:

Table 9.5 Base case steady-state solution

Mole fraction of ethanol in distillate	0.8165
Mole fraction of ethanol in bottoms	0.2214
Reboiler duty, kJ/h	4.2×10^6

9.5.2 RGA CALCULATION

For this exercise, the steady-state values for the compositions will be considered to be the desired set points for the controllers. The best control pairings must be determined for maintaining the distillate and bottoms product compositions. The RGA will be calculated for each pairing of the three possible pairings, Cases 4, 10, and 18 from Table 9.1.

Pairing comparison for cases 4, 10, and 18

At this point, the steady-state gain between the distillate flow rate and the distillate composition, k_{11}, will be calculated using the steady-state distillation column model. Perform a step input in the distillate flow rate from 30 to 40 kmol/h. Change one of the column specifications from reflux ratio to reboiler duty, specifying a reboiler duty equal to the base case steady-state solution of 4.2×10^6 kJ/h. The new specifications for the column should be the same as those given in Table 9.6.

Table 9.6 Specifications for case 4 open loop

Specification	Value
Reboiler duty, kJ/h	4.2×10^6
Distillate rate, kmol/h	30

Run the column and determine the new mole fractions for ethanol in the distillate and bottoms. The results should be very close to those shown in Table 9.7.

Table 9.7 Steady-state solution for case 4 open loop

Mole fraction of ethanol in distillate	0.7890
Mole fraction of ethanol in bottoms	0.1407
Reboiler duty, kJ/h	4.2×10^6

The open loop gain is then calculated as follows:

$$k_{11} = \frac{\Delta x_D}{\Delta D} = \left(\frac{0.7890 - 0.8165}{40 - 30}\right) = -2.75 \times 10^{-3} \tag{9.19}$$

The closed loop gain, k^*_{11} can also be calculated from the steady-state solution. One can close the bottoms composition control loop by making the desired bottoms composition a steady-state specification. The closed loop specifications are shown in Table 9.8.

Table 9.8 Specifications for case 4 closed loop

Specification	Value
Mole fraction of ethanol in bottoms	0.2214
Distillate rate, kmol/h	30

Run the column and determine the new mole fractions for ethanol in the distillate and bottoms. The results should be very close to those shown in Table 9.9.

Table 9.9 Steady-state solution for case 4 closed loop

Mole fraction of ethanol in distillate	0.6680
Mole fraction of ethanol in bottoms	0.2214
Reboiler duty, kJ/h	2.6×10^6

Now, the closed loop gain is calculated as follows:

$$k_{11}{}^* = \frac{\Delta x_D}{\Delta D} = \left(\frac{0.6680 - 0.8165}{40 - 30}\right) = -1.48 \times 10^{-2} \tag{9.20}$$

At this point, the RGA matrix can be calculated because this is a 2×2 system.

$$\lambda_{11} = \frac{k_{11}}{k_{11}{}^*} = \left(\frac{-2.75 \times 10^{-3}}{-1.48 \times 10^{-2}}\right) = 0.185 \tag{9.21}$$

The RGA matrix is then:

$$\begin{bmatrix} \lambda_{11} & 1-\lambda_{11} \\ 1-\lambda_{11} & \lambda_{11} \end{bmatrix} = \begin{bmatrix} 0.185 & 0.815 \\ 0.815 & 0.185 \end{bmatrix} \tag{9.22}$$

The step changes to calculate λ_{11} for Cases 10 and 18 are shown in Table 9.10. The resulting RGA matrix for Case 10 is:

$$\begin{bmatrix} \lambda_{11} & 1-\lambda_{11} \\ 1-\lambda_{11} & \lambda_{11} \end{bmatrix} = \begin{bmatrix} 0.985 & 0.015 \\ 0.015 & 0.985 \end{bmatrix}$$

Table 9.10 Open and closed loop results with the corresponding relative gain

Case	Steady state specifications	New value	New distillate X_{EtOH}	New distillate x_{EtOH}	Steady state gain	Relative gain λ_{11}
	L = 60	L = 70				
10 Open loop	B = 70	L = 70	0.8239	0.2180	$k_{11} = 7.4x10^{-4}$	0.9487
Closed loop	B x_{EtOH} = 0.2214	L = 70	0.8243	0.2214	$k^*_{11} = 7.8x10^{-4}$	
	L = 60	L = 70				
18 Open loop	Reboiler duty = 4.2×10^6		0.8388	0.2880	$k_{11} = 2.23 \times 10^{-3}$	2.8590
Closed loop	B x_{EtOH} = 0.2214	L = 70	0.8243	0.2214	$k^*_{11} = 7.8 \times 10^{-4}$	

and for Case 18 is:

$$\begin{bmatrix} \lambda_{11} & 1-\lambda_{11} \\ 1-\lambda_{11} & \lambda_{11} \end{bmatrix} = \begin{bmatrix} 2.859 & -1.859 \\ -1.859 & 2.859 \end{bmatrix}$$

Using the three RGA matrices calculated from the steady-state model of the distillation column, it can be concluded that Case 10 would be the best pairing of measured variables to manipulated variables. The RGA matrix associated with Case 10 has elements that approach unity, indicating very little interaction.

The Niederlinski Index can be calculated from the full steady-state gain matrix. Using the steady-state model of the distillation column, the remaining elements for the steady-state gain matrix can be calculated for Case 10. The resulting matrix is:

$$K = \begin{bmatrix} 7.4\times 1^{-4} & -3.4\times 10^{-4} \\ 1.7\times 10^{-3} & 7.0\times 10^{-3} \end{bmatrix}$$

The Niederlinski Index can then be calculated from:

$$N = \frac{|K|}{\prod_{i=1}^{n} k_{ii}}$$

$$= \frac{|(7.4\times 10^{-4})(7.0\times 10^{-3}) - (-3.4\times 10^{-4})(1.7\times 10^{-3})|}{(7.4\times 10^{-4})(7.0\times 10^{-3})} = 1.11$$

Because the Niederlinski Index is not < 0, the control pairing can not be ruled out because it is definitely unstable. In this 2×2 case, the Niederlinski Index indicates that the system is stable. However, as mentioned earlier, a positive index value for higher order systems would indicate only that the system is not definitely unstable. The positive index value does not indicate stability for higher order systems: the system may or may not be unstable. Therefore one should also test the selected control scheme extensively via dynamic simulation before adoption.

9.6 References

1. Bristol, E.H., "On a New Measure of Interactions for Multivariable Process Control". *IEEE Trans. Auto. Con.*, 1966, **AC-11**: 133.
2. Niederlinski, A., "A Heuristic Approach to the Design of Linear Multivariable Interacting Control Systems". *Automatica*, 1971, **7**: 691.
3. Grosidideier, P., Morari, M., and Holt, R.B., "Closed Loop Properties from Steady State Information". *I&EC Fund.*, 1985, **24**: 221.
4. HYSYS. Process™, Version 1.5.2. Hyprotech Ltd., AEA Technology Engineering Software, Calgary, Alberta, Canada, 1998.

10 PLANT-WIDE MODELLING AND CONTROL

The fundamental questions in plant-wide control are whether the feed rates can simply be set for a process and left unattended and whether the process is meeting the desired purity and quality specifications [1,2]. What happens when common disturbances occur such as feed composition changes, production rate changes, product mix or purity specification changes, ambient temperature changes, or when measurement sensors either fail or are in error? This chapter covers some of the most common problem areas encountered when designing a plant-wide control scheme.

10.1 Short term versus long term control focus

When applying a plant-wide control scheme, it is important to be aware of the propagation of variation and the transformation that each control system performs. Management of that variation is the key to good plant-wide operation and control. A healthy variation management strategy should have both a short term and a long term focus. The short term focus is to use control strategies to transform the variation to less harmful locations in the plant. The long term focus should concentrate on improvements which reduce or eliminate either the variations or the problems caused by variations.

To better illustrate the idea of short and long term control focus, consider an acid recovery plant [3,4]. An example of a short term focused control scheme for the plant is shown in Figure 10.1. In this system acid feeds of varying concentrations are pumped to four storage tanks. Tank A contains high concentration acid that varies greatly in concentration. Tank B is fed with slightly less concentrated acid and a feed which varies noisily. Tank C is fed by streams which are similar to Tank B but vary to a lesser degree. Tank D is fed by a stream that has a much gentler but increasing variance.

The acid feed is then sent from the tanks to a separation system. This separation system removes water and other impurities to produce the final anhydrous grade product.

The control scheme shown in Figure 10.1 attempts to minimise variance but unfortunately passes along much of the disturbance to the extraction

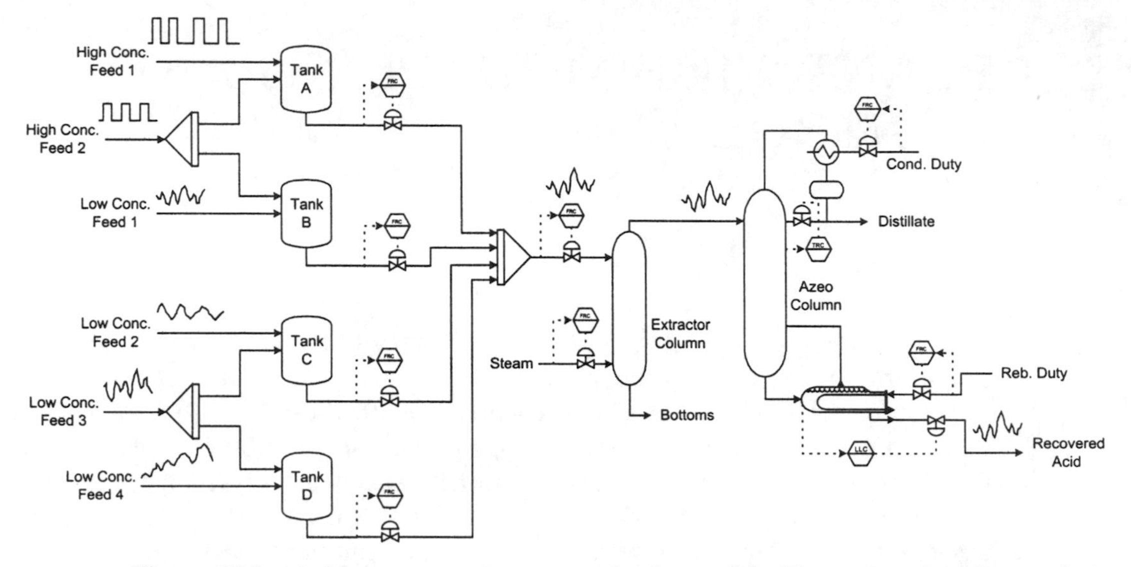

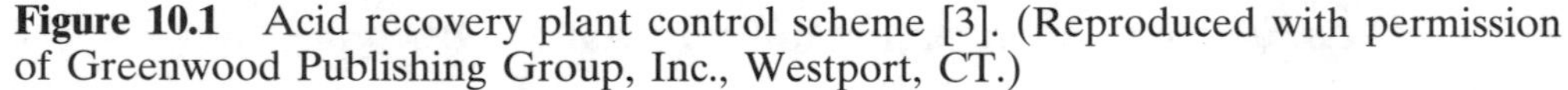
Figure 10.1 Acid recovery plant control scheme [3]. (Reproduced with permission of Greenwood Publishing Group, Inc., Westport, CT.)

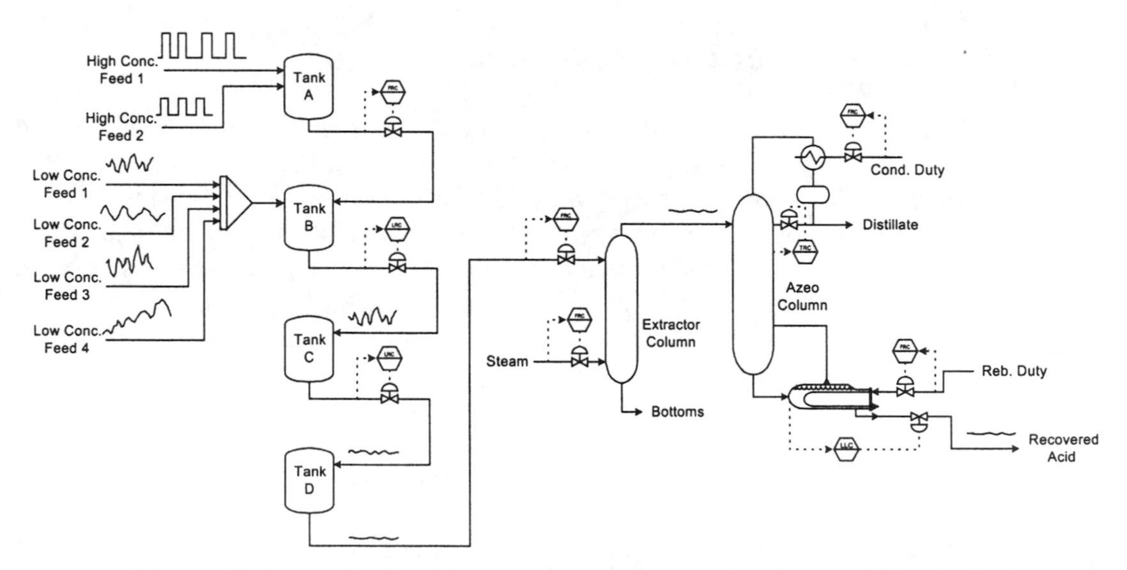

Figure 10.2 Revised acid recovery plant control scheme [3]. (Reproduced with permission of Greenwood Publishing Group, Inc., Westport, CT.)

column and the azeotropic column. Consequently, the desired product, which is the bottoms of the azeotropic column, varies significantly in quality.

The long term focus strategy for the plant involves redesigning the feed inventory system to filter out the high frequency variations. Figure 10.2 shows the same acid recovery system with a different configuration that helps to achieve this long term focus. All the high concentration feeds are collected in Tank A, while Tank B gathers the low concentration feeds. The feed from Tank A is then sent to Tank B at a constant rate, thus eliminating some of the problems in variation seen in the short term focus scheme.

The other major change to the control scheme is in the control systems used for the feed inventory. Level controllers are used on Tanks B and C. These level controllers use the capacitance of the tanks to attenuate the fluctuations in feed flow. The feeds to the extraction column and azeotropic column are considerably dampened, resulting in a much more consistent end product.

10.2 Cascaded units

The dynamics and control of continuous process units that operate as a cascade of units, either in parallel or in series, have been studied extensively for many years [5,6,7]. A wealth of knowledge is available to help design effective control systems for a large number of unit operations when these units are run independently [6,8]. This knowledge can be directly applied to the plant-wide control problem if a number of process units are linked together as a sequence of units. Each downstream unit simply sees the disturbances coming from its upstream neighbour.

The design procedure was proposed almost three decades ago [5] and has since been widely used in industry. The first step of the procedure is to lay out a logical and consistent "material balance" control structure that handles the inventory controls, i.e. levels and pressures. This hydraulic structure provides gradual and smooth flow rate changes from unit to unit. Thus, flow rate disturbances are filtered so that they are attenuated and not amplified as they work their way down through the cascade of units. Slow acting, proportional only level controllers provide the simplest and most effective way to achieve this flow smoothing.

Then product quality control loops are closed on each of the individual units. These loops typically use fast proportional-integral controllers to hold product streams as close as possible to specification values. Since these loops are considerably faster than the slow inventory loops, interaction between the two is generally not a problem. Also, since the manipulated variables used to hold product qualities are often streams that are internal to each individual unit, changes in these manipulated variables have little effect on the downstream process. The manipulated variables frequently are utility streams that are provided by the plant utility system, i.e. cooling water, steam, refrigerant, etc. Thus, the boiler house will be disturbed, but the other process units in the plant will not see disturbances coming from upstream process units. Of course, this is only true when plant utilities systems have effective control systems that can respond quickly to the many disturbances that they see coming in from units all over the plant.

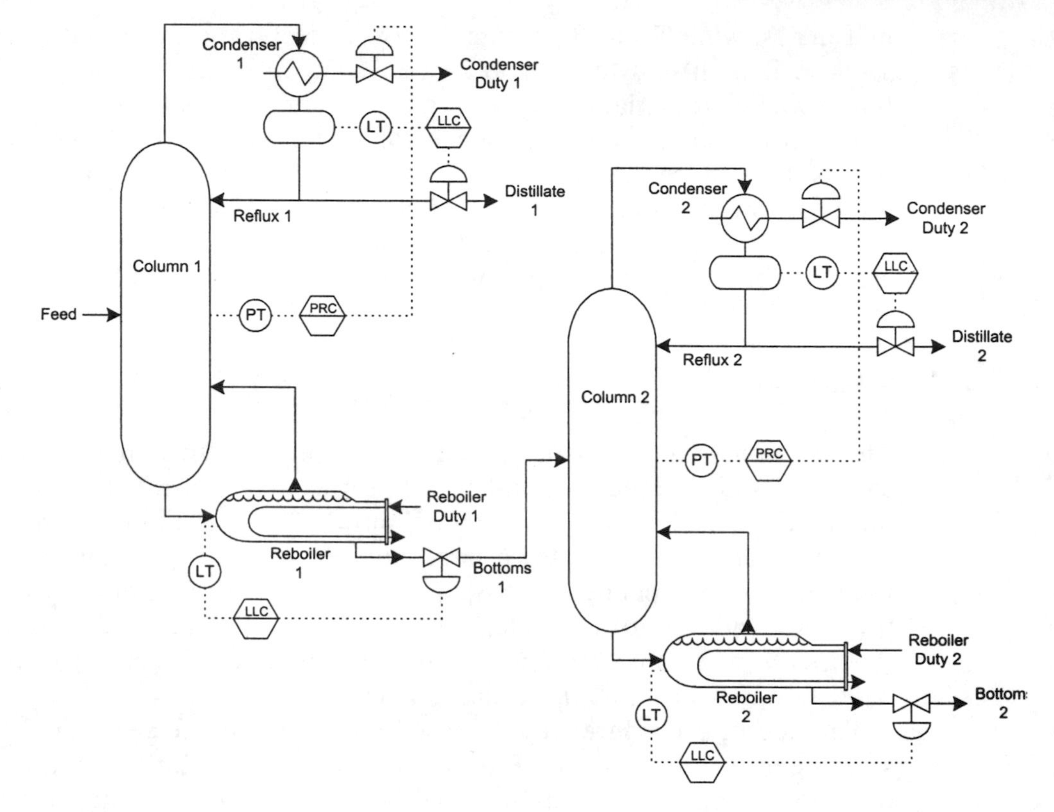

Figure 10.3 Cascade system with two distillation columns in series.

As an example of a cascade system, consider a sequence of distillation columns in which the bottoms of the first column feeds the downstream column. This is shown in Figure 10.3.

Figure 10.3 shows the column with the inventory loops closed. Now that the inventory loops have been closed the product quality loops can be chosen. Each column has two degrees of freedom remaining, reflux and vapour boilup, so some combination of two variables can be controlled in each column, i.e. two compositions, two temperatures or one temperature and one flow. Vapour boilup changes require changes in steam flow to the reboiler and also in cooling water flow indirectly through the pressure controller. Both vapour boilup and reflux changes affect the two liquid levels and therefore the distillate and bottoms flow rates, but proportional level controllers usually provide effective filtering of these disturbances. Based on these guidelines and the information provided in Chapter 8, the product quality loops can be closed with the specifics of the loops depending on the control objectives.

Since the propagation of the disturbances in such a system is sequential down the flow path, the use of feedforward control in each unit can also help to improve product quality control [7].

It should be noted that the inventory controls can be in the direction of the flow, i.e. products come off due to level control, or in the opposite direction, i.e. feed is brought in on level control. The same design procedure applies.

10.3 Recycle streams

If recycle streams exist in the plant, the procedure for designing an effective plant-wide control scheme becomes more complicated. Processes with recycle streams are quite common, but their dynamics are poorly understood at present.

The typical approach in the past for plants with recycle streams has been to install large surge tanks. This isolates sequences of units and permits the use of conventional cascade process design procedures. However, this practice can be very expensive in terms of capital costs and working capital investment. In addition, and increasingly more important, the large inventories of chemicals can greatly increase safety and environmental hazards if dangerous or environmentally unfriendly chemicals are involved.

To demonstrate the principles of plant-wide control for a recycle system, consider the ethylene glycol plant shown in Figure 10.4. Equivalent amounts

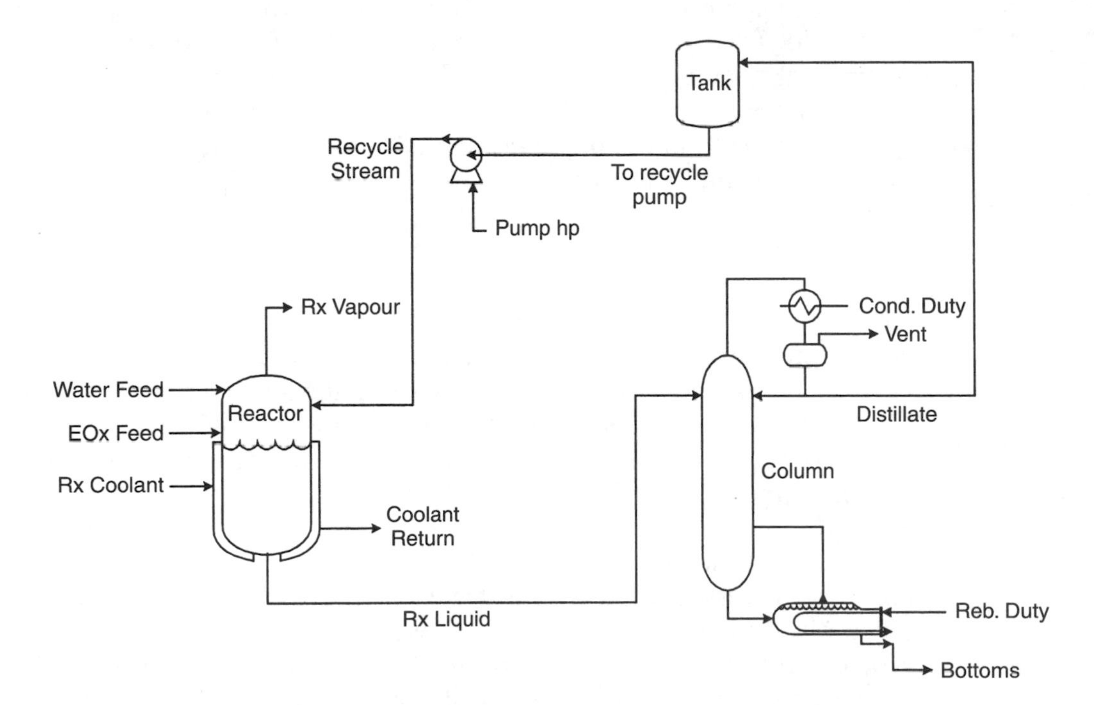

Figure 10.4 Ethylene glycol plant.

of water and ethylene oxide are fed to a reactor, as dictated by the reaction stoichiometry, to produce ethylene glycol. The liquid product stream is sent to a distillation column to separate unreacted water and ethylene oxide from the ethylene glycol. The unreacted feed is sent back through a recycle loop to the reactor.

The reactor control problem is a problem of heat management. The reactor is modelled as a continuously stirred tank reactor (CSTR) with a cooling jacket. As such, the reactor temperature can be measured and controlled by adjusting the rate of cooling flow through the jacket until a desired reactor operating temperature is reached.

The problem of distillation control was addressed in Chapter 8. The issue now is how to control the reactor liquid level, the recycle tank liquid level, the recycle flow rate, the ethylene oxide feed flow rate and the water feed flow rate.

The biggest danger in the operation of the whole plant is the "Snowball Effect" in the recycle [9]. This effect occurs when material accumulates within the recycle loops and cannot be removed. As a result, the plant shuts down. A comparison of two plant-wide control schemes will be made to demonstrate their respective advantages and disadvantages.

The first control scheme involves controlling the level of the reactor by manipulating the flow rate of the reactor effluent. The flow rates of the reactor feed streams are controlled through a ratio controller to meet the required feed ratio. Finally, manipulating the flow rate of the stream to the recycle pump controls the level of the recycle tank. This control scheme is shown in Figure 10.5.

To test the weakness or robustness of this first control scheme, a measurement error is introduced to the flow controller manipulating the water feed flow rate. The water feed flow controller receives a signal that is too low. It adjusts the flow to meet the current set point, when in fact it is supplying excess water. The ethylene oxide flow controller moves to match the water feed flow rate through a cascaded ratio controller. The ratio is 1:1 to supply equal amounts of water and ethylene oxide to the reactor.

When excess water is added to the system, the level of the reactor increases. The level controller increases the liquid flow rate leaving the reactor to compensate. Assuming that the distillation column separates the ternary mixture almost perfectly, the unreacted ethylene oxide and the excess water are driven overhead into the distillate stream. This stream feeds the recycle tank and thus increases the level. The flow rate of the stream to the recycle pump is increased to compensate. This increased flow is recycled to the reactor and increases the level. The cycle begins again which results in the accumulation of water in the system. The recycle stream "snowballs".

To better illustrate this concept of snowballing, a strip chart was recorded for the appropriate variables in the plant using dynamic simulation (see

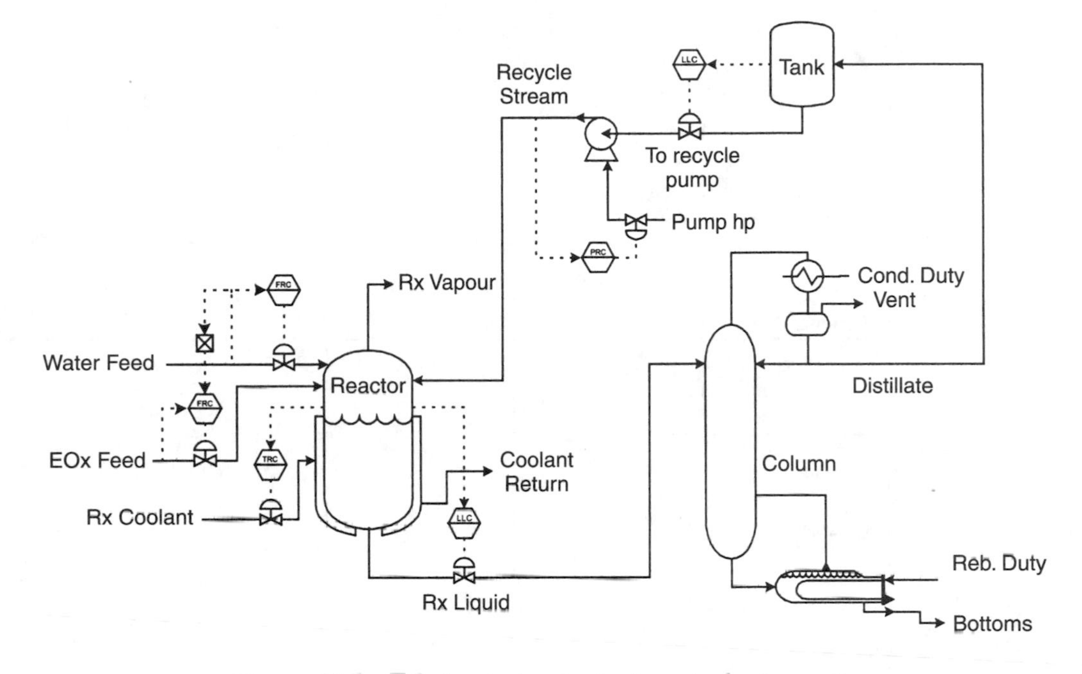

Figure 10.5 Ethylene glycol plant control scheme 1.

Figure 10.6). The ethylene glycol plant was set with a recycle tank level of 95% and with a valve size on the stream to the recycle pump which results in almost saturated flow, i.e. the valve is almost fully open. The excess water increases the level in the recycle tank, thus opening the valve on the stream to the recycle pump even further until it saturates. The level in the recycle tank then continues to increase past the 100% point where the tank begins to overflow. The plant must shut down.

If a positive measurement error is supplied, the flow sensor transmits a flow that is too large. Since the ethylene oxide controller is set up so that the set point is in a 1:1 ratio to the water feed flow rate due to the reaction stoichiometry, too much ethylene oxide enters the system. If this is the case, the excess ethylene oxide reacts with the surplus water in the recycle loop, thus consuming the water and producing ethylene glycol. This reduces the material inventory within the plant until there is only ethylene oxide remaining.

While there are a number of different ways to control this plant, it is helpful to keep in mind two fundamental rules of plant-wide control. These are affectionately known as "Luyben's Rules", referring to the original author [10].

Luyben's Rules:

1. Only flow control a feed if it is sure to be fully consumed in the reaction.
2. Always put one stream in the recycle path on flow control.

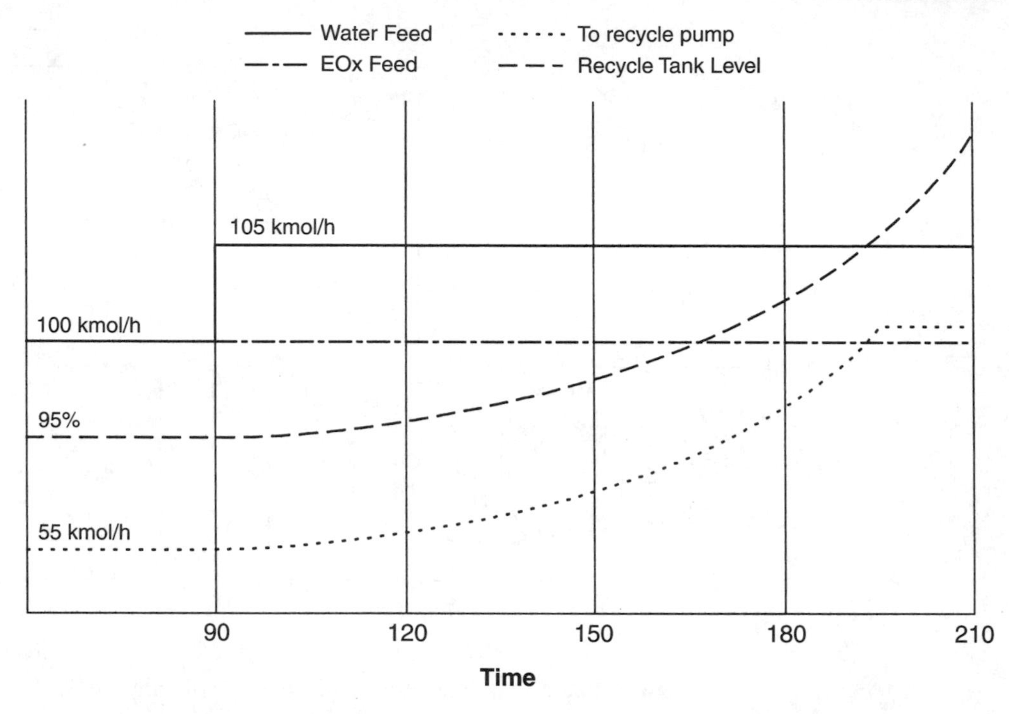

Figure 10.6 Control scheme 1 response to a measurement error.

With these rules in mind, a new control scheme can be proposed, illustrated in Figure 10.7. This time the stream to the recycle pump is under flow control. In order to control the level of the recycle tank, it is necessary to manipulate the water feed rate. However, with the water feed introduced to the reactor, a considerable amount of dead time is unnecessarily introduced to the system. To overcome this dead time, the water feed is introduced directly into the recycle tank and is used to control the liquid percent level.

In order to introduce a similar transmittor error to the one used in the previous control scheme, the recycle tank level controller output is cascaded to provide a set point for a flow controller on the water feed stream. The water feed flow controller uses the measured flow, complete with an error. The ethylene oxide stream is controlled using a composition controller that manipulates the ethylene oxide flow rate to meet a specified composition in the liquid stream leaving the reactor. The liquid level of the reactor will be controlled using the same controller manipulating the flow rate of the liquid leaving the reactor. With these controllers in place, the process flow schematic has been modified. The updated schematic is shown in Figure 10.7.

This control scheme is more robust in the event of a disturbance. The flow rate of the stream to the recycle pump is controlled, preventing any increases in the recycle flow rate. The recycle tank level is controlled by the water

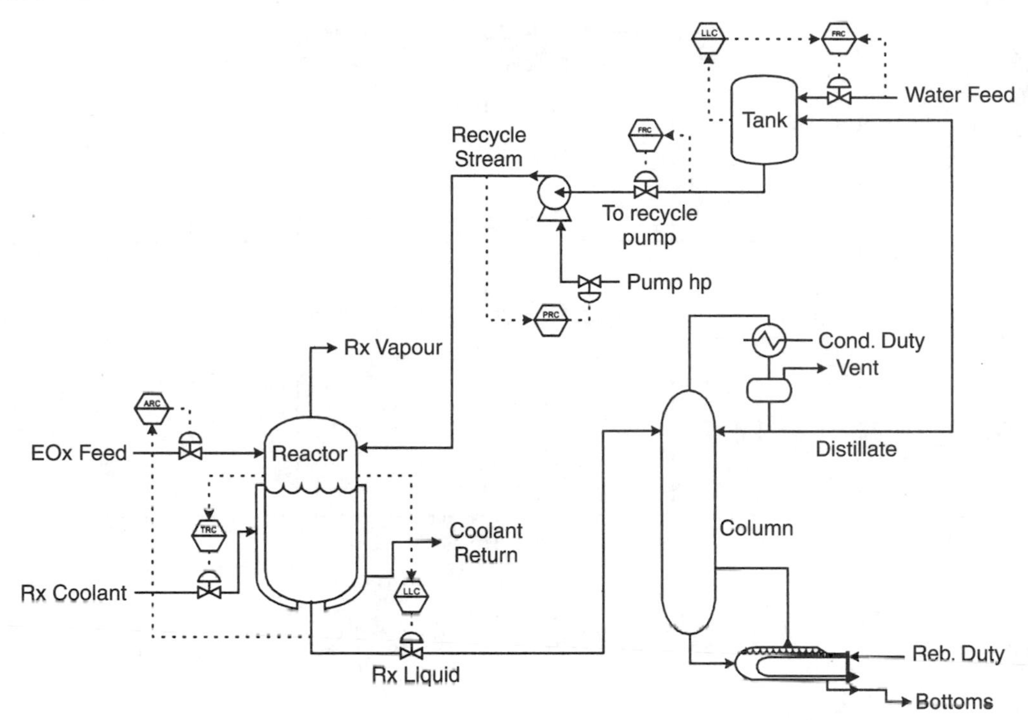

Figure 10.7 Ethylene glycol plant control scheme 2.

feed flow rate. The reactor is still under level control by manipulating the flow rate of the liquid stream leaving the reactor. The ethylene oxide feed flow rate is also manipulated by a composition controller, which measures the exit composition of ethylene oxide from the reactor. This control scheme does not allow for excess ethylene oxide or excess water in the system, and hence this system cannot snowball.

To demonstrate this system's robustness, the same measurement error can be introduced to the water feed flow rate. The following strip chart shows an introduction of −5 kmol/h error into the sensor transmitting a flow measurement to the water feed flow controller.

With an increase in the amount of water fed to the recycle tank, the level controller adjusts the set point of the water feed flow controller and reduces the amount of water being introduced to the plant. The recycle tank level attains a new operating level. Remember, there will always be offset from the set point due to using only proportional control for the level control. More importantly, notice that there is no accumulation within the system.

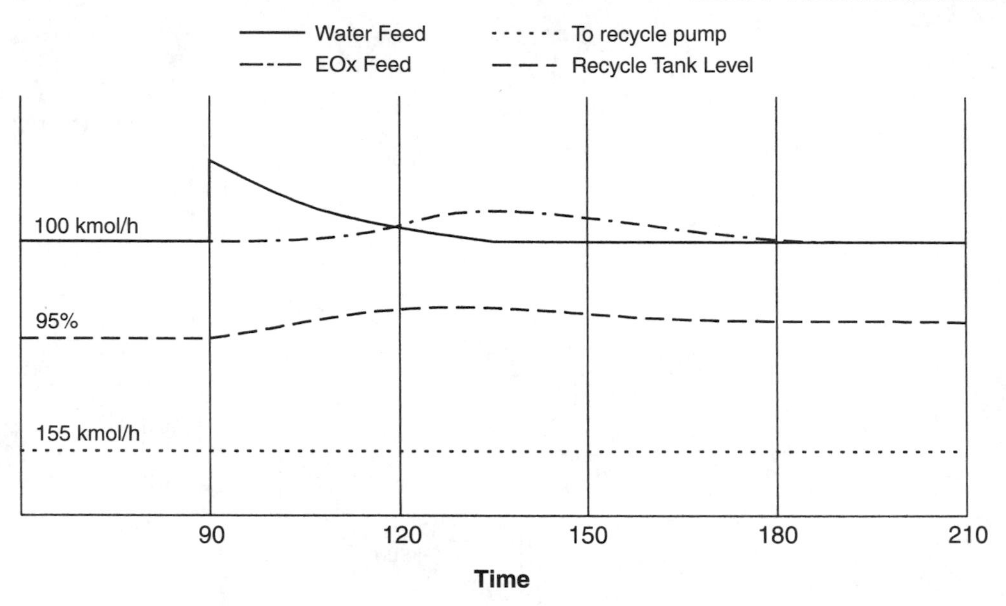

Figure 10.8 Control scheme 2 response to measurement error.

10.4 General considerations for plant-wide control

When considering plant-wide control, a number of questions must be answered:

- What are the primary objectives of the plant?
- Where are the production bottlenecks and constraints?
- Where should the production rate be set?
- Where are the bulk inventories and how should they be controlled?
- Will an additional inventory improve the operation and control?
- Will changes in the process design improve the operation and control?
- Where should recycle streams be placed?
- How are the component inventories controlled in these recycle systems?
- Will small changes in a feed cause a very large change in the recycle rate around the system ("snowball effect")?
- What are the primary sources of variation?
- What can be done to reduce or eliminate variation at its source?
- How does variation propagate through a plant-wide system?
- What can be done to transfer the variation to less harmful locations?
- How much of the plant-wide operation should be automated and how much should be left for the operator?

Plant-wide design, operation and control is a fast developing area for research. As such, it cannot be summarised simply in one paragraph. For a more in-depth discussion of this topic, refer to the series of papers authored by W.L. Luyben, B.D. Tyreus and M.L. Luyben [9,11,14], and *Plant Wide Process Control* by the same group of authors [10].

10.5 References

1. Downs, J.J., "Plant-Wide Control Fundamentals – Analysis of Material Balance Systems". Presentation at Plant-wide Control Course, Lehigh University, May 3–7, 1993.
2. Vogel, E.F., "Plant-Wide Process Control". In *Practical Distillation Control*, Luyben, W.L. (ed.), Van Nostrand Reinhold, New York, 1992, p. 86.
3. Moore, C.F., "A New Role for Engineering Process Control Focused on Improving Quality". In *Competing Globally Through Customer Value: The Management of Strategic Super Systems*, Stahl, M. and Bounds, G. (Eds), Greenwood Publishing Group, Inc., Westport, CT, 1991.
4. Vogel, E.F., "An Industrial Perspective on Dynamic Flowsheet Simulation". Proceedings, CPC IV, Padre Island, Texas, Feb. 17–22, 1991. Published by CAChE, AIChE, New York, 1991, pp. 181–208.
5. Buckley, P.S., *Techniques of Process Control*, John Wiley & Sons, New York, 1964.
6. Shinskey, F.G., *Process Control Systems: Application, Design, and Tuning*, 2nd edn. McGraw-Hill, New York, 1988.
7. Wilson, H.W. and Svrcek, W.Y., "Development of a Column Control Scheme: Case History", *Chem. Eng. Prog.*, 1971, **67** (2): 45.
8. Considine, D.M. (ed.), *Process Industrial Instrument and Controls Handbook*, 4th edn. McGraw-Hill, New York, 1993.
9. Luyben, W.L., "Dynamics and Control of Recycle Streams. 1. Simple Open-Loop and Closed-Loop Systems". *Ind. Eng. Chem. Res.*, 1993, **32**: 466–75.
10. Luyben, W.L., Tyreus, B.D. and Luyben, M.L., *Plant-Wide Process Control*. McGraw-Hill, New York, 1999.
11. Luyben, W.L., "Dynamics and Control of Recycle Streams. 2. Comparison of Alternative Process Designs". *Ind. Eng. Chem. Res.*, 1993, **32**: 476–86.
12. Luyben, W.L., "Dynamics and Control of Recycle Streams. 3. Alternative Process Designs in a Ternary System". *Ind. Eng. Chem. Res.*, 1993, **32**: 1142–53.
13. Tyreus, B.D. and Luyben, W.L., "Dynamics and Control of Recycle Streams. 4. Ternary Systems With One or Two Recycle Streams". *Ind. Eng. Chem. Res.*, 1993, **32**: 1154–62.
14. Luyben, M.L., Tyreus, B.D. and Luyben, W.L., "Plant-Wide Control Design Procedure". *AIChE Journal*, 1997, **43** (12): 3161–74.

APPENDIX 1: PROCESS AND INSTRUMENTATION DIAGRAM SYMBOLS

SYMBOL	DESCRIPTION
	Control valve
	Valve
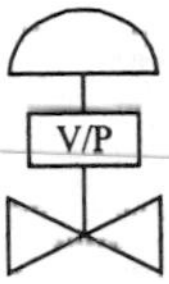 V/P	Control valve with valve positioner
	Check valve
	Pressure relief valve
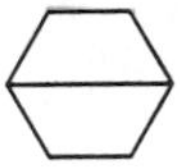	Controller
	Transmitter/Sensor
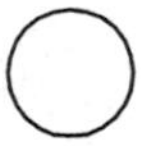	Transmitter/Sensor
 Σ	Controller summer
	Summer or multiplier
 ÷	Divider
 <	Selector

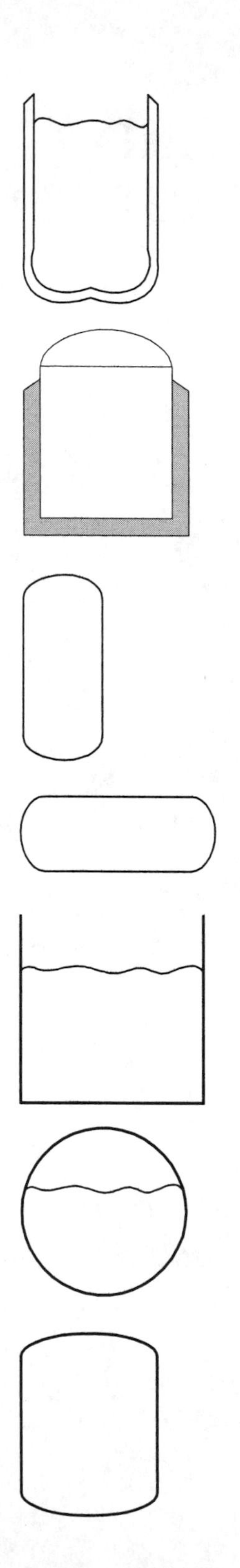

Reactor with cooling jacket

Reactor with cooling jacket

Knock out drum

Reflux drum

Tank

Horizontal tank

Vertical tank

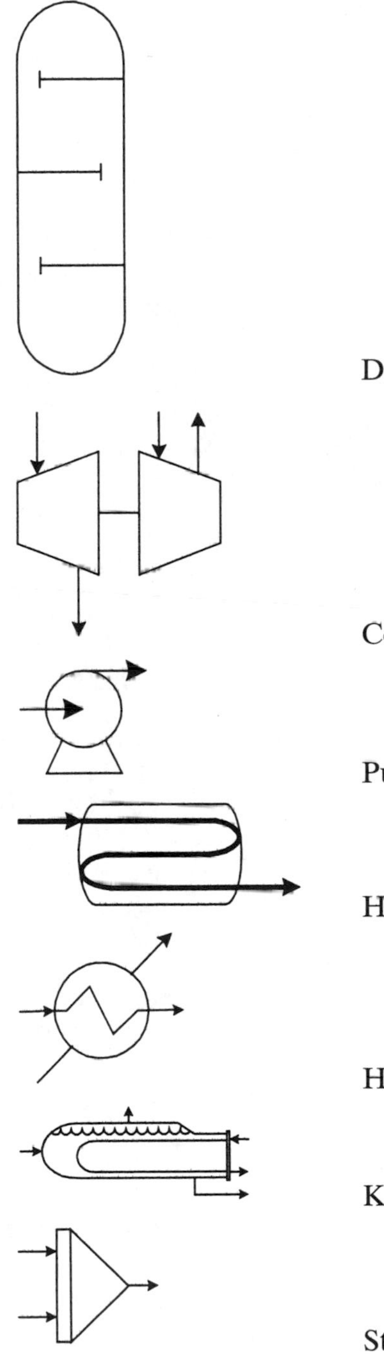

Distillation column tray section

Compressor

Pump

Heat exchanger

Heat exchanger

Kettle reboiler

Stream mixer

APPENDIX 2: GLOSSARY OF TERMS

Amplitude	The difference between the average value of a sinusoidal variation and the maximum or minimum value.
Amplitude ratio	The ratio of the amplitude of a system's response to its forcing function's amplitude when the forcing function is a continuous sinusoid; a form of dynamic gain.
Analog controller	A controller that operates on continuous signals such as voltages, pressures or currents; this is a common older type of controller to be distinguished from a digital controller.
Attenuation	A decrease in the strength of a signal by a system component.
Automatic controller	A device which operates to correct or limit the deviation of a variable from a reference value.
Automation	The use of automatic control devices in a process so that human supervision is minimised.
Capacitance	The amount of energy or material which must be added to a closed system to cause unit change in potential; the partial derivative of content with potential.
Cascade	A series of stages in which the output of one is the input to the next.
Cascade control	Automatic control involving "cascading" of controllers; that is, having one controller's output as the input to the next controller instead of directly manipulating a process variable.
Comparator	The portion of the control element which determines the difference between the set point and the measured feedback variable.
Compensator	A component added to a system to improve characteristics of its response.
Control element	The portion of the control system which relates the error between the desired value and the manipulated variable.

Controlled variable	That quantity or condition of the controlled system which is controlled.
Controller	A device that receives the set point and feedback signals, computes the difference and uses adjustable parameters to produce an output signal to eliminate the difference between the set point and feedback signals.
Critical gain	A value of system gain beyond which closed loop operation is unstable.
Cycling	Periodic changes in the controlled variable. Also known as *oscillation*.
Damping ratio	Also known as *damping coefficient* and *damping factor* (ζ) which characterises the nature of damping of the transient response.
Dead band	The largest range of values of the input variable to which a component does not respond.
Dead time	An interval of time between an input to a component and the beginning of a response to the input.
Derivative action	A controller mode in which there is a continuous linear relationship between the derivative of the error signal and the controller output signal. Also known as *rate action*.
Derivative time	The time difference by which the output of a proportional plus derivative controller leads the controller input when the input changes linearly with time.
Desired value	The value of the controlled variable which is desired. Also known as *set point*.
Deviation	The difference at any instant between the value of the controlled variable and the set point.
Digital controller	A controller which operates on signals, usually electronic, which have only discrete numerical values.
Distance-velocity lag	The effect resulting from a signal being transmitted over an appreciable distance at a finite velocity; a type of dead time. Also known as *transportation lag* and *time delay*.
Disturbance	An input signal other than the set point which directly affects the controlled variable. Also known as *load*.
Error	In measurement, the difference between the value found and the true value; in control, the set point minus the measured value of the feedback variable.
Feedback	The signal to the controller representing the condi-

	tion of the controlled variable; a control system in which corrective action is based on such signals.
Feedback elements	The portion of the control loop which establishes the primary feedback variable in terms of the controlled variable.
Feedforward	A control mechanism in which corrective action is based on measurements of inputs to the process; a form of *predictive control.*
Final control element	The controlling means or element which directly changes the manipulated variable.
First order system	A system whose dynamic behaviour is described by a first order linear differential equation.
Frequency response	The amplitude ratio and the difference in phase of a system or element's output with respect to a sinusoidal input. The frequency of the input and the system or element's differential equation determine the frequency response.
Gain	The proportionality constant in a transfer function.
Impulse	A sharp increase or decrease in a variable immediately followed by a return to its original value.
Input	A variable that is dependent only on conditions outside the system.
Input elements	The portion of the control system which provides a reference input to a comparator in response to the set point.
Integral action	A controller mode in which there is a continuous linear relation between the integral of the error signal and the output signal of the controller.
Integral time	For a step input this is the time required for the output of a proportional plus integral controller to change an amount equal to the proportional response alone.
Interacting	Two or more consecutive stages whose combined transfer function is not the product of the transfer functions of the preceding stages if they appeared alone.
Lag	The retardation of one condition with respect to another.
Linear	A relationship showing output proportional to input; a system whose behaviour is adequately described by such equations; a system that follows the principle of superposition.

Linearise — To substitute an approximate linear function for a nonlinear function.

Load or load variable — Any outside input to a control system except the set point. Also known as *disturbance*.

Load change — A change in input conditions such that a change in manipulated variable is necessary to maintain the controlled variable at the set point.

Loop — A series of stages forming a closed path.

Lumping — An assumption that the effects of two or more aspects of a system can be considered together as a single quantity; an assumption that a parameter distributed over space may be considered at a single point in space.

Manipulated variable — A process variable that is changed by the controller to eliminate error.

Mode — The classification of a controller by the manner in which the manipulated variable responds to the error signal.

Model — A conceptual approximation of a physical system that is usually mathematical in nature.

Natural frequency — The frequency of oscillation that a system would have if there were no damping.

Noise — Accidental and unwanted fluctuations in a variable that tend to conceal it.

Nonlinear — An equation that contains a term not conforming to linearity; a system whose behaviour is not described by linear equations.

Offset — The steady-state deviation in the controlled variable caused by a change in the load variable.

On-off control — A system of regulation in which the manipulated variable has only two possible values, high and low, maximum and minimum, or on and off. Also known as *two-position control*.

Output — The variable that is chosen to describe the condition of a system; the dependent variable in the dynamic equation.

Oscillation — See *cycling*.

Overdamped — Said of a system of second or higher order whose transient response has no tendency to oscillate or overshoot.

Overshoot — In a step response, the difference between the final steady-state value and the value of the first maximum (or minimum if the response is downward);

	often expressed as a fraction; only defined for underdamped systems.
Parameter	A constant coefficient in an equation that is determined by the properties of the system.
Period	The amount of time between consecutively recurring conditions; the reciprocal of frequency.
Predictive control	A control scheme that predicts the effect of a load change and takes corrective action before the controlled variable is affected e.g. *feedforward control.*
Primary element	That portion of the measuring means which first senses a change in the controlled variable. Also known as a *sensor/transmitter.*
Process	The system being controlled.
Proportional action	A controller mode in which there is a continuous linear relation between the value of the error signal and the value of the controller output.
Proportional band	The range of the controlled variable that corresponds to the full range of the final control element.
Proportional sensitivity	A proportional action; the steady-state ratio of the controller output to the error signal.
Rangeability	The ratio of maximum flow to minimum controllable flow in a final control element.
Rate action	See *derivative action.*
Reset rate	The inverse of integral time; usually expressed as repeats per unit time.
Resistance	The potential required to produce change; the partial derivative of driving force with flow rate.
Response	A system's output due to a change in its input.
Response time	The time required for an output to increase from one specified percentage of its final value to another, based on a step input.
Self regulation	The inherent characteristic of a system that produces a steady state without the aid of automatic control.
Sensor/transmitter	See *primary element.*
Set point	See *desired value.*
Settling time	The time required for the absolute value of the difference between the output of a component or system and its final value to become and remain less than a specified amount.
Signal	Information in transmission.
Stable	A system whose response to a bounded input is also bounded.

Steady-state	The condition when all properties are constant with time, the transient response having died out.
Steady-state error	A control error at steady-state.
Time constant	The time required for the output of a first order system to change 63.2% of the amount of total response to a step forcing function.
Transducer	Any device that transmits, amplifies or changes a signal.
Transfer function	A mathematical relationship that describes the ratio of an output of a system to the input to the system.
Transient response	That part of a system's response that approaches zero as time proceeds.
Transportation lag	See *distance/velocity lag.*
Two-position control	See *on-off control.*
Undamped	Oscillatory transient response of constant amplitude.
Underdamped	Oscillatory transient response of diminishing amplitude.

References

1. Murrill, P.W., *Automatic Control of Processes*, International Textbook Company, Scranton, PA, 1967, pp. 451–9.

WORKSHOPS

"Do, or do not, there is no try."
Yoda

WORKSHOP 1 – LEARNING THROUGH DOING

Course philosophy: Learning through doing
In conjunction with this workshop, you should review Chapter 1.

This course consists of a set of learning modules or workshops, each of which is intended to enhance your understanding of process control and simulation theory and application through hands-on experience using the latest simulation technology and actual laboratory experiments.

Key learning objectives

1. Develop an understanding of the general organisation of the course and what is expected.
2. Understand how to proceed through the course.
3. Develop a working knowledge of the HYSYS Steady-State Dynamic Simulation Package.
4. Understand the fundamentals of steady-state and dynamic process simulation.

Course coverage

This course will deal with fundamental and underlying principles of automatic process control and simulation. The course covers the theory associated with single input/single output (SISO) loops and how these are configured into multiloop schemes to control complex unit operations and entire plants. It will not discuss in detail the hardware or individual measurement techniques such as flow, temperature, pressure, etc., except as the measurement affects the control loop. In the course notes, Chapter 2 provides a simplified summary of control loop hardware.

Prerequisites

No elaborate prerequisites are required; however, an understanding of unit operation modelling is assumed. It is inevitable that modelling involving differential equations will, by necessity, be involved in parts of the theory and workshops. Quite often mathematics is a barrier that prevents a clear understanding of control concepts and implementation of process control theory. It is anticipated that the "Real Time Approach" will remove or at least minimise these barriers.

Study material

The course notes and the HYSYS Process Simulator are the only materials required for the course. The notes are independent, unique, stand-alone, and specifically designed for the tutorial/workshop approach.

The references that are provided at the end of each chapter detail additional selected study and reference reading. The available literature on the subject of process dynamics and control is massive. Additional literature is readily available from instrumentation vendors such as Honeywell, Foxboro, Fisher, etc. This vibrant area of chemical engineering is represented by the Instrument Society of America (ISA), PO Box 12277, Research Triangle Park, NC 27709–2277.

Organization

The course consists of eight workshops and three laboratory sessions. Each workshop is associated with a specific portion of the book that provides the necessary theoretical background. During the workshop a specified assignment using the dynamic process simulator will be completed. The results achieved during each workshop, along with an explanation written in Microsoft Word, should be submitted in the form of a diskette at the end of each tutorial session.

Total course objectives

1. Understand the basic theoretical concepts of feedback and SISO loops.
2. Understand the components of a control loop and how they interact.
3. Understand process control terminology.
4. Have an appreciation of process dynamics.

5. Know how to develop the fundamental models for first order plus dead time processes.
6. Know how to tune controllers.
7. Know how and where to implement such techniques as cascade, feed-forward, ratio, dead time, and multiloop control.
8. Appreciate the use of process simulation in the development and validation of control strategies.
9. Develop an understanding of the unit operation control schemes.
10. Understand what is meant by plant-wide control and be able to implement a plant-wide control strategy.
11. Familiarise yourself with the appropriate simulation software.

WORKSHOP 2 – FEEDBACK CONTROL LOOP CONCEPTS

"What we have to learn to do, we learn by doing."
Aristotle

Introduction

Prior to attempting this workshop, you should review Chapter 2.

Process systems respond to various disturbances (or stimuli) in many different ways. However, certain types of responses are characteristic of specific types of processes. The characteristic response of a process can be described as its personality. Process control engineers have developed a range of terms and concepts to describe different process personalities and they use this knowledge to develop effective control systems.

Two of the most common personalities are those for first and second order systems. First order systems may also be called first order processes or first order lags and can be mathematically modelled through the use of a first order differential equation. Figure 2.1 shows the typical step response of a first order process. The time constant, τ, was discussed in Chapter 3 and is related to the speed of the process response; the slower the process the larger the value of τ.

Unlike first order processes, second order processes can have several different types of responses. Second order processes are more complex than first order, and hence the mathematical models used to describe these processes are also more complex. There are three types of second order systems to consider. The key parameter in determining the type of system is the damping coefficient, ξ. When $\xi < 1$, the system is underdamped and has an oscillatory response as shown in Figure 2.2. An underdamped system overshoots the final value and the degree of overshoot is dependent upon the value of ξ. The smaller the value, the greater the overshoot. If $\xi = 1$, the system is deemed critically damped and has no oscillation. A critically damped system provides the fastest approach to the final value without the overshoot that is found in an underdamped system. Finally, if $\xi > 1$ the system is overdamped. An overdamped system is similar to a critically damped system in

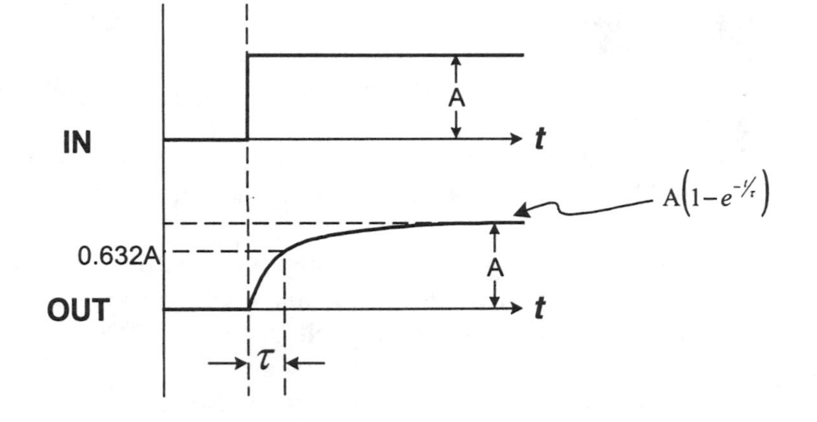

Figure W2.1 First order process response to a step change.

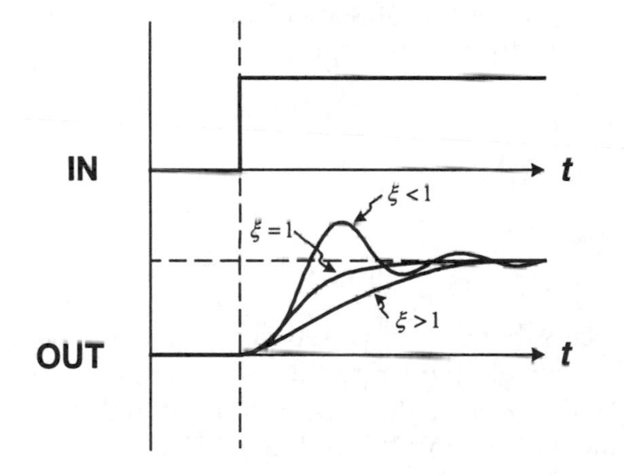

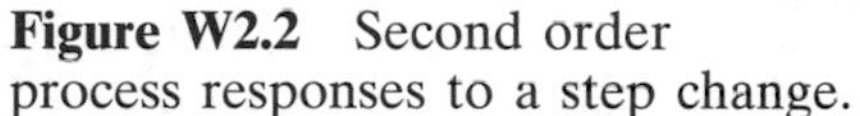
Figure W2.2 Second order process responses to a step change.

that the response never overshoots the final value. However, the approach for an overdamped system is much slower and varies depending upon the value of ξ. The larger the damping coefficient is, the slower the response.

There are two main differences between first and second order responses. The first difference is obviously that a second order response can oscillate while a first order cannot. The second difference is the steepness of the slope for the two responses. For a first order response, the steepest part of the slope is at the beginning, whereas for the second order response, the steepest part of the slope occurs later in the response.

First and second order systems are not the only two types of systems. There are higher order systems, such as third or fourth order systems. However, these higher order systems will not be discussed here.

Key learning objectives

1. Understand the components of the loop and how these components interact.
2. Become familiar with the terminology associated with process control.
3. Be able to explain the components in a single input/single output (SISO) block diagram.
4. Be able to develop the underlying mathematical models and relationships for each component of the feedback control (FBC) loop.
5. Understand the effect of capacitance.
6. Understand the effect of resistance.
7. Understand the concept of a response and the metaphor of process personality.
8. Understand the effect of self-regulation on process response.
9. Recognise the open-loop response of second order processes.
10. Recognise the open-loop response of capacity dominated processes, with and without dead time.

Tasks

1. LEVEL RESPONSE

Capacity dominated process behaviour can best be studied using a very common process element, namely the surge tank or separator. The ordinary differential equation (ODE) for a single tank was presented in the notes in Chapter 3 and is implemented in HYSYS as the TANK unit operation.

Build a simulation in HYSYS using water with a flow of 20 kmol/h at 15°C and 1 atm as the only component and the Peng-Robinson Equation of State as the fluid property package. Use the default tank volume of 2 m^3 and specify liquid flow control on the Liquid Valve page of the tank unit operation. Calculate the flow out of the tank using Equation W2.1, which describes a linear valve, and then Equation W2.2, which describes a non-linear valve. In both cases, the outlet flow rate is a function of the liquid head on the tank only.

$$Flow = K \times head \tag{W2.1}$$

$$Flow = K \times \sqrt{head} \tag{W2.2}$$

In Equations W2.1 and W2.2, SI units are flow in m^3/h and head in metres. Use the spreadsheet unit operation to incorporate Equations W2.1 and W2.2 into the simulation. Import the tank liquid level into the spreadsheet, assume a tank geometry, and use the resulting value of head to calculate the liquid outlet flow. A K value of 20 should work well with SI units. Then export the calculated flow back to the worksheet as the outlet flow rate. This last action can only be performed in the dynamic mode otherwise the material balance will be violated.

System identification is the term used to define a procedure to characterise the process response. In this case, system identification can be accomplished by adjusting the feed rate to the tank in steps, up and down, and then observing the tank level response on a strip chart. This is termed step response testing.

Set up strip charts, recording the important variables, to study the open-loop response of capacity dominated processes consisting of:

1. a single tank,
2. two tanks in series,
3. two tanks in series with a pipe segment between them, *[Hint: Use the PFR unit operation to simulate a pipe segment and calculate a volume that should give a dead time of around 10 minutes.]*
4. three tanks in series (use only linear valves).

- What is the open-loop response of the tank level to a step change in the feed rate in a process with a single tank only?
- What effect does the addition of a second tank have on the level response in Tank 1? And in Tank 2?
- What effect does the addition of a third tank have on the level response in Tank 1? And in Tank 2? What is the open-loop response of Tank 3?
- What effect does the volume of the tank have on the personality of the response?
- How does the valve type affect the process personality?
- What effect does the addition of the pipe segment between the two tanks have on the response of the level in Tank 1 and Tank 2? Did the PFR properly simulate dead time in this situation? Why or why not?

2. TEMPERATURE RESPONSE

The next exercise in this workshop requires that you set up a mixing tank to heat water directly using live steam, as illustrated in Figure W2.3. Use the default tank volume and set the default tank control to liquid level with a 50% set point. The feed water stream to the tank enters with a flow of 100 kmol/h at 15°C and 1 atm.

Perform a series of steady-state runs to determine the amount of steam required to raise the temperature of the feed water stream to about 200°F. Then switch to the dynamic mode of operation and perform step response testing by varying the inlet flow rate and feed temperature to determine the

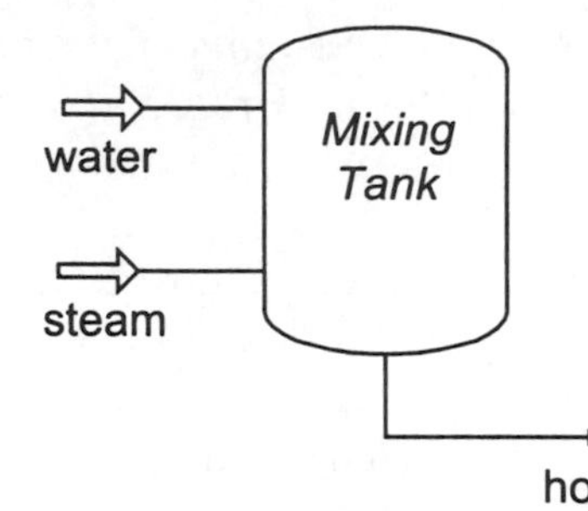

Figure W2.3: Mixing tank process.

process response. Remember to use the strip charts to observe the important process variables.

Add a pipe segment to the system on the outlet of the tank. Simulate the pipe segment using a PFR, as in the previous exercise. Calculate a PFR volume to approximately 10 minutes of dead time. Make sure that the number of segments in the PFR operation is set to at least 50 instead of the default of 20 (under "Reactions"). Repeat your analysis of step disturbances noting the relationship between the tank temperature and the temperature at the outlet of the pipe segment. Repeat the analysis again but with a different volume for the PFR, and note any differences on the response.

- What type of response does the process produce?
- How does the pipe segment affect the response?
- What effect does the pipe segment volume have on the response? Did the PFR properly simulate dead time in this situation? Why or why not?

Present your findings on diskette in a short report using MS-Word. Also include on the diskette a copy of the HYSYS files which you used to generate your findings.

WORKSHOP 3 – PROCESS CAPACITY AND DEAD TIME

"Knowledge is a treasure but practice is the key to it."
Thomas Fuller

Introduction

Prior to attempting this workshop, you should review Chapter 3.

This workshop will illustrate the effect on the process response of the three key process dynamic parameters: process gain, process time constant and process dead time. You will also explore the impact that capacitance or "lag" has on these process parameters.

Key learning objectives

1. Process gain is the key process parameter affecting the extent (magnitude) of the response of a process or process element.
2. The time constant determines the personality of the response for a process or process element.
3. The time constant is the key dynamic parameter that determines the ability of a process to reject, or attenuate, disturbances.
4. The period and the amplitude of the disturbance will determine the amount of attenuation/rejection.
5. Capacitance is good for disturbance rejection, but the downside is that it results in very slow and long response times.
6. Dead time has no effect on the filtering capability of the process.
7. Dead time has no redeeming features and can make the control loop unstable.
8. Tight process control can only be achieved if the loop dead time is small compared with the smallest time constant of a disturbance of significant amplitude.

Tasks

1. SYSTEM IDENTIFICATION

The process used for this workshop is shown in Figure W3.1. Build a simulation of this system using the Wilson activity model as the fluid package. The feed is a 50/50 mixture of water and methanol (100 kmol at 30°C and 200 kPa) which is heated in a steam heater to about 70°C. The hot mixture is then stored in a surge tank for future use. Note that you do not need to enter any further information about vessel volumes, etc. Simply use the default values for hold-ups for dynamic runs.

This very simple example has many analogies in process plants. Whenever material or energy enters a piece of plant equipment that can accumulate some of the material or energy, the process has capacitance. Virtually all process equipment has the potential to store mass or energy and hence create a capacity dominated process.

System identification is the term used to define a procedure to characterise the process response. In this case, system identification can be accomplished by setting the default level controller set point at 50% (under "Liquid Valve"), adjusting the steam flow to the heater in steps, up and down, and then observing the temperature response on a strip chart. This is termed step response testing and is the same as was done in the previous workshop.

Figure W3.2 illustrates a typical step test response for a first order system. The relevant process parameters of gain, K_p and time constant, τ, for this first order process are shown and can be calculated as follows:

$$K_p = \frac{\text{(Tank temperature change/temperature transmitter span)}}{\text{(Steam rate change/steam valve span)}} \quad \text{(W3.1)}$$

τ = the time it takes for the tank temperature to reach 63.2% of its final value

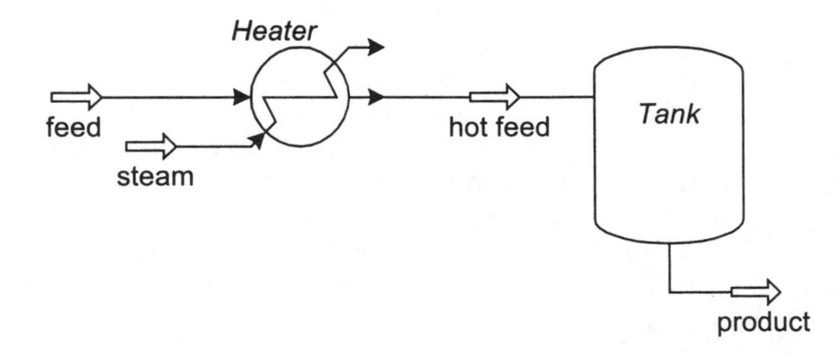

Figure W3.1 Capacity dominated process.

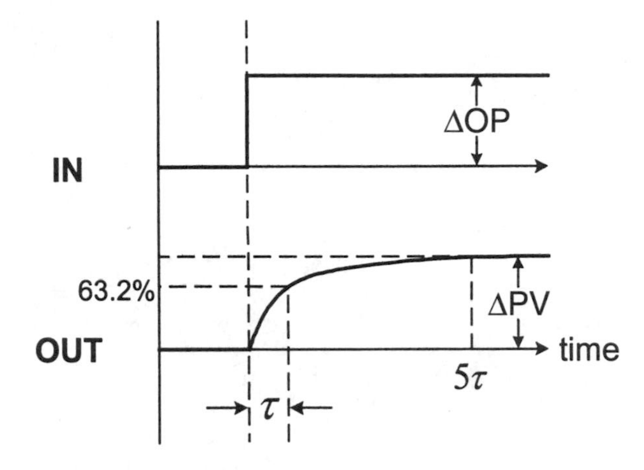

Figure W3.2 System step response.

- In order to check how linear the process is, it is necessary to determine if the gain is the same regardless of the steam rate and to see if the magnitude of the gain is unchanged for increases and decreases in steam. Do this by using the step testing method described above and Equation W3.1.

2. CAPACITANCE

Now we will examine how the gain, time constant and dead time vary for different tank levels or different amounts of capacitance in the process shown in Figure W3.1.

- Make step changes in the steam rate for three different tank level set points of 5%, 50%, and 95%. Calculate the gain, time constant, and dead time for the three different process capacities. Present your results in Table W3.1 and then plot the time constant and gain in Figure W3.3.

Table W3.1 Summary of process parameters

Tank level (%)	Process gain	Time constant (min)	Dead time (min)
5			
50			
95			

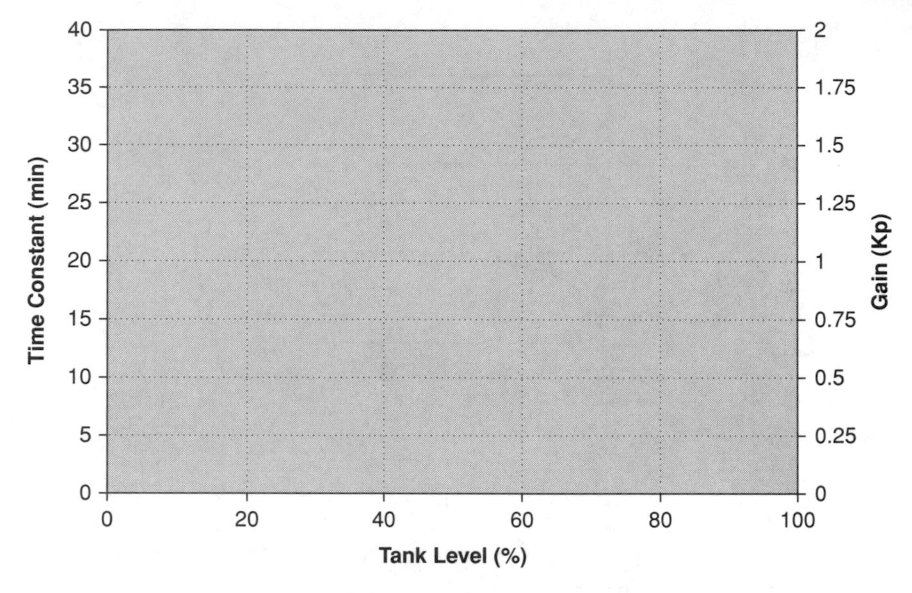

Figure W3.3 Process gain and time constant versus tank level.

3. ATTENUATION

As has been indicated in the notes, the object of both the process and control system is to reject or at least minimise the effect of disturbances. In order to quantify disturbance rejection the term attenuation has been borrowed from electrical and mechanical engineers and is defined as follows:

$$\text{Attenuation} = 1 - \frac{\text{Disturbance Amplitude Out}}{\text{Disturbance Amplitude In}}$$

For example, if the tank temperature varies with an amplitude of 5°C and the input temperature disturbance has an amplitude of 25°C, the attenuation is $(1 - 5/25) = 0.8$ or 80%.

In HYSYS, a transfer function block is used to generate a sinusoidal feed temperature. From the main simulator builder, or through the PFD, select the transfer function block. On the Connections page attach the output of the transfer function block to the feed "Object" and select the process variable (Select PV) as temperature. Move to the Parameters page and set the nominal (User Input PV) temperature to 30°C and the span of the PV output to vary between 0°C and 100°C. To select the actual wave type move to the page labelled "Lead/2nd Order" and select "Sine Wave". Under Sine Wave Parameter, specify an amplitude of 25 and a period of 10 minutes to start with. It is important to note that the amplitude is entered as a percentage of the PV span, i.e. 25% of 100°C gives an amplitude of 25°C. The distur-

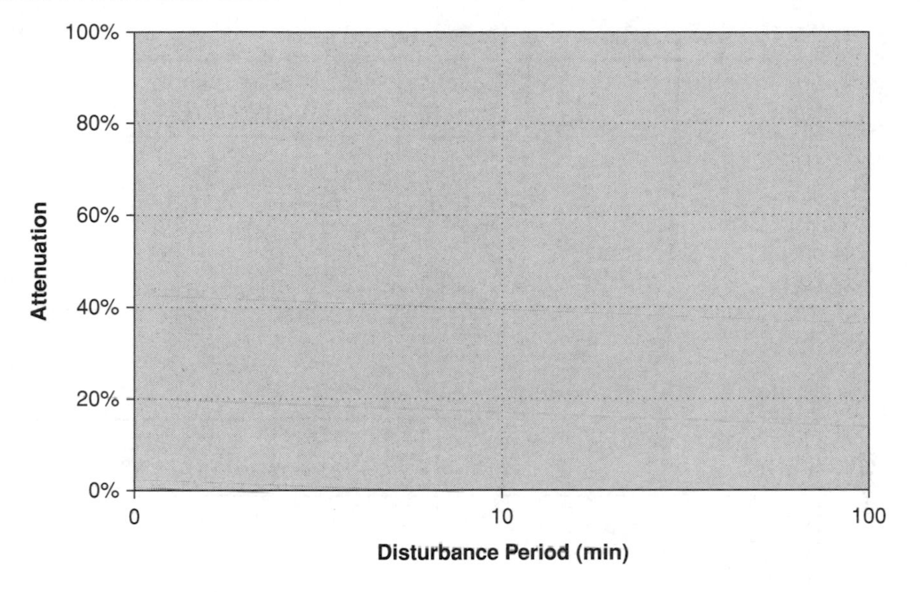

Figure W3.4 Attenuation of feed temperature disturbance – first order process.

bance period will be varied depending on the dynamic test being run. Finally, open the faceplate to complete the set-up.

- Complete the following tables for each tank level and then plot the results with a curve for each level, on Figure W3.4. [*Note: You will need your results from this section of the workshop for later workshops so remember to save a copy of the results for yourself.*]

Table W3.2 Attenuation for level at 5%

Disturbance period (min)	Frequency (1/min)	ΔT – product temperature (°C)	Attenuation
5			
10			
20			
30			
50			

Table W3.3 Attenuation for level at 50%

Disturbance period (min)	Frequency (1/min)	ΔT – product temperature (°C)	Attenuation
10			
20			
30			
40			
100			

Table W3.4 Attenuation for level at 95%

Disturbance period (min)	Frequency (1/min)	ΔT – product temperature (°C)	Attenuation
10			
20			
30			
40			
100			

4. DEAD TIME

In this section of the workshop the dynamic characteristics of processes with capacitance and appreciable dead time will be studied. The process you will work with is the simple feed heater and storage tank, shown in Figure W3.1, except that additional equipment will be added between the heater and the surge tank. The additional equipment will be a plug flow reactor (PFR) with a volume of 0.3 m^3 and length of 2 m, which gives a dead time of 6 minutes. Figure W3.5 shows an example of what this process should look like. The objective is to see how dead time affects the temperature response of the warm solution leaving the surge tank.

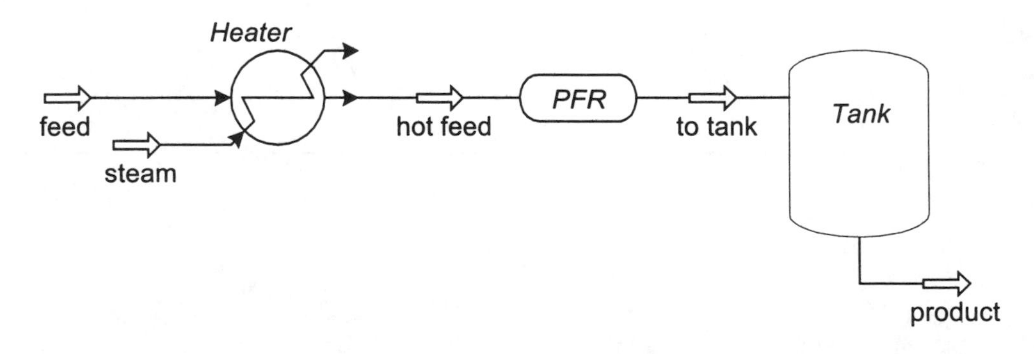

Figure W3.5 A process containing dead time.

Again, this simple example is illustrative of many real plant situations. Even the time it takes for fluid to move through pipes between connecting items of equipment is an example of dead time. A sensor located a distance from a vessel such as a reactor introduces process dead time. The time it takes for a process analyser to sample a process stream and measure a particular property or the time it takes for a manual sample to be taken to the laboratory for analysis are also both examples of dead time. This is a time during which there no knowledge of what is happening in the process.

- Using the built-in tank level controller with a set point of 50%, increase and decrease the steam rate to the heater and record the tank temperature response. What is the dead time for the process shown in Figure W3.5? Is there a difference between the dead time predicted and the actual dead time from the simulation? If the answer is yes, why is there a difference between the two values?
- Vary the tank level between 5% and 95%. How does this affect the process dead time?

An important indication of the effect of dead time on a process is the dead time to time constant ratio (t_{DT}/τ). If this ratio is less than 0.3, the dead time has little or no effect on the process response. However, if the ratio is greater than 0.3, the process becomes dead time dominated and thus is virtually uncontrollable.

- Repeat the step response testing done above to determine the time constant of the new system and use these results to calculate the dead time / time constant ratio. You will need to stop the sine wave feed temperature input and use a constant feed temperature. Record your results in the table provided.

Table W3.5 Dead time/time constant ratio

Tank level	Dead time (min)	Time constant (min)	Dead Time/ time constant
5			
50			
95			

- To test the hypothesis that dead time has no effect on the open loop process attenuation for a capacity dominated process, again perform a frequency response test using the sine wave input. Run the 40 minute (frequency = 0.025/min), 25°C amplitude disturbance through the process of Figure W3.5 with the tank level set to 50%. Has the attenuation changed from the value you calculated earlier?

Present your findings on diskette in a short report using MS-Word. Also include on the diskette a copy of the HYSYS files which you used to generate your findings.

WORKSHOP 4 – FEEDBACK CONTROL

"Nothing comes from doing nothing."
Shakespeare

Introduction

Prior to attempting this workshop, you should review Chapters 3 and 4.

The previous workshop introduced the concepts of capacitance and attenuation. These are "natural" characteristics of a system, as are dead time and the process time constant. Now that we have a basic understanding of the way processes behave, we can apply this knowledge to control the process response.

Once the process personality is understood, we can manipulate process flows to maintain a desired variable at constant conditions, which are called set points. This is known as feedback control, where the value of a variable is "fed back" to a controller which manipulates another variable according to the difference between the controlled variable and its set point.

Key learning objectives

1. Feedback control is easiest and most successful for low capacity processes without dead time.
2. Dead time reduces the ultimate gain of a process.
3. A large time constant decreases the responsiveness of a process and reduces the achievable control performance.
4. Proportional only control suffers from offset, which can be eliminated through integral action.
5. Derivative action can only be used where there is no significant process noise and relatively little dead time.
6. Buffer tanks and surge drums can help smooth out changes in flow and thereby isolate equipment from upstream disturbances. This can only result from loose level tuning where the primary interest is flow smoothing, not level control.

7. Proportional only level control with a controller gain (K_c) of 2.0 is generally sufficient. When the manipulated variable is the outlet flow, this implies that the valve is fully shut at 25% level and fully open at 75% level. K_c less than 1.0 won't hold the level between 0% and 100%.
8. For averaging level control:
   - Attenuation decreases as the gain increases.
   - Adding integral action to an integrating process (level control) can become a disturbance generator rather than a disturbance smoother if not properly tuned.
   - Level loop tuning is always dependent on the system characteristics.
   - If the hold-up is too small to get good flow smoothing, reduce K_c and add integral action to ensure that the level stays in the tank ($K_c \times T_i = 4.0$).
   - If the hold-up time is long, you do not need any integral action.
   - If the level is cycling, increase K_c and decrease T_i (this is the opposite of other loops).

Tasks

1. LOW CAPACITY, NO DEAD TIME PROCESSES

In the previous workshop, you should have built the system shown in Figure W4.1. Check that you still have the Wilson property package specified. The only components required are water and methanol. The inlet temperature should be 30°C, while the heater outlet should be fixed at 70°C. Initially, we wish to analyse a process without dead time so that you will need to delete the PFR that you added in the previous workshop. Retain the strip charts

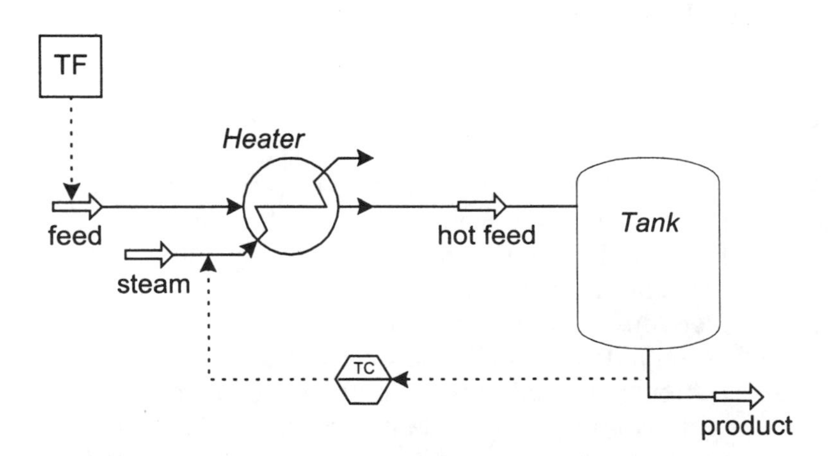

Figure W4.1 Low capacity, no dead time process.

that you set up previously. If they have been deleted, rebuild them to contain the following variables: feed temperature, tank outlet temperature, steam heat flow and feed molar flow. Select suitable ranges for each variable.

The results from the previous workshop indicated that processes with low capacitance and a relatively long disturbance period have the lowest attenuation and are in greatest need of process control. Hence, for this portion of the workshop you will need to use a low capacity process so adjust the tank level accordingly. Your simulation should still contain a transfer block operation, which adds noise to the system. Set the feed water temperature disturbance period to 30 minutes and its amplitude to 25°C.

Now add a controller for the tank outlet temperature. The controller should manipulate the steam rate to the heater between 0 and 1×10^6 kJ/hr (direct Q). The PV range should be 0 to 100°C, and the set point should be 70°C to match the steady-state conditions. Set the controller gain (K_c) to 1.0, but leave the integral time and derivative time blank. Make sure that you correctly specify whether your controllers are direct acting or reverse acting so that the controller will open/close the valve when required. Finally, set the controller to automatic.

The previous workshop demonstrated that capacity dominated processes have significant disturbance rejections (attenuation) properties without requiring any form of process control. This is called open-loop attenuation. Controllers can usually increase the attenuation of process systems. When operated in "automatic", the additional attenuation is called closed-loop attenuation. When in "manual" the system behaves as it would without the controller present.

- Vary the gain to achieve the maximum closed-loop disturbance attenuation. How effective is the controller in rejecting disturbance not already rejected by the natural attenuation of the process?
- What happens when the controller gain gets very high? Is there a limit to how much you can increase the gain?

2. PROCESSES WITH DEAD TIME

Capacity dominated processes are relatively easy to control. However, the presence of dead time makes the control problem more difficult. We can demonstrate this by adding a PFR to the system shown in Figure W4.1 in order to simulate dead time. The PFR should have a length of 2.0 m, a total volume of 0.5 m^3 (dead time = 10 min) and a pressure drop of 0 kPa. Remember that you want to work with a capacity dominated process, so ensure that the tank level is set accordingly.

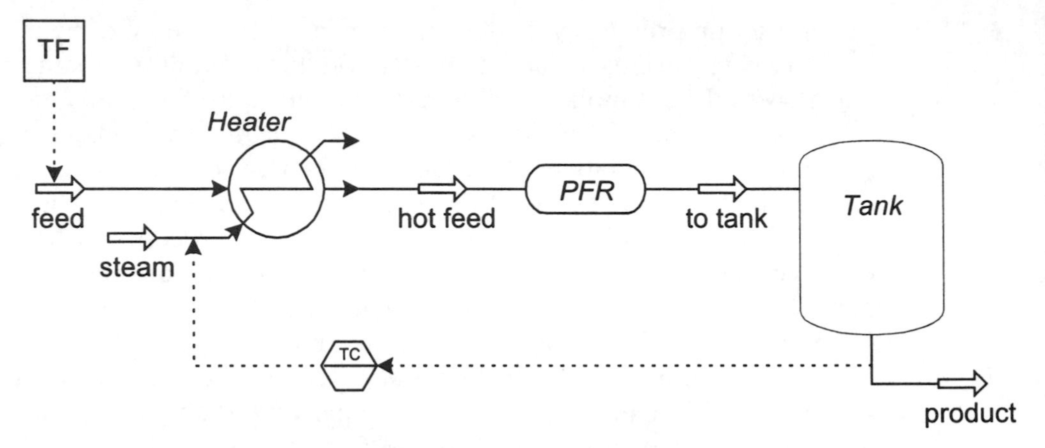

Figure W4.2 High capacity process with dead time.

- What is the maximum attenuation for the process with dead time?
- Is there an optimum/maximum gain that maximises process attenuation?
- Fix the controller gain at $K_c \approx 10$ and calculate attenuation for dead times of 2, 5, 10 and 20 minutes. Vary the size of the PFR to change the amount of dead time in the system. Record the results in Table W4.1.

Table W4.1 Process attenuation with dead time

Dead time (min)	ΔT – product temperature (°C)	Attenuation
2		
5		
10		
20		

3. PROPORTIONAL ONLY CONTROL

We have found that feedback control can provide good attenuation of process disturbances provided that the dead time is not too great. The ability to provide attenuation to a process is sometimes called disturbance rejection. However, disturbance rejection is only one of the requirements of an effective controller. The other main requirement is that we should be able to change the set point whenever we want and have the controller manipulate the process so that the controlled variable continues to match the set point. This is sometimes referred to as controller performance. These two parameters, disturbance rejection and controller performance, are used to assess the effectiveness of a controller.

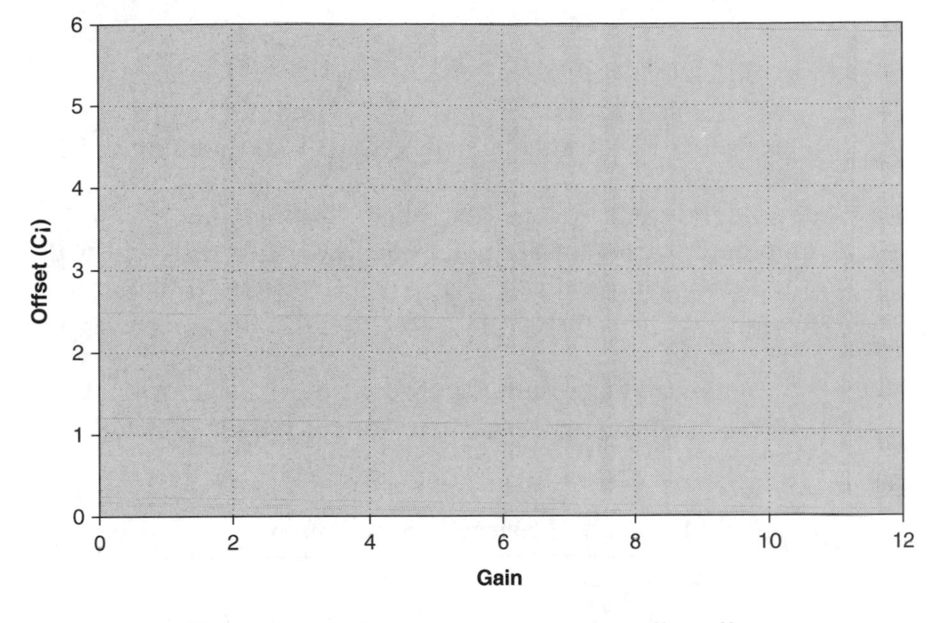

Figure W4.3 Proportional only controller offset.

Eliminate dead time from the system by deleting the PFR or setting its volume to a very low value, and pause the feed temperature disturbance to remove the noise from the process. Starting with a gain of 1.0, interactively change the tank outlet temperature set point to 80°C.

- Where does the tank outlet temperature stabilise? The difference between this value and the set point is called offset.
- Plot the relationship between offset and the controller gain on Figure W4.3.
- Can you explain the relationship shown on Figure W4.3 in terms of the proportional-only controller equation given in Equation W4.1?

$$\text{Output} = \text{Gain} \times (SP - PV) + \text{Bias} \qquad \text{(W4.1)}$$

4. PI AND PID CONTROL

The feedback controller we have employed to this point has only contained one term: controller gain. As suggested above, this type of controller is called a proportional only controller, which suffers from the problem of offset. Offset can be reduced with high values of gain, but sometimes this makes the controller unstable, particularly when there is dead time in the system. We can eliminate offset by introducing another term to the controller equation: integral time. The control equation for a proportional integral (PI) controller is:

$$\text{Output} = \text{Gain} \times \left((SP - PV) + \frac{1}{T_i} \int_0^T (SP - PV)dt \right) \qquad \text{(W4.2)}$$

- How does PI control eliminate offset? Use Equation W4.2 to help explain.

Add integral time to your controller, starting with T_i = 1.0, and check that it eliminates offset for set point changes, both with and without dead time in the system. Interactively change the feed rate to determine how effective the controller is at disturbance rejection, i.e. step response testing.

- Summarise your results for PI control in Table W4.2. Record the details of each type of step test performed, i.e. 30–40 kmol/h under the Test column.

Table W4.2 PI controller optimisation

Test	Gain	Integral time (min)	Offset	Time to steady-state (min)

Integral action can be slow since it relies on the integral of the error being large, where the error is the difference between the set point and the process variable. Proportional action usually provides the "muscle" for the controller. However, too much proportional action creates instability. In some circumstances, PI controllers are not sufficiently fast, making a third controller action necessary. This term is called derivative time and can sometimes be introduced to speed up the response time of the controller. The control equation for a proportional-integral-derivative (PID) controller is:

$$\text{Output} = \text{Gain} \times \left((SP - PV) + \frac{1}{T_i} \int_0^T (SP - PV)dt + T_d \frac{d(SP - PV)}{dt} \right) \text{(W4.3)}$$

Derivative time increases the controller response when the controlled variable is moving away from its set point most quickly, i.e. straight after a disturbance has affected the system. Apart from increasing the responsiveness of the controller, derivative action also reduces oscillation. Derivative action can be very effective under some circumstances but very damaging under others. For example, if the system is essentially stable but there is a small amount of process noise (usually very high frequency), the derivative action will interpret the noise as being the start of a large disturbance and will make large changes in the manipulated variable which are clearly not required.

Add derivative action, starting with $T_d = 1.0$, to your system to determine whether or not it improves the controller effectiveness for this example.

- Optimise controller performance by varying the three controller parameters. Consider the responsiveness to process disturbances and the ability to track a set point. Record your results in Table W4.3. Under the "Test" column, record the type of step change test performed.

Table W4.3 PID controller optimisation

Test	Gain	Integral time (min)	Derivative time (min)	Offset	Time to steady-state (min)

5. AVERAGING LEVEL CONTROL

Surge drums and intermediate product tanks are critical parts of any process system. Their principal purpose is to provide hold-up and capacitance to smooth out flow disturbances so they do not carry through to downstream process units. This function must be recognised and it is frequently overlooked in many operating plants. A consequence of this function is that the level in surge drums and intermediate tanks should **not** be tightly controlled. Tight level control will transmit flow disturbances to downstream units and negate the effectiveness of the surge volume. The level controller must only control the level between the low limit (when the tank/drum approaches empty and thereby risks damaging the outlet pump) and the high limit (when the tank overflows). Intentional loose level control is called averaging level control.

One exception to the rule of averaging level control for surge drums is in the case of distillation column hold-ups. Averaging level control should not be used to control the reflux drum level or the reboiler sump level. Tight level control is required for these vessels to maintain the integrity of the column material balance so that changes in the reflux rate and reboiler duty will have the desired effect on product compositions and yields without introducing additional lag to the system.

In order to better understand how averaging level control works, build a simple system consisting only of a 2 m^3 surge drum. The feed to the surge drum should be 250 kmol/h of water at 25°C and 100 kPa. Add a level controller and enter a set point of 50%. Finally add a feed disturbance using a transfer function unit operation set up to vary the feed rate sinusoidally with an amplitude of 25 kmol/h and a period of 4 minutes. The control valve range should be 0–500 kmol/h. Remember that the amplitude is entered as a percentage of the PV span for the transfer function operation.

- Test the following combinations of PI control for the surge drum level controller for the range of disturbance periods given in Table W4.4:

  1. $K_c = 1.0$, $T_i = 50$ minutes

Table W4.4 Averaging level control

Disturbance period (min)	$K_c = 1.0$ $T_i = 50$ min	$K_c = 2.0$ $T_i = 25$ min	$K_c = 0.5$ $T_i = 100$ min
5			
10			
20			
30			

  2. $K_c = 2.0$, $T_i = 25$ minutes
  3. $K_c = 0.5$, $T_i = 100$ minutes

- Which combination of gain and level control provides the best disturbance attenuation?
- Are there any problems with using a very low gain?
- Are there any problems with using a proportional only level controller?

Present your findings on diskette in a short report using MS-Word. Also include on the diskette a copy of the HYSYS files which you used to generate your findings.

WORKSHOP 5 – CONTROLLER TUNING FOR CAPACITY AND DEAD TIME PROCESSES

"A little experience often upsets a lot of theory."
Samuel Parks Cadman

Introduction

Prior to attempting this workshop, you should review Chapter 5.

This workshop will illustrate that HYSYS may be used to determine the appropriate parameters for a PI controller that is controlling a capacitive process with significant dead time. You will learn that controller tuning is determined by the desired load or set point response as well as the type of process and the values of the process parameters, which include process gain, time constant and dead time. A review of the two tuning techniques used in this workshop is provided below.

AUTO TUNE VARIATION (ATV) TUNING TECHNIQUE

The auto tune variation or ATV technique of Åström is one of a number of techniques used to determine two important system constants, called the ultimate period and the ultimate gain. Tuning values for proportional, integral and derivative controller parameters may be determined from these two constants. All methods for determining the ultimate period and ultimate gain involve disturbing the system and using the disturbance response to extract the values of these constants.

In the case of the ATV technique, a small limit cycle disturbance is set up between the manipulated variable (controller output) and the controlled variable (process variable). Figure W5.1 shows the typical ATV response plot with critical parameters defined. It is important to note that the ATV technique is applicable only to processes with dead time. The ultimate period will just equal the sampling period if the dead time is not significant.

General ATV tuning method for a PI controller

1. Determine a *reasonable* value for the valve change, h. This value is arbitrarily chosen, but typically 0.05 is reasonable, i.e. 5%.
2. With the controller in the off position, manually move the valve $+h$ units.
3. Wait until the process variable (PV) starts to move and then move the valve $-2h$ units.
4. When the process variable crosses the set point, move the valve $+2h$ units.
5. Repeat until a limit cycle is established, as illustrated in Figure W5.1.
6. Record the value of the amplitude, a, by picking it off the response graph.
7. Perform the following calculations:

$$\text{ultimate period} = P_u = \text{period taken from the limit cycle}$$

$$\text{ultimate gain} = K_u = 4 * \text{h} / 3.14 * \text{a}$$

$$\text{controller gain} = K_c = K_u / 3.2$$

$$\text{controller integral time} = T_i = 2.2 * P_u$$

ZIEGLER-NICHOLS CLOSED LOOP TUNING TECHNIQUE

The closed loop technique of Ziegler-Nichols (Z-N) is another technique that is commonly used to determine the two important system constants, ultimate period and ultimate gain. It was one of the first tuning techniques to be widely adopted.

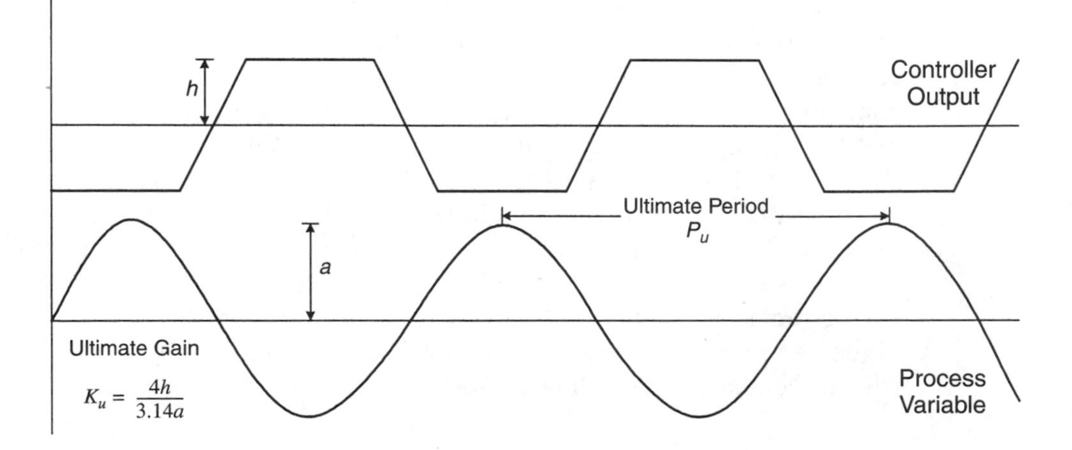

Figure W5.1 ATV critical parameters.

In Z-N closed loop tuning, as for the ATV technique, tuning values for proportional, integral and derivative controller parameters may be determined from the ultimate period and ultimate gain. However, Z-N closed loop tuning is done by disturbing the closed loop system and using the disturbance response to extract the values of these constants.

Z-N closed loop tuning method for a PI controller

1. Attach a proportional only controller with a low gain (no integral or derivative action).
2. Place the controller in automatic.
3. Increase proportional gain until a constant amplitude limit cycle occurs.
4. Perform the following calculations:

ultimate period, P_u = period taken from limit cycle

ultimate gain, K_u = controller gain that produces the limit cycle

controller gain = $K_c = K_u / 2.2$

controller integral time = $T_i = P_u / 1.2$

Key learning objectives

1. Controller tuning is determined by the desired controller response.
2. Controller tuning is determined by the type of process.
3. Controller tuning is affected by the value of the process gain.
4. Controller tuning is affected by the value of the time constant.
5. Controller tuning is affected by the value of the dead time.
6. The auto tune variation (ATV) tuning technique is a powerful method for many loops.
7. The Ziegler-Nichols closed loop technique is also useful, but more aggressive than ATV.
8. HYSYS can be used to find appropriate tuning parameters for a PI controller.

Tasks

1. TUNING CONTROLLERS

The process used for this workshop is shown in Figure W5.2. A 50/50 feed mixture of water and methanol (T = 30°C, P = 200 kPa, F = 100 kmol/h) is heated

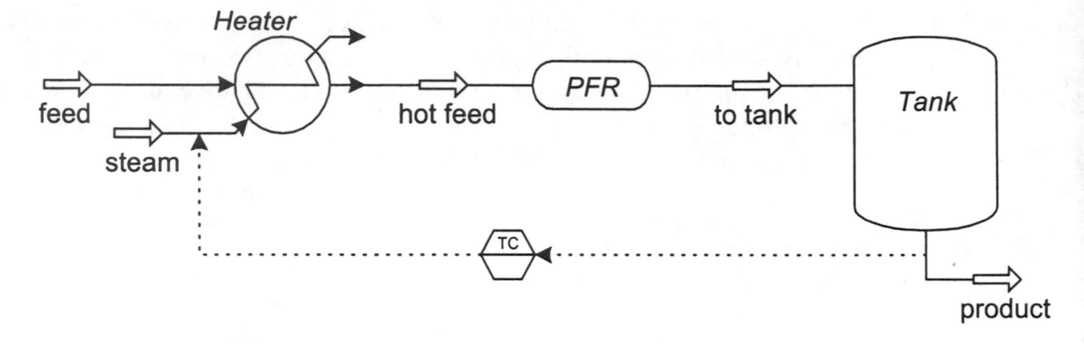

Figure W5.2 Illustrative capacity plus dead time process.

in a steam heater to approximately 70°C. The hot stream passes through a dead time leg before being stored in a tank for future use. Use a PFR unit operation to simulate the dead time with volume = 0.3 m^3 and length = 2 m. This was the process you worked on in the latter part of Workshop 3.

Set the tank level to 50% with no incoming disturbances. With the temperature controller in manual, adjust the steam valve to get a tank temperature of approximately 70°C. Bring up the temperature controller faceplate. Set the mode to auto-tune. The controller will bring the process into a limit cycle.

- Determine the period of this limit-cycle in minutes. Use this limit-cycle to determine the amplitude of the temperature cycle of the stream exiting the tank and make this dimensionless by dividing by the temperature transmitter span.
- Now determine the fractional amplitude of the controller output (h).
- Calculate the ultimate gain and use this with the ultimate period to compute the controller settings.
- Determine the controller settings at two more tank levels (5 and 95%).

Now use the Z-N tuning technique to determine the controller settings.

- Compare the results of using both the ATV and Z-N tuning techniques.

2. CONTROLLER CONTRIBUTIONS TO ATTENUATION

We have seen in Workshop 3 that the process itself is able to attenuate with no control, i.e. open loop. We have just tuned our feedback controller for various levels of capacitance and can now determine what the process plus control (closed loop) is able to attenuate. By subtracting the open loop attenuation from the total attenuation we can determine what the controller itself contributes to the overall process attenuation.

- Determine the total closed loop attenuation of the tank operating at the 50% level for sinusoidal disturbances of periods of 10, 20, 30, 40 and 100 minutes with an amplitude of 25°C.
- Compute the controller contribution to attenuation for these disturbances.
- At 5% level determine the controller attenuation for sinusoidal disturbances of periods of 5, 10, 20, 30 and 50 minutes and amplitude 25°C.
- At 95% level determine the controller attenuation for sinusoidal disturbances of periods of 10, 20, 30, 40 and 100 minutes and amplitude 25°C.
- Plot attenuation versus the logarithm of the disturbance period. Compare the curves using their dead time to time constant ratios that you calculated in Workshop 3.

Present your findings on diskette in a short report using MS-Word. Also include on the diskette a copy of the HYSYS files which you used to generate your findings.

WORKSHOP 6 – ADVANCED TOPICS IN CLASSICAL CONTROL

"Theory without experience is sterile, practice without theory is blind."
George Jay Anyon

Introduction

Prior to attempting this workshop, you should review Chapters 6 and 7.

This workshop will show how the response of feedback control loops can be improved through the use of other control methods. These other methods include measuring common disturbances and taking action before they affect the controlled variable (feedforward control) and using a faster responding loop to decrease the response time of a system with a large time constant (cascade control). You will determine what conditions are necessary for feedforward or cascade control to be useful and identify which parameters reduce the effectiveness of these control methods.

Key learning objectives

FEEDFORWARD CONTROL

1. Feedforward controllers can respond faster than feedback controllers can since they react to process disturbances immediately without waiting for them to affect the process.
2. Feedforward control can only compensate for disturbances that are measured. Its effectiveness is reduced if unmeasured disturbances are significant.
3. Feedforward control is less effective for nonlinear processes, where nonlinearities exist between the disturbance measurement and controlled variable.

CASCADE CONTROL

4. Cascade control can significantly improve the control performance if a secondary variable can be found in the system that directly affects the primary loop and is faster responding than the primary loop.
5. The inner control loop helps to reject disturbances to the primary control variable.
6. The ultimate period of the inner (slave) loop should be at least four times smaller than the outer (master) loop in order for cascade control to be effective.
7. The most frequently used slave loop is a flow loop but other types of fast responding loops, such as pressure loops, can also be used.

RATIO CONTROL

8. Ratio control is a type of simple feedforward control that is most effective for low frequency disturbances.

Tasks

1. BASIC PROCESS CONFIGURATION

Build the system shown in Figure W6.1. The feed is pure water with a temperature of 20°C, atmospheric pressure, and a flow rate of 1.5 m^3/h. The outlet temperature of the heater should be set to 55°C with steam as the heating medium. Assume that the pressure drop is negligible, i.e. set equal to 0. Incorporate dead time into the process by adding a PFR with a length of 1.0 m, a total volume of 0.2 m^3 and a pressure drop of 0 kPa. Finally, add a tank with a volume of 1.2 m^3 and set the liquid level set point at 50%. Alternatively, you might be able to modify the simulation you used for Workshop 4.

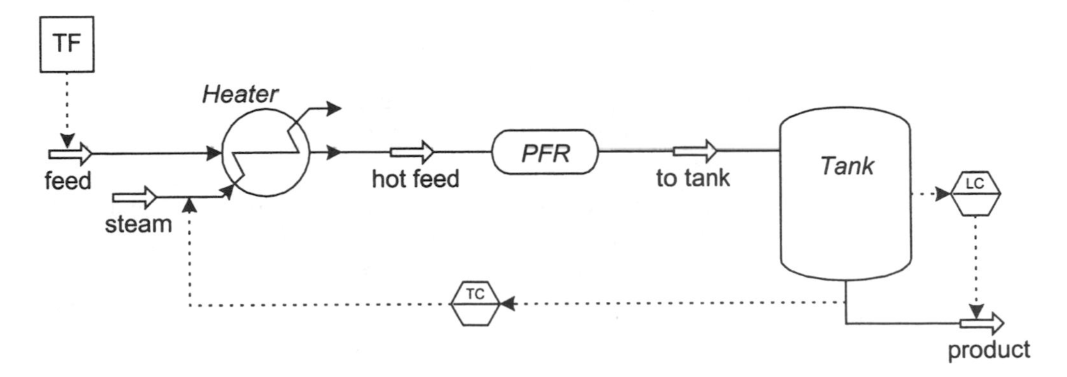

Figure W6.1 Simple heating system.

Use strip charts to view your results, monitoring the following variables: feed temperature, product temperature, steam heat flow and product molar flow. Select suitable ranges for each variable and iconise the strip chart view for later use.

Add two controllers to the system to set up feedback control of the process:

1. The first controller should manipulate the steam rate to the heater between 0 and 5×10^5 kJ/h (direct Q) to control the product temperature between 0 and 100°C. The set point should be 55°C initially, with tuning constants of $K_c = 0.5$ and $T_i = 10$.
2. The second controller should manipulate the product flow from the tank between 0 and 200 kmol/h to control the tank level between 0 and 100%. The set point should be 50% with tuning constants of $K_c = 10$ and $T_i = 10$.

Make sure that you correctly specify whether your controllers are direct acting or reverse acting. Set both controllers to automatic.

Finally add feed noise to the system using the Transfer Block unit operation. The transfer block *PV* target is the feed temperature. Create sine wave noise with an amplitude of 10°C and a period of 10 minutes. Remember that the amplitude is entered as a percentage of the PV span, i.e. PV max. − PV min. The feed temperature should still oscillate around a mean of 20°C.

2. DETERMINE BASELINE CONTROL PERFORMANCE

The heater-tank system, shown in Figure W6.1, has a large capacitance, which should provide good attenuation of process disturbances and help reject high frequency process noise. However, it will make the system slow to respond to set point changes or permanent disturbances, i.e. feed rate changes. The significant dead time in the system will compound any control problems and make it difficult to achieve tight control of the system when only feedback control (FBC) is used.

- Vary the period of the disturbance to the feed temperature and fill in Table W6.1 to demonstrate this characteristic of the system.
- Should the magnitude of the feed temperature disturbance affect the attenuation?
- From your results, identify any deficiencies of the feedback control system simulated above. If necessary, vary the temperature controller tuning constants to try to improve the performance of the control loop. [*Hint: How long does it take the controller to respond to a change in the feed temperature? Can the warm water temperature be stabilised by manipulating the tuning constants? How does the controller respond to changes in the feed rate, i.e. step response testing?*]

Table W6.1 Baseline control performance

Disturbance period (min)	Frequency (1/min)	ΔT – product temperature (°C)	Attenuation
5			
10			
20			
40			
60			

3. FEEDFORWARD CONTROL

Feedforward control can be used to combat the control problems associated with processes containing significant dead time. This is achieved by measuring process disturbances and compensating for them before they affect the controlled variables. Ideal feedforward control is realised if pre-emptive control action is taken to completely cancel out the effect of measured disturbances before they enter the process. Sometimes the ideal feedforward controller is not realisable because disturbances affect the system more quickly than the manipulated variable. However, feedforward control can still be useful in these scenarios when teamed with feedback control because the feedforward control reduces the duty on the master controller and improves the overall system response. Clearly, no action can be taken if the disturbances are not sensed or measured.

Build a feedforward controller for the heater tank system, using the built-in spreadsheet function in HYSYS, to compensate for changes in the feed temperature before they become apparent in the warm water temperature. This feedforward control will be combined with the feedback control to see if process response can be improved.

Set up the following titles in cells A1–A6:

A1 Actual Feed Temp
A2 Nominal Feed Temp
A3 Temp Difference
A4 Process Gain 2
A5 Steam Valve Span
A6 Process Gain 1
A7 Feedforward Duty

Complete the spreadsheet as follows:

B1 Drag and drop the feed temperature from the process feed stream.

B2 Input the nominal feed temperature which is equivalent to the feed temperature set point. [*Hint: Refer to the Transfer Function operation.*]

B3 + B2 − B1

B4 Input the value of the process gain between the product temperature and the feed temperature. [*Hint: How much does the product temperature rise for a 1°C step increase in the feed temperature?*]

B5 Drag and drop the span of the heater duty valve.

B6 Input the value of the process gain between the product temperature and the heater duty. [*Hint: How much does the warm water temperature rise if the heater duty changes from 0% to 100%?*]

B7 The feedforward duty can be calculated from Equation W6.1. Incorporate this equation into the spreadsheet.

$$\Delta Q = \frac{\text{nominal feed temp} - \text{actual feed temp}}{\text{process gain 1}} \times \frac{\text{steam valve span}}{\text{process gain 2}} \qquad \text{(W6.1)}$$

ΔQ represents the changes in heater duty required to produce a 1°C change in the product temperature. Equation W6.1 is not necessarily exact at all values of the feed temperature due to the process nonlinearity. An exact expression is not necessary for successful feedforward control; even if the calculated duty is incorrect by 50%, the controller will still perform better than with no feedforward action.

Create a new energy stream named FF Duty. Export the result in cell B7 to the new energy stream. Add a mixer to the flowsheet and combine the new FF Duty with the original heater duty stream, called master signal. Attach the output to the heater duty energy stream, called steam. Your process should be similar to the one shown in Figure W6.2.

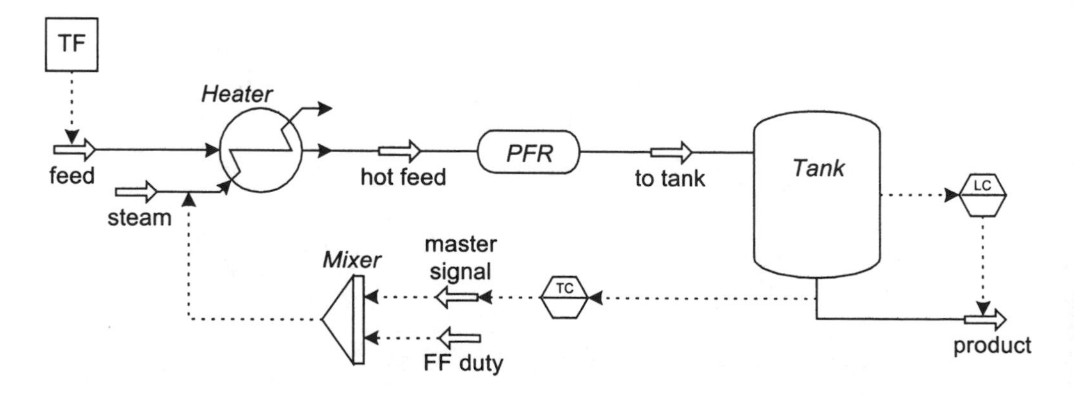

Figure W6.2 Feedforward control system.

- Test the feedforward controller for the same range of feed temperature disturbances that you analysed for FBC in the previous task. Record the results in Table W6.2.

Table W6.2 Feedforward control performance

Disturbance period (min)	Frequency (1/min)	ΔT – product temperature (°C)	Attenuation
5			
10			
20			
40			
60			

- How effective is the feedforward controller?
- What are its major deficiencies? [*Hint: Test the effectiveness of the feed-forward controller for changes in the feed rate.*]
- Briefly comment on any implementation issues that might be relevant with feedforward control. [*Hint: How is the feedforward gain calculated? How can dynamics be incorporated into the feedforward controller? How important is tuning of the feedforward controller?*]

4. CASCADE CONTROL

Cascade control is an alternative way to manage processes that contain large time constants and/or significant dead time. It is not necessary to sense or measure disturbances but a secondary variable must exist that directly affects the primary (master) loop and is faster responding than the primary loop. The secondary variable is usually, but not necessarily always, a flow or a pressure that is directly controlled via a control valve. Normally, this secondary (slave) controller is a flow controller, a pressure controller or a fast responding temperature controller. The time constant of the slave loop should be less than 25% of the time constant of the master loop for cascade control to be effective. Also, the secondary loop should contain little or no dead time. This allows the secondary variable to be controlled tightly, which provides attenuation for the primary loop.

Build a cascade controller for the heater tank system using an inner loop that manipulates the steam rate based on the heater outlet ("hot feed") temperature. The master controller (i.e. tank temperature) should provide the set point for the slave loop, and the slave controller should manipulate the steam rate directly. The process is shown below in Figure W6.3.

To implement cascade control into your existing simulation, delete the feedforward controller but retain the tank level controller and the original

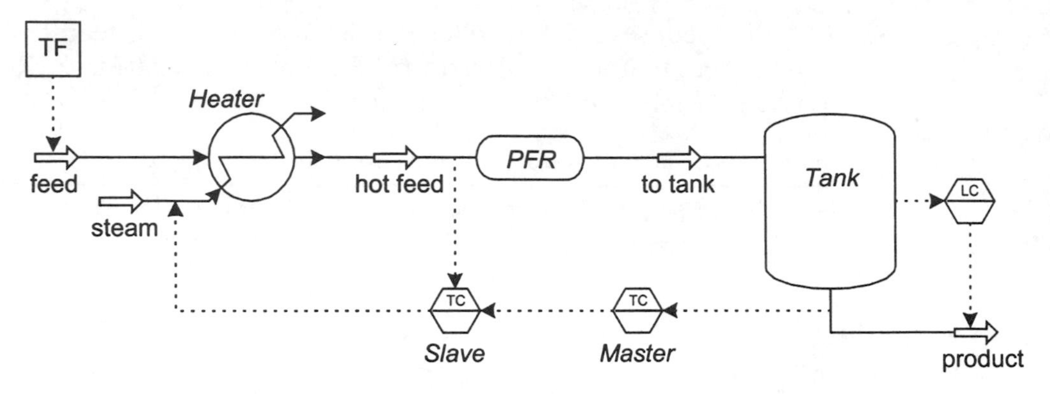

Figure W6.3 Cascade control system.

tank temperature controller. The tank temperature controller will now be the master controller for the cascade loop.

Add another controller unit operation to the PFD. Connect the *PV* point to the "hot feed" temperature. Connect the *OP* point to the heater energy stream ("steam") and specify "Direct Q" between 0 and 5×10^5 kJ/h. Connect the *SP* point (cascaded set point source) to the *PV* point of the master controller. The slave loop should be tuned tightly; specify a gain of 10 and an integral time of 10 minutes. The master loop can be tuned more loosely; specify a gain of 1.0 and an integral time of 10 minutes. Note that the gain of the master loop is not numerically comparable to the gain of the temperature loop from the previous simulation (without cascade control) because a different variable is being manipulated in the two cases. Your process should now be similar to the one shown in Figure W6.3.

- Test the cascade controller for the same range of feed temperature disturbances that you analysed for the previous two systems, which contained FBC only and feedforward control. Record the results in Table W6.3.

Table W6.3 Cascade control performance

Disturbance period (min)	Frequency (1/min)	ΔT – product temperature (°C)	Attenuation
5			
10			
20			
40			
60			

- Try varying the tuning constants for both the slave loop and the master loop. Which combination(s) of tuning constants work best?

- What comments can you make about the slave and master controller settings? How sensitive is the overall control performance to the slave loop tuning?
- Overall, how effective is the cascade controller? [*Hint: How does the controller respond to changes in the feed rate? How does the controller respond to changes in the master controller set point? Does the duty control valve open and shut excessively, i.e. is there too much control action?*]
- What are the major advantages and shortcomings of cascade control?
- How does cascade control compare with feedforward control?

5. RATIO CONTROL

Ratio control is a simple form of feedforward control that is commonly employed in controlling reactor feed compositions and in blending operations. It is also used to control the fuel to air ratio in heaters and boilers and to control the reflux ratio in distillation columns. The flow rate of one stream is used to provide the set point for another stream so that the ratio of the two streams is kept constant even if the flow of the first stream varies. Alternatively, the actual ratio between two flows can be used as the input to a controller.

Build a new system consisting of two streams, two tanks and a mixer using the Wilson thermodynamic package. Pick any two components that are liquid phase at ambient temperatures. The first stream should be pure component "A" at 25°C and 100 kPa. The second stream should be pure component "B" at the same temperature and pressure. Set the flow of the first stream to 400 kg/h and the second stream to 100 kg/h. These flows are consistent with the desired ratio of 4:1 between component A and B. The tanks are used to simulate dead time in the system so choose relatively small volumes for the tanks and locate them in series with the first stream. Both tanks should be on level control rather than liquid flow control. Simulate process noise with a sine wave input to the first stream using an amplitude of 50 kg/h and a period of 10 minutes. The system should resemble the one shown in Figure W6.4.

Incorporate ratio control via a spreadsheet. Import the flow on the first stream into cell A1. Put the ratio of 0.25 in cell A2 and add a formula to give the flow of the second stream in cell A3. Export the result of cell A3 to the flow of the second stream.

Run the simulation in dynamic mode with several values for the disturbance period. Watch how the combined flow rate and concentration changes with different conditions. You may need to reduce the integrator step size to see the effects of very high frequency disturbances (period < 5 minutes).

- How effective is the ratio controller at filtering out low frequency noise?
- How effective is the ratio controller at filtering out high frequency noise?

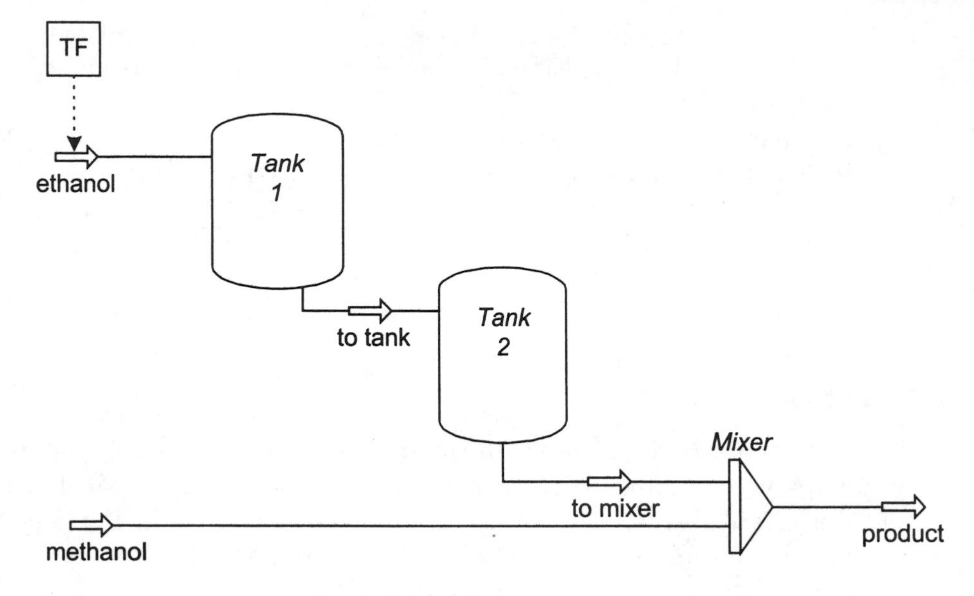

Figure W6.4 Ratio control system.

- What are some of the advantages and limitations of ratio control?
- What is the significance of the dead time in the system? [*Hint: How would your answers change to the above questions if the dead time was the same for both streams?*]

Present your findings on diskette in a short report using MS-Word. Also include on the diskette a copy of the HYSYS files which you used to generate your findings.

WORKSHOP 7 – DISTILLATION CONTROL

"Have confidence that if you have done a little thing well,
you can do a bigger thing well, too."
Joseph Storey

Introduction

Prior to attempting this workshop, you should review Chapter 8.

Distillation is one of the most important unit operations in chemical engineering. It forms the basis of many processes and is an essential part of many others. It presents a more difficult control problem compared with many other unit operations as at least five variables need to be controlled simultaneously and there are at least five variables available for manipulation. Thus a distillation column provides an example of a multiple input/multiple output (MIMO) control problem. It is critical that variable pairing is done appropriately between controlled and manipulated variables. The overall control problem can usually be reduced to a 2×2 composition control problem since the inventory and pressure loops frequently do not interact with the composition loops. This workshop will highlight some fundamental rules of distillation control and show how a basic distillation control scheme can be selected.

Key learning objectives

1. There are five degrees of freedom present in a simple distillation column with no side draws and a total condenser. The degrees of freedom increase by one with each side draw and for a partial condenser, which has two overhead products.
2. The degrees of freedom equal the number of controlled variables, the number of manipulated variables, and the number of control valves: $DF = N_{cv} = N_{mv} = N_{valves}$.
3. Steady-state analysis can be used to develop a basic control strategy that

provides good sensitivity. Dynamic analysis is required to develop a control scheme that provides good responsiveness.

4. Vapour and liquid flows have different dead times and response times.
5. Two-point composition control involves the control of both the distillate and bottoms compositions simultaneously. One-point composition control involves operating one of the composition control loops in manual or using a non-composition related variable as the controlled variable.
6. Distillation control schemes are usually described by the two variables which are manipulated for composition control (i.e. the LV, DV or LB configuration) and may work by manipulating either the mass balance (i.e. the DV and LB configurations) or the energy balance (i.e. the LV configuration).

Tasks

1. STABILISER CONFIGURATION

The distillation column shown in Figure W7.1 is typical of a stabiliser, which is found in most refineries. The column is designed to remove volatile components from potential gasoline blendstocks. The feed is usually a mixture of

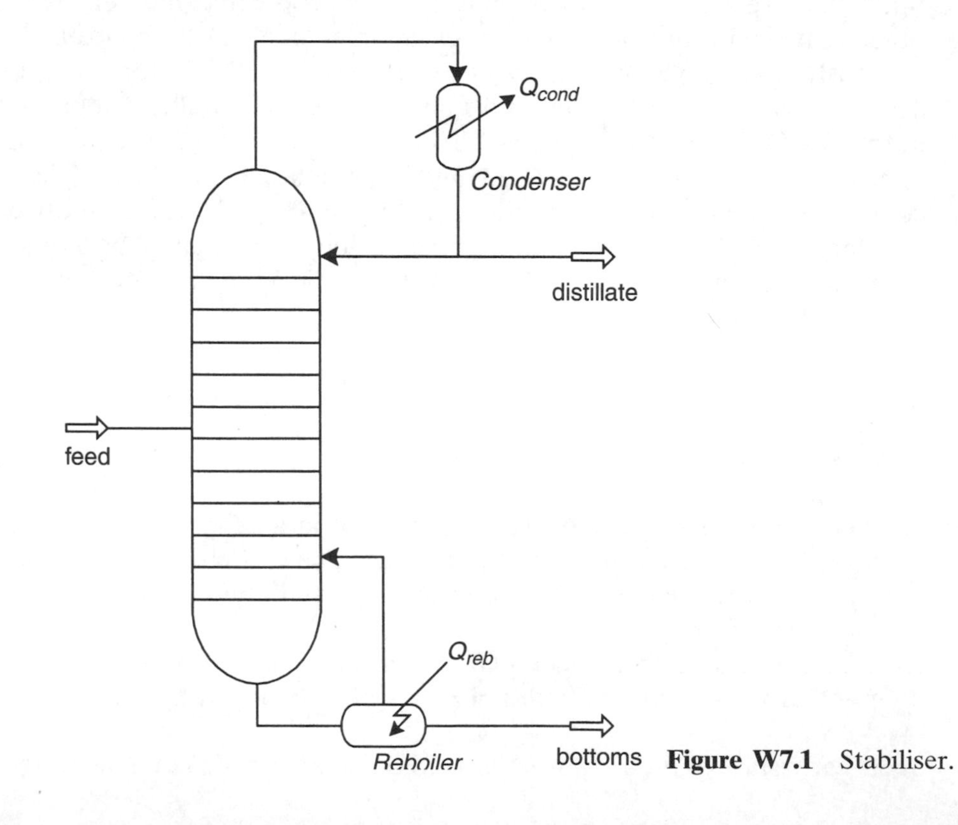

Figure W7.1 Stabiliser.

C_3, C_4, and C_5. In this case, the feed contains 5% propane, 40% isobutane, 40% n-butane, and 15% isopentane. The total flow rate is 40,000 bbl/day at 720 kPa and 30°C.

The stabiliser contains 20 trays and a total condenser. Feed enters at tray 10. The normal column overhead pressure is 700 kPa and there is a 20 kPa pressure difference that is evenly distributed between the condenser and the reboiler. Each tray is 2.0 m in diameter with a 0.10 m weir, which is 1.6 m long. The bottoms product specification is 0.01 wt% C_3, and the column is normally operated with a reflux ratio of 1.0.

Use the Peng-Robinson property package and the above information to build a simulation model of the stabiliser. View the results to get a feel for the column in steady-state mode. [*Hint: The column has two degrees of freedom after the inbuilt inventory and capacity controls have been considered. These should initially be taken up by a bottoms composition specification and a reflux ratio specification.*]

2. CONTROL SYSTEM OBJECTIVES AND DESIGN CONSIDERATIONS

Ultimately the importance of process control is seen through increased overall process efficiency allowing the plant engineer to get the most from the process design. This is especially true of distillation control. Most distillation columns are inherently flexible and a wide range of product yields and compositions can be obtained at varying levels of energy input. A key requirement of any control system is that it relates directly to the process objectives. A control system that does not meet the process objectives or produces results that conflict with the process objectives does not add value to the process.

The first step of control system design is to identify the process objectives, which may not always be obvious. One process objective of the system shown in Figure W7.1 is to control the propane content of the bottoms at 0.01 wt%. However, other process objectives are not clear from the information given above. Controlling the reflux ratio at 1.0 may be desirable for the given conditions, but this is not a key process objective as it does not directly fix any property of the products or the energy consumption.

Some possible process objectives for this system include:

- minimum energy input;
- maximum product yield;
- minimum isopentane in the overhead product;
- maximum recovery of C_4+ components;
- overhead temperature to meet utility requirement.

In the stabiliser described above, we will initially use the bottoms composition to set the feed split and the energy input to set the fractionation. As yet, we have not paired these controlled variables to control valves (manipulated variables).

- Define the control objectives for the stabiliser built in the previous task.

3. DISTILLATION CONTROL DESIGN VIA STEADY-STATE ANALYSIS

The feed to the stabiliser described above is expected to vary in both rate and composition. The preferred control system design should be able to handle the design feed conditions and any extremes that might be expected. The data in Table W7.1 should be used to test any control system design.

Simulations allow essentially all types of variables to be used as controlled variables. However, a control scheme must be implementable in an operating plant. This requires that the variable being controlled can be accurately measured to provide feedback in a control loop. Examples of variables that can be easily measured include flows (especially liquid flows), temperature and pressures. Some compositions, mainly mass and volumetric fractions, can be measured using analysers but these instruments are generally expensive and often introduce considerable dead time to processes. Consequently, they are excluded from many control systems.

Preliminary control system design has traditionally been conducted using steady-state data only. Steady-state simulations are performed to gain an understanding of the process and the way it responds to certain changes (or disturbances). This information is used to select a candidate control system, which is then either tested with dynamic simulations or immediately implemented in a plant environment in the hope that it provides adequate control with correct tuning. A possible control system design strategy, using only steady-state simulations, is given below.

Control strategy selection using steady-state analysis

1. Select two potential controlled variables (i.e. temperature on tray 3 and the reflux rate). Note the values of these variables from the base case

Table W7.1 Expected feed variance

	Base case	Minimum flow	Maximum flow
Feed Rate (bbl/day)	40,000	20,000	50,000
C_3 wt%	5	1	12
iC_4 wt%	40	44	33
nC_4 wt%	40	45	37
iC_5 wt%	15	10	18

Table W7.2 Control strategy selection using steady-state analysis

Primary controlled variable	Set point	Secondary controlled variable	Set point	Control objectives					
				Base case		Min. flow		Max. flow	
				Prim. *mv*	Sec. *mv*	Prim. *mv*	Sec. *mv*	Prim. *mv*	Sec. *mv*

solution (see Task 1) and list them under the "Set Point" column in Table W7.2.

2. Use these two potential variables as the new specifications for the column in the steady-state, base case simulation. The value for these two new specs will be the values noted as the set points in the previous step.
3. Solve the column with the new specifications and record the value of the two manipulated variables under the "Prim. *mv*" and "Sec. *mv*" columns. These manipulated variables are based on the control objectives, i.e. propane content in the bottoms product and energy input.
4. Repeat Step 3 for the expected extremes in flow rate and composition. Again, record the values of the manipulated variables for each case.
5. Repeat starting from step 1 for two new potential controlled variables and/or different set points. Determine which combination of controlled variables produces the best overall control performance for all the feed variances by finding the pair with corresponding set points that meet the control objectives for all variances tested.
6. After the control variables have been established to control the feed-split and column fractionation, set the other variables to control inventory and capacity of the column.

Often the variable pairings required for inventory and capacity control will be immediately evident after the composition control variables have been selected. If not, the following guidelines can help:

- Control the pressure with the condenser cooling duty.
- Control levels with an outlet stream from the vessel (i.e. reflux or distillate for the reflux accumulator) or an energy stream that affects the inlet flow to the vessel (i.e. the reboiler duty for the reboiler sump).

- Where more than one stream is available, choose the largest stream for level control.

Use the method listed above to build a control system for the stabiliser given in Figure W7.1 with the following control objectives:

- 0.01 wt% propane in the bottoms.
- fixed energy input.

- You should consider the range of feed variance given in Table W7.1 and use only variables that can be easily measured. Record your results in Table W7.2 and provide a diagram of the candidate control system in your report write-up.

- Are there any particular advantages or disadvantages associated with your preferred combination of controlled variables?

Once the controlled variables have been chosen, determine what set points will allow the control objectives to be met at all operating conditions (i.e. all three cases from Table W7.1) using the following steps.

1. Modify the column specifications to control the process objectives (i.e. 0.01 wt% propane in the bottoms and fixed heat input) directly.
2. Note the values of the selected controlled variables for all three cases in Table W7.1.
3. Identify which value of each controlled variable is the worst case (i.e. most conservative).
4. Re-solve the column with the "conservative" values of the controlled variables to confirm that the control objectives are met or exceeded for each case, meaning that the concentration of propane in the bottoms is equal to or lower than the specification and the energy input is equal to or lower than the target you have specified.

4. DYNAMIC RESPONSIVENESS

Once you have completed the steady-state design, use the results to size all the valves in the system. Add the control loops (two) determined from the steady-state design, with the set points which allow the control objectives to be met at all times. Inventory and pressure control will be available automatically but you can add your own control loops if desired. Solve the column in dynamic mode. You are free to change set points or create disturbances to the system to examine how the control system performs. You may need to tune the control loops in order to produce an adequate response.

- Does your candidate control scheme provide satisfactory control? [*Hint: Test the control schemes for a wide range of disturbances, including some outside the bounds of the steady-state design. Are the system dead times and lags dominating the control behaviour? Are both set point changes and disturbances handled adequately?*]

5. DISTILLATION COLUMN CONTROL CONFIGURATIONS

We noted previously that a simple distillation column with a total condenser normally has five degrees of freedom. Each degree of freedom corresponds to a control valve and a controlled variable. Three of these degrees of freedom must be used to control the inventory and capacity variables, i.e. levels and pressures. The remaining two degrees of freedom are used for composition control. The condenser duty (or a related variable) is usually reserved for pressure control. However, any of the remaining four variables can be used for composition control. The following notation is often used for the four degrees of freedom:

L = liquid flow down the column, reflux rate
V = vapour flow up the column, boilup or reboiler duty
D = distillate rate
B = bottoms rate

The relationship between boilup and reboiler duty is not exact but it is usually sufficiently close so that the two variables can be used interchangeably.

Distillation control configurations are frequently described by the two variables that are used for composition control or not used for inventory/capacity variable control. For example, the LV configuration uses the reflux rate and reboiler duty to control the product compositions. By inference, the condenser duty is used for pressure control, the distillate rate is used to control the reflux accumulator level, and the bottoms rate is used to control the reboiler sump level.

- Complete Table W7.3 by listing the manipulated variables (*mv*) for each of the controlled variables.

Table W7.3 Distillation column control configurations

Control configuration	*mv* for reflux accumulator level control	*mv* for reboiler sump level control	*mv* for primary composition control	*mv* for secondary composition control
LV				
DV				
LB				

- Why is the LD or BV configuration not likely to produce satisfactory control?
- Why is the DB configuration not likely to produce satisfactory control?
- For the DV and LB configurations, which variable should be used as the primary composition control variable and which should be used as the secondary?

The LV control configuration is often described as an energy balance configuration while the DV and LB configurations are material balance configurations. This is because the DV and LB configurations manipulate the feed-split or material balance directly by changing one of the product rates. However, the LV configuration only affects the feed-split indirectly through the level controllers.

The basic distillation control configurations have been listed above. However, there are many other configurations which use linear or even nonlinear combinations of the basic manipulated variables. One common example, which is sometimes called Ryskamp's Scheme [1], manipulates the reflux ratio (L/D) via ratio control, and the reboiler duty (V). Another relatively common scheme is the double ratio configuration which manipulates the reflux ratio and the boilup ratio. This scheme has been widely recommended as it results in relatively small interactions between the two control loops.

- The principal control objectives for the stabiliser were previously listed as 0.01 wt% propane in the bottoms and fixed energy consumption. Describe how each of the three control configurations listed in Table W7.3 could be set up to satisfy these control objectives.
- Test these configurations via simulation. Are there any significant differences between the three schemes? [*Hint: You may have to vary the tuning constants to get equivalent performance.*]

Present your findings on diskette in a short report using MS-Word. Also include on the diskette a copy of the HYSYS files which you used to generate your findings.

References

1. Ryskamp, C.J., "New Strategy Improves Dual Composition Column Control (also effective on thermally coupled columns)". *Hydrocarbon Processing*, June 1980, p. 51–9.

WORKSHOP 8 – PLANT OPERABILITY AND CONTROLLABILITY

"There's a better way to do it. Find it!"
Thomas Edison

Introduction

Prior to attempting this workshop, you should review Chapters 8, 9, and 10.

Traditionally, process design has been performed using steady-state analysis only. Simple rules of thumb have been used to size vessel hold-ups and to set other variables that affect the dynamic performance of a plant. This can sometimes lead to operability and controllability problems as a design might look good in the steady state but be very difficult to operate or control due to the presence of dead time or insufficient capacitance.

A key consideration for plant operability and controllability is variable interaction. We have learned that dead time is one of our enemies as it always makes tight control more difficult to achieve. Variable interaction places similar restrictions on the way we can control a process and can significantly reduce the overall control system performance. Three common sources of variable interaction are the nature of the process (i.e. distillation), the combination of multiple unit operations and heat integration. Each of these points can be highly advantageous in the steady state but they can also create operability and controllability problems that may not be evident without considering the process dynamics at the design phase.

This workshop will investigate several examples where variable interaction is significant and will introduce an analytical technique for finding the best variable pairings in multiple input/multiple output (MIMO) systems. The potential trade-off between capital savings and plant operability will also be demonstrated. The problems in this workshop are more open-ended than other workshops. You are encouraged to work more freely and continue your analysis until you are satisfied that you have pursued all paths.

Key learning objectives

1. Distillation control can usually be reduced to a 2×2 control problem. Interaction between variables plays a key role in control strategy selection and performance.
2. Inventory control (i.e. reflux accumulator and reboiler sump) should always be via the largest outlet stream.
3. The relative gain array (RGA) is a control system design technique that can be used to minimise control loop interaction.
4. Process understanding and a clear understanding of the key process objectives is essential to the development of a good control scheme.
5. Tight process control requires that the equivalent dead time in a loop should be small compared with the smallest time constant of a disturbance with significant amplitude.
6. There is often a trade-off between steady-state cost savings and dynamic operability.
7. Steady-state minimum-cost designs utilise very small hold-ups and high levels of heat integration. Both of these factors reduce the dynamic operability and controllability of a process.
8. Too much hold-up provides good attenuation but makes the process too slow to respond to disturbances.

TASKS

1. TWO-POINT COMPOSITION CONTROL FOR DISTILLATION

Most industrial distillation columns are operated similarly to the stabiliser that we studied in the previous Workshop. One degree of freedom is used to control a product composition and the second available degree of freedom is used to control fractionation or energy consumption. This mode of operation is often called one-point or single composition control.

Sometimes both distillation products are equally important or equally valuable and both the bottoms composition and the distillate composition need to be controlled. This is called two-point or double composition control, and results in a much more difficult control problem than one-point composition control [1]. The primary cause of the extra difficulty is the interaction that exists between composition loops in a distillation column. This property of distillation columns (inherent interactions between two or more control loops) is called ill-conditioning.

One problem created by ill-conditioning is that there usually exists only a very narrow operating range that satisfies both composition control loops. Essentially both the manipulated variables being used for composition control need to be adjusted together to produce the required results. Many newer control schemes are based on this principle, including model based control and dynamic matrix control (DMC) [2, 3].

A second problem with two-point composition control is that there are no degrees of freedom left to operate around equipment constraints such as a reboiler duty limitation or flooding limitation. One-point composition control schemes have one degree of freedom which is not used for composition control and is available for this purpose (cf. our stabiliser which is operated at a fixed heat input). This can create operational difficulties for many industrial columns.

Reconfigure the controllers on the stabiliser from Workshop 7 to provide two-point composition control. Assume that each product is equally important and that the control objectives are 0.01% propane in the bottoms and 0.1% isopentane in the distillate. Also assume that you have two perfect analysers (i.e. no dead time, no error) available so that the two compositions can be controlled directly.

Select one of the basic distillation control configurations (i.e. the LV, DV, or LB configurations) and tune composition controllers for both the distillate and bottoms products. Test the responsiveness of your candidate control structure using dynamic simulations of the stabiliser.

- Can you configure the system to give tight control of both the bottoms and distillate compositions and give good responsiveness to set point changes?
- Do the controllers interact? [*Hint: If one loop is stable, does a set point change in the other loop disturb the first loop?*]

- Using steady-state analysis techniques, find an improved control configuration as follows:
  1. For each of the three feed conditions given in Table W7.1 of the previous Workshop, record the distillate rate, bottoms rate, reflux rate and reboiler duty when the two composition specifications are satisfied simultaneously. This information should be already available from your previous steady-state simulation.
  2. Calculate ratios of the manipulated variables and/or ratios to the feed rate (i.e. B/F, L/D, Q_R/B and others, if necessary).
  3. Find the combination of two manipulated variables (ratios) that shows the smallest variability for the whole range of feed conditions. This is done since it will maximise the natural disturbance attenuation of the control system.
  4. Outline a candidate composition control structure using the two ratios chosen, making sure that you apply the normal rules of steady-state sensitivity and dynamic responsiveness for distillation control from

Table W8.1 Manipulated variables and calculated ratios for steady-state analysis technique

Manipulated variables and ratios	Base case	Minimum flow	Maximum flow
D (bbl/d)			
B (bbl/d)			
L (bbl/d)			
Q_R (MW)			
B/F			
L/D			
Q_R/B			

the previous Workshop. [*Hint: You will need to use the spreadsheet operation and cascade controllers. Using the ratios from Table W8.1 for direct control is not feasible. The correct approach for the control structure is outlined in Ryskamp [1].*]

Test the responsiveness of your candidate control structure using dynamic simulations of the stabiliser. Do not forget that you will have to retune the two composition control loops.

- Has loop interaction been reduced?
- Is the overall control better than the basic control configuration that you tested above?
- If you didn't have perfect analysers available or didn't want to introduce dead time or error that would be present with real analysers, is there a combination of easily measured temperatures that you could successfully use to infer the distillate and bottoms compositions for the whole range of feed variance given in Table W7.1?

2. RELATIVE GAIN ARRAY (RGA)

The relative gain array is a steady state analysis tool that can be used to select an appropriate control structure from several candidate structures in a MIMO system. The relative gain is the ratio between the open loop gain and closed loop gain in a system. In a distillation column, only the composition control variables are normally considered. The open loop gain is the process gain between the controlled and manipulated variables with the secondary manipulated variable held constant. The closed loop gain is the process gain between the controlled and manipulated variables with the secondary controlled variable held constant.

The relative gain array, Λ, has a property which makes the calculation of all the elements of the array unnecessary. This is shown in Equation W8.1

and applied to a 2×2 system in Equation 8.2. λ_{ij} refers to the relative gain between the ith controlled variable and jth manipulated variable.

$$\sum \lambda_{ij} = \sum \lambda_{ji} = 1 \quad \text{(W8.1)}$$

$$\Lambda = \begin{bmatrix} \lambda_{11} & \lambda_{12} \\ \lambda_{21} & \lambda_{22} \end{bmatrix} = \begin{bmatrix} \lambda_{11} & 1-\lambda_{11} \\ 1-\lambda_{11} & \lambda_{11} \end{bmatrix} \quad \text{(W8.2)}$$

- Determine the λ_{11} element of the relative gain array for each of the basic distillation control configurations (i.e. the LV, DV and LB configurations). Consider only the composition control loops so that the problem reduces to a 2×2 system.
- Which configuration appears to be most suitable for the stabiliser? Does this agree with your dynamic simulation results? [*Hint: The gains you need can all be calculated via simulation.*]

3. REACTOR TEMPERATURE CONTROL

A key design consideration with exothermic reactions is the utilisation of the heat of reaction. This energy optimisation is often critical for plant profitability. The obvious heat integration technique is to use the hot reactor product to heat the cold reactor feed. The alternative would be to employ a hot utility to heat the reactor feed to near the reactor temperature and then a cold utility to cool the reactor product to the desired level.

Build the simple reactor system shown in Figure W8.1. The reactor should be modelled as a separator with a volume of 2 m^3. The reaction, given below, is equilibrium limited. The relationship between K_{eq} (in terms of activities) and the reaction temperature (in °K) is also given below. The system is very non-ideal, so an activity model should be used, i.e. the UNIQUAC model, and the reaction equilibrium should be measured in terms of activities rather than molar concentrations.

$$CH_3OH + CH_2 = C(CH_3) \Leftrightarrow CH_3OC(CH_3)$$

$$\text{methanol} + \text{isobutene} \Leftrightarrow \text{MTBE}$$

$$K_{eq} = -10.54 + 4870 / T\ (°K)$$

The feed is composed of two streams. The first stream is a hydrocarbon stream that contains 30 mol% isobutene and 70 mol% 1-butene. The second stream, consisting of pure methanol, is in 5% molar excess of the reaction stoichiometry. The hydrocarbon feed rate is 1000 kg/h. Both streams are at

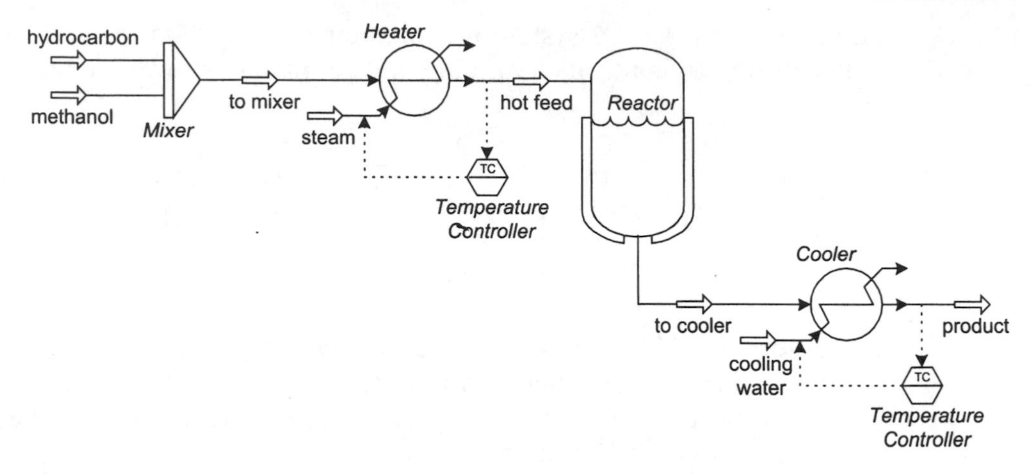

Figure W8.1 MTBE reaction system without heat integration.

30°C and 1500 kPa. The reactor inlet temperature should be controlled at 70°C. The reactor outlet temperature will be higher than the inlet since the reaction is exothermic and a considerable amount of heat is released. This has the effect of limiting the conversion of isobutene in the reactor. The reactor product should be cooled to around 40°C so that a second reaction stage can increase the isobutene conversion to around 99%. The reactor pressure drop is 140 kPa while the pressure drops through the exchangers are 70 kPa. The exchanger volumes can be estimated at 0.1 m^3 each.

Set up temperature control loops on both heat exchangers, then tune PI controllers for both of these temperature control loops.

- Test your control system for disturbances in the feed rate, feed temperature and feed composition. The following disturbances are suggested as starting points for your analysis.
- Which disturbances are most difficult to control?

Table W8.1 MTBE reaction system disturbances

	Base case	Test case 1	Test case 2	Test case 3
Hydrocarbon feed rate (kg/h)	1000	750	1000	1000
Feed temperature (°C)	30	30	45	30
Hydrocarbon feed composition (mol%)	30% i-but 70% 1-but	30% i-but 70% 1-but	30% i-but 70% 1-but	40% i-but 60% 1-but

- How does the isobutene conversion vary during disturbances? [*Hint: You may need to use a spreadsheet to calculate the isobutene conversion*

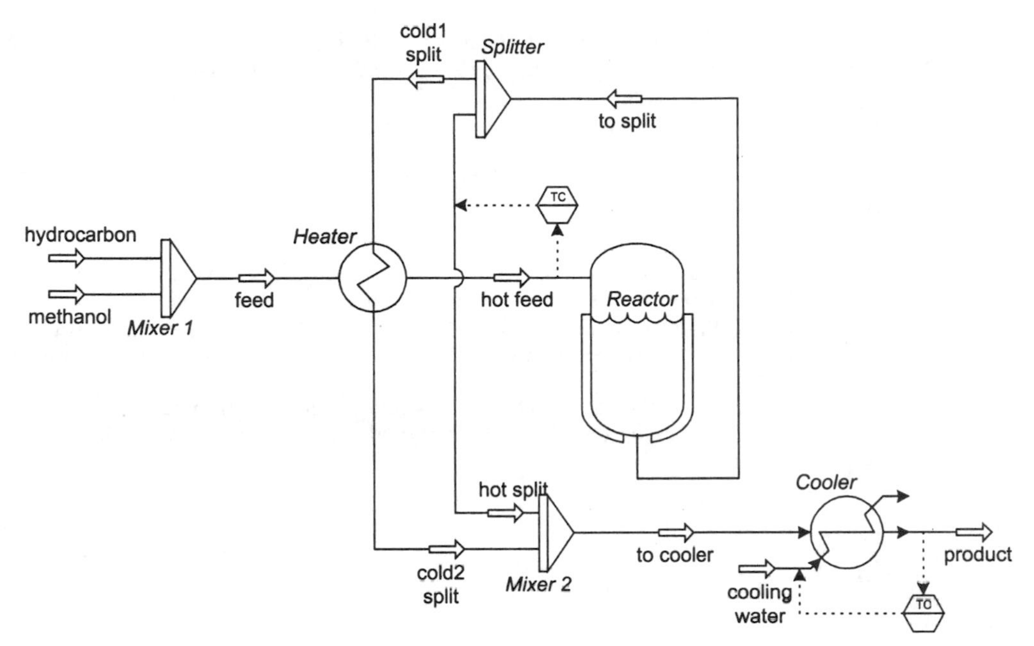

Figure W8.2 MTBE reaction system with heat integration.

continuously.]

Modify the system to incorporate heat integration between the reactor outlet (hot) and the reactor feed (cold), as shown in Figure W8.2. Again, assume pressure drops of 70 kPa in both sides of the exchangers and a volume of 0.1 m^3. This should also reduce the load on the existing cooler.

Retune both temperature controllers. Test the new control structure against a similar range of disturbances.

- How much energy is saved by using heat integration in the process? Consider both heating and cooling duties.
- Does the system with heat integration still provide adequate control?
- What implications does the controllability or lack of controllability have for safety?
- Overall, which process configuration would you prefer?
- How could you modify the system to incorporate elements of both process designs to minimise utility consumption without compromising operability, controllability and safety?

Present your findings on diskette in a short report using MS-Word. Also include on the diskette a copy of the HYSYS files which you used to generate your findings.

References

1. Ryskamp, C.J., "New Stategy Improves Dual Composition Control (also effective on thermally coupled columns)". *Hydrocarbon Processing*, 1980, **June**: 51–9.
2. Cutler, C.R. and Ramaker, B.L., "Dynamic Matrix Control – A Computer Control Algorithm". AIChE National Meeting, Houston, 1979; Joint Auto Control Conference, San Francisco, 1979.
3. Prett, D.M. and Gillette, R.D., "Optimization and Constrained Multivariable Control of a Catalytic Cracking Unit". AIChe National Meeting, Houston, 1979; Joint Auto Control Conference, San Francisco, 1979.

INDEX